Creative Understanding

CREATIVE UNDERSTANDING

Philosophical Reflections on Physics

Roberto Torretti

The University of Chicago Press
Chicago and London

Roberto Torretti is professor of philosophy at the University of Puerto Rico and editor of the journal *Diálogos*. Among his several books is *Relativity and Geometry*.

The University of Chicago Press, Chicago 60637
The University of Chicago Press, Ltd., London

Printed in the United States of America

99 98 97 96 95 94 93 92 91 90   5 4 3 2 1

Library of Congress Cataloging-in-Publication Data

Torretti, Roberto, 1930–
  Creative understanding: philosophical reflections on physics / Roberto Torretti.
    p.   cm.
Includes bibliographical references (p.) and index.
ISBN 0–226–80834–3. — ISBN 0–228–80835–1 (pbk.)
1. Physics—Philosophy. 2. Physics—Methodology. I. Title.
QC6. T653 1990   90–11007
530′.01—dc20   CIP

♾ The paper used in this publication meets the minimum requirements of the American National Standard for Information Sciences—Permanence of Paper for Printed Library Materials, ANSI Z39.48.1984.

For Carla

CONTENTS

PREFACE

ὁ τοιοῦτος νοῦς . . . εἴη ἄν καὶ ταύτῃ ποιητικός, ᾗ
αὐτὸς αἴτιος τοῦ εἶναι πᾶσι τοῖς νοουμένοις.

Moreover, such an understanding is creative
inasmuch as it brings every notion into being.

ALEXANDER OF APHRODISIAS†

The understanding of natural phenomena developed by mathematical physics since the 17th century stands today as one of the most remarkable achievements of human history. Rightly or wrongly, many regard it as a paradigm for every scientific endeavor. And its practical applications have, for good or ill, drastically altered the fabric of life.

In a lecture on the method of theoretical physics delivered at Oxford in 1933, Albert Einstein noted that the concepts and fundamental laws of physics are not derived by abstraction from experience, nor can they be justified by appealing to the nature of human reason, for they are "free inventions of the human mind" (Einstein 1934, p. 180). Clearly then, if we take our cue from him, the understanding wrought from those concepts and laws must be termed 'inventive' or 'creative'.

The aim of this book is to elucidate, at least in some important respects, the workings of that creative understanding. No attempt is made to unveil the mysteries surrounding intellectual creativity. But through the examination and interpretation of examples from the history of science, and the critical discussion of good and not so good ideas from recent philosophical literature, it seeks to throw light on the means and ends of the intellectual enterprise of physics.

The book is divided into five chapters, labelled "Observation," "Concepts," "Theories," "Probability," and "Necessity." The conventional starting-point with observation was chosen, not to pay lip service to the philosophy of inductivism, but to underscore its inadequacy. Observation without understanding is blind. We must grasp phenomena under universal concepts in order to make them out, and so make them into facts. This Kantian thesis is rehearsed and illustrated in Chapter 1. A strong argument for it follows from the preferential status of instrumental observation in science. The recorded modification of an instrument can yield information about the state of the observed object only to the extent that the latter is a necessary condition of the former. However, a necessary connection between them cannot be found by inspecting the instrument's dial, but must be read into it in the light of an overall understanding of the physical situation. In modern physics, the required understanding is supplied by the richly articulated grasp of physical systems afforded by physical theories.

Physical theories are in effect the unifying theme of the remaining four chapters. The formal examination of their typical (idealized) structure and mutual relations in Chapter 3 is preceded and prepared by the informal presentation and discussion, in Chapter 2, of a problem that has much exercised philosophers of science since the publication of Thomas S. Kuhn's influential essay on *The Structure of Scientific Revolutions* in 1962; namely, the alleged incomparability or, as the saying goes, "incommensurability" of succeeding theories. The problem arises if the identification of the objects of scientific discourse depends on the concepts employed for describing them, and there is no fixed set of concepts shared by all scientific theories. After discussing and dismissing with familiar arguments the view that such a fixed set of concepts is available, I embark on a lengthy criticism of Hilary Putnam's doctrine of reference without sense, according to which one can single out a determinate physical object, e.g., a physical magnitude, independently of how one conceives it. Although Putnam himself has discarded this doctrine, it is still favored by several authors.

My own approach to the problem of the incommensurability of physical theories is explained and defended in Chapter 2, especially in Sections 2.5 and 2.7. The problem can be posed in earnest only with respect to the theories of fundamental physics, which can, in turn, supply the requisite conceptual bridges between successive theories of narrower scope. The problem does not arise when a new theory of fundamental physics is reached—as Special Relativity was by Einstein in 1905—through internal criticism of the preceding theory. Yet even when that is not quite the case, a shared tradition of mathematical thought makes it possible to read the earlier theory in terms of the new one, or to devise an ad hoc common framework for the description and assessment of experimental data. But apart from such internal means of comparison, the fundamental theories of mathematical physics communicate across the common ground of understanding from which they grow and which they serve: the loose, unpretentious grasp of things and events in everyday life. This is not to say that science is accountable to common sense, or that scientific discourse should be translatable into ordinary language. But since the latter has not been—and presumably never will be—replaced by the former, it continues to supply the murky global perspective within which each particular theory of physics discerns the facet it seeks to conceive with clarity and precision.

In Chapter 3 I take the view that a physical theory conceives an open-ended host of physical situations as instances of a mathematical concept. This view was put forward by Joseph Sneed in *The Logical Structure of Mathematical Physics* (1971), and has recently been presented in a clearer and more elaborate form by Balzer, Moulines, and Sneed in *An Architectonic for Science* (1987). Sneed and his collaborators explicate the mathematical concept at the heart of a physical theory in set-theoretical terms, as a Bourbakian species of structure. This approach is also adopted here, not due to any sympathy for Bourbaki's ideas, but because they are good enough for the present job, and after several decades of dominance over the teaching of mathematics, they have become fairly well known. Sneed's analysis of the structure of a physical theory is illustrated and motivated with some examples from history and then formally explained. There follows some criticism of important aspects of Sneed's doctrine. It is shown that his distinction between the models and the potential models of a physical theory, although useful for explicating the familiar contrast between the *concepts* of physics and its *laws*, is in each case relative to the peculiar way chosen for reconstructing the central concept of the theory in question as a Bourbakian species of structure. Such relativity undermines Sneed's use of so-called partial potential models in the solution of his problem of theoretical terms. This, however, is a pseudoproblem, stemming from a refusal to countenance a genuinely creative understanding of natural phenomena, and nothing is lost by forgoing its purported solution.

Sneed and his associates have developed the means of conceiving both the links that can be established between several closely related or widely divergent theories and the constraints which bind together the different applications of a single theory. They readily account for the fact that the number and variety of applications of a physical theory can change without prejudice to its conceptual identity. But they do not tell us how a fragment or aspect of experience is turned into an application of a physical theory; how scientific thinking takes hold of a domain of reality, fills in the intelligibility gaps, and articulates it as a domain of objectivity. In Section 3.5 this problem is elucidated in the light of the work of Günther Ludwig. Ludwig has also originated the best available formal treatment of approximation in physics. His contribution and its implications for the coexistence and joint employment of seemingly incompatible theories are explained in Section 3.6. Sections 3.7 and 3.8 methodically examine the different types of relations that exist between physical theories and contain the book's last word on the problem of Chapter 2.

In Chapter 4 I interrupt the discussion of physical theories in general, to grapple with a difficult but rewarding illustration: the physicomathematical concept of probability or, more precisely, of a probability space. The choice of this example may seem questionable, for the species of structure *probability space* is not by itself the central mathematical concept of any physical theory. It does, however, occur as a constituent in several such concepts (just as the *real number field* occurs in all), and due to its comparative simplicity, it lends itself better for the didactic purpose of this chapter than the conceptual core of a full-fledged theory. A deeper motivation for choosing it lies in the controversies that surround it. What sort of physical reality is represented in physics by the concept of a probability space? There is no general agreement on this point, and a powerful school professes that there is no such reality at all. The issue had to be discussed before dealing with physical necessity and determinism in the final chapter. It is a welcome opportunity for considering the creative understanding at work as a source of objectivity.

That necessary connections between physical events are involved in the current understanding and utilization of scientific observations was the main result of Chapter 1. Chapter 5 shows how physical theories use mathematical concepts to embed natural phenomena in a tissue of such connections. The conception of a physical theory as a species of structure modelled by the theory's applications finds its ultimate vindication here. The chapter also includes some reflections on the sources of the notion of physical necessity in ordinary human experience (§5.1) and on the uneasy coexistence of physicomathematical determinism with commonsense causality (§5.4).

*****

The notes serve several purposes. A few of them explain technical terms, especially from mathematics. Others pursue special questions or try to ward off possible objections. Others provide illustrative quotations, mention sources, or make suggestions for further reading. I suppose that the main text can be understood without referring to the notes, but I expect that some readers will find them useful.

References to the literature are usually identified by the author's name followed by the year of publication. Exceptionally, when I refer to an edition published much later than the original work, I substitute a word or a few identifying letters for the year of publication. (The sole purpose of this is to avoid vexing anachronisms such as Einstein 1987 or Heraclitus 1855). Titles are given in References, at the end of the book. That list contains only works which have been mentioned in the main text or the notes and does not fully reflect my debt to other writers. I have tried my best to indicate the provenance of my ideas, but I do not always remember.

I have indulged in a minor deviation from standard English usage. I let the pronoun *she* stand for the noun *person* when the latter refers to an indeterminate human being. I see this as a tame—and etymologically justifiable—gesture against linguistic *machismo.* Of course, in my own language a person—*una persona*—is always referred to by the feminine third person pronoun *ella* even if she happens to be a male.

ACKNOWLEDGMENTS

In writing this book I have been much assisted and encouraged by others, to all of whom I express here my warmest thanks.

I conceived the book and began working on it during a six-month visit to the Center for Philosophy of Science at the University of Pittsburgh in 1983-84. I am grateful to Larry Laudan and Nicholas Rescher, who, as directors of the Center, invited me to go there, and to Adolf and Thelma Grünbaum, who did much to make my stay intellectually fruitful and in every sense enjoyable. While in Pittsburgh I benefited from long conversations with David Malament, John Norton, and Richard Healey.

An earlier version of Chapter 1 was published in the *British Journal for the Philosophy of Science*, vol. 37, pp. 1–23 (1986). The editor, Professor G. M. K. Hunt, has kindly authorized me to print a modified version here. The first version of Chapter 3 appeared in *Diálogos*, Nº 48, pp. 183–212 (1986), and Nº 49, pp. 147–88 (1987), and I have permission to republish the portions that have not been rewritten.

John Stachel read the first draft of Chapters 1 and 2 and made some incisive comments which I hope I have duly taken into account. C. Ulises Moulines commented on the previously published text of Chapter 3 in two long letters which were most helpful to me when I rewrote Sections 3.3 and 3.4. Ronald Giere kindly clarified to me a point in one of his writings. Georg Fromm detected a mistake in Chapter 5. Jorge López instructed me about a basic fact of measure theory which I had not properly grasped. Two referees for the publisher pointed out quite a few errors which I have since corrected. They also made broad proposals for rewriting the book (in different ways) which, however, I did not venture to follow. Pamela Bruton edited the manuscript with unfaltering precision, greatly improving my use of relative pronouns and introducing some very welcome stylistic changes.

I typed the book myself, so I have no one to thank for that, but I am glad to record here my grateful admiration for the creators of the personal computer and the software that goes with it. I am especially grateful to Joe Carroll, for allowing me to use his equipment and teaching me how to print with it, and to Marc Cogan, for his prompt and friendly replies to my questions regarding the Greek font Kadmos.

I thank the staff of the University of Chicago Press for their pleasant and efficient handling of the book's publication.

The University of Puerto Rico granted me leave to go to Pittsburgh and has since repeatedly freed me from some of my teaching duties so that I could work on the book. I am, of course, deeply grateful for such continued support of my research; but above all I have to thank the University and the people of Puerto Rico for twenty years of complete academic freedom and almost complete tranquillity at a time when these prerequisites of scholarly work could no longer be taken for granted in my homeland.

My greatest debt is to Carla Cordua, who has long been to me a tireless source of light and strength. I am very happy to be able to dedicate this book to her.

1 Observation

"Begin at the beginning!"—said the King of Hearts to the White Rabbit.[1] Many philosophers—call them *foundationists*—have been at pains to do so. They feared with good reason that if they let their thoughts take root right where they stood, in medias res, their quest for certainty would run afoul of a host of uncertified presuppositions. Reaching for the "unhypothesized First"[2] foundationists conjured up such fancy creatures as the bodyless ego and the unconnected simple sense datum and burdened their successors with thankless tasks, like proving the existence of the "external" world or "justifying" the way we ordinarily try to figure out the future from the past. I shall not go along with them. The understanding currently alive in language and other standard practices is the unfinished, unpredictably evolving outgrowth of partly random events, but we cannot do without it. We can work on it from within to broaden it and refine it, to fill in its gaps and to tie its loose ends, but there is no way that we can step outside it and build a better understanding from scratch. I must therefore rely on my own and the reader's command of some words and phrases and assume some truisms into which, so to speak, they fall of themselves. Of course, ordinary language—especially where it is not being used for survival—is not free of fog, and the line between incontestable truisms and disposable dogmas is often elusive. But the leeway that philosophy gains from such uncertainties should be husbanded with care. In particular, one ought not to cauterize the fuzzy edges of traditional meanings for the sake of clarity and precision, lest one wind up in yet another prim philosophical house of cards.

Science grows by observation. This commonplace I take for granted. For the foundationist who accepts it, its implications are plain enough: Observations must convey independent bits of knowledge, or else they would not deliver increments to science; hence, the systematic generalizations of science must be distilled from the aggregate of such particulars by some sort of logical alchemy; to establish its rules and to vindicate them is the main business of the philosophy of science. There is, however, a different and, to my mind, more promising philosophical approach to scientific observation: Just take it as it is currently practiced and understood, and try to tell from it what a science feeding on it can be. This approach can be traced back to Norwood Russell Hanson (1958). In the last few years it has been developed and applied in several excellent philosophical and historical studies.[3] In a similar spirit we shall now briefly review the main features of observation, to brace ourselves for the inquiries of the following chapters.

In Section 1.1 I bring to the fore two facets of ordinary observation, viz., physical interaction and awareness, and note that the former, but not the latter, is involved in every scientific observation. I introduce the expressions *personal* and *impersonal observation* to name observation with and without awareness and stress the role of general concepts in both forms of observation. Our conceptual grasp of physical objects will be the main subject of this book. With regard to it, foundationists traditionally raise two questions which I shall deliberately neglect: Where do the concepts come from? How can one be certain that a given concept is appropriate for a particular use? I comment briefly on these questions in Section 1.2. In Section 1.3 I consider another traditional question that we ought to come to terms with before proceeding with our inquiry: When personally observing a physical object, is one immediately aware of that object, or must one infer its presence and its features from directly perceived mental objects? Although this question cannot be satisfactorily dealt with in the space I can devote to it here, our short discussion should help to set, so to speak, the philosophical tone of the book and to get the reader attuned to it. The next three sections broadly examine our current understanding of observation as a physical process. Having made some very general considerations about it in Section 1.4, I compare personal with impersonal observation from this viewpoint in Section 1.5. Finally, in Section 1.6 I draw the conclusions I have been reaching for in this chapter: If every observation involves physical interaction between the observed object and a living or inanimate receiver, and every difference observed in the former must correspond to a difference recorded in the latter, then, obviously, a state of the receiver can supply information about the presence of a certain feature in the object only if—or insofar as—this feature is judged to be a necessary condition of that state. Consequently, the information we obtain by observation depends on our understanding of necessary connections in nature.

1.1 Two main forms of observation

Consider a few examples. A plainclotheswoman observes a slim clean-shaven young man come out of a house suspected of being a terrorist den. A nurse observes a mercury column rise and then slowly descend while she attends to the heartbeats she hears through a stethoscope, as she takes a patient's blood pressure. A tourist observes a small green caterpillar crawl slowly yet confidently on the piece of Danish pastry she has ordered for breakfast. In all these cases, a person, the *observer*, pays attention to something, the observed *object*—typically, a physical process or state of affairs, involving one or more things and their changing properties and relations—of which she is distinctly aware. Moreover, in all these cases the object physically interacts with the observer in a manner characteristic of the type of observation performed.

Some characteristic form of physical interaction is a necessary condition of the observation of physical objects, as it is currently understood. If the nurse in my example saw the mercury column fall with her eyes shut, one would say that she hallucinates it, not that she observes it.[4] Whether any physical interaction is required for the observation of one's own mind is a moot question which cannot be tackled here. However, to avoid cumbersome exceptions to what I shall be saying, I propose to use the word 'observation' in such a way that it does not cover the introspection of mental states and processes. (If the reader feels unhappy about this admittedly conventional restriction of the term's extension, he may add a constant prefix of his own choosing to all further occurrences of 'observation', 'observe', 'observable', etc., in this book.)

On the other hand, I should not say that every observation is made by a human observer who thereby becomes personally aware of the observed object. Indeed, most of the observations on which science thrives are carried out by artifacts designed to interact in a specific way with objects of some sort and to record the effects of the interaction so that human beings can eventually make use of them. The record may consist in a passing event (e.g., a click) or a lasting state (e.g., a pattern of black and white spots, as in an X-ray picture), of which a person can become aware by observing it in turn. But the record can also be inputted into a computer which combines many observation records, according to some preset program, in a computer output that a person can read. Now, I do not think that reading—unless it is *proof*reading—can be properly said to be a form of observation. You might observe a computer printout to see whether it is neatly printed or whether you ought to change the ribbon or make some other adjustment, but not to learn what the printout says. Consider, for instance, a scientist who peruses the output of a computer programmed to make weather forecasts from the data supplied by a widely distributed array of diverse meteorological instruments. Could one say that he is indirectly observing today's weather conditions? No more than one could say that he indirectly observes the weather conditions of July 11, 1953, when he rereads the record of the calculations he made himself with data gathered by personally reading the appropriate meteorological instruments on that day. And, of course, observation records can be used as feedback in an automated industrial or laboratory process without anybody taking notice of them. Evidently, many observations are now being performed in science and also in medicine, manufacturing, etc., that do not, and never will, involve observational awareness of their outcome, and hence should be described as observations without an observer.[5]

Let us therefore distinguish between *personal* and *impersonal* observation. In both forms of observation a result of a physical process is recorded in a distinct physical system I shall call the *receiver*. In personal observation, the receiver is the body—or should we say the nervous system?—of a human being who, upon reception of the record, becomes aware of the object of observation. In impersonal observation the receiver can be any of a wide variety of things—even a live human body on which, for instance, one tests a new drug or the tyrant's dinner. In all observations, the record of the physical interaction may last for years—as is the case with photographs—or just for a fraction of a second—as the neuronal excitation caused by a flash of lightning we see with our eyes—but it is never instantaneous. Even if, like everything else in nature, the physical state constitutive of an observation record is continually changing and is thus never quite the same, it is only by virtue of its more or less lasting sameness—as far as it goes, and however it may have to be understood—that it is the record of just that observation. Instantaneous states and events can no doubt be detected and measured, but only through more or less stable receiver records.

The awareness involved in personal observation is a peculiar mode of consciousness that differs, say, from fear or recollection, but also from other forms of perception, such as watching a movie or basking in the sun on a beach. Like other deliberate, attentive modes of consciousness, observational awareness is *self*-conscious: the observer must be aware of observing, or else he cannot properly be said to observe. He may, indeed, be so absorbed by his task that he becomes oblivious of himself; but even then he will be aware of the observed object as something that is being *observed*—not, say, imagined or merely thought about. Observational awareness is always, of course, awareness *of* something, the object of the observation, a complex of changing and unchanging features singled out from life's flux. Characteristically, the observer grasps the object as a particular instance of a universal. I do not mean to say that it is impossible to pay attention to an individual as such, and not as a member of a class. When talking to a close friend, or laughing together, or holding her in my arms, I am often aware only of the unique individual who is with me, and not of any universals that she instantiates. Awareness of individuality in its irreplaceable uniqueness is, I should say, a necessary ingredient of any genuine personal relationship. But such awareness is not observational, and as soon as one begins to observe one's partner, her teeth, her accent, her syntax, one does in effect subsume her under a general concept. Thus alone can the observer establish what is being observed. As the observation proceeds, the concept is further specified and articulated and may also be revised or replaced. For, far from being incorrigible as it has sometimes been suggested, a personal observation undergoes continual revision while it lasts.

In impersonal observation general concepts function in a rather different, but no less decisive, manner. Conceiving the object is not here an integral part of the act of observation. But in order to constitute an *observation*, a physical process must be conceived as such by the people who set it up or who intend to use its results. Recorded receiver states cannot disclose anything about a purported object of observation unless this object and the receiver are appropriately distinguished and the interaction between them is somehow understood. We shall have more to say on this later. For the time being, I conclude that, since concepts go into every observation, empirical knowledge is intellectual through and through. Kant said it bluntly: sense awareness without concepts is blind.

1.2 Conceptual grasp of the objects of observation

In the remaining chapters of this book we shall in one way or another be dealing with the conceptual grasp of physical objects, especially as it is practiced in mathematical physics. Here I only wish to mention—and to dismiss—two problems regarding that grasp which have greatly exercised philosophers of the foundationist persuasion.

One is the problem of noogony,[6] or the origin of concepts. They cannot *all* be obtained *from* observation by the standard procedures of comparison, reflection, and abstraction. In each observation some concepts must be at work from the very outset. Each time a judgment is revised, some concepts must remain stable. Thus, every exercise of our understanding involves concepts which are, one may say, locally a priori—i.e., presupposed and taken for granted for the occasion. Clearly, then, the doctrine that all concepts proceed from observation would entangle us in an infinite regress. This does not imply that any concepts are permanent, or global in scope. But some of them, at least, must be born ex nihilo. Whether we lay them at the door of Mother Nature or of the Muse of inquiry or of God Himself is presumably a matter of taste. Whatever one's choice may be, it appears that in dealing with noogony one cannot avoid mythology.

The second problem concerns the appropriateness of any given concept for a particular use. Every conceptual grasp of an object of observation is liable to revision and correction in the light of other observations; indeed, as I noted above, *self*-correction is the very soul of personal observation. What justifies our preference for one concept over another? How can we judge that we have achieved a better conceptual grasp? Foundationists miraculously solve—or rather dissolve—this problem by appeal to the dogma of Immaculate Perception. According to it, we observe virgin data, unpolluted by our deceptive intelligence, and can adjust our concepts and judgments to them. Now, if one could ever make a perfectly self-contained observation, not signifying anything beyond itself, there would certainly be no means—and no motive—for revising it. One would just blow away the conceptual chaff and leave the observational grain alone in its splendid isolation—and irrelevance. But in fact no observation is thus self-contained.[7] Each one of them is constitutively linked by concepts to other observations welded into a complex network of assumptions and beliefs, together with which it gives rise to a wealth of expectations.

Failure of expectations is perhaps, in the end, the main inducement for revising and correcting our observations. But earlier observations can assist in the correction—and even lead to the rejection—of subsequent ones, if they are more detailed or more careful or more consonant with one another. Consonance and detail furnish unquestionable, more or less unambiguous criteria of preference. But to say that an observation is more careful than another one would seem to presuppose the very choice that we seek to vindicate. However, some observational procedures may well be deemed more careful than others if they normally lead to more successful expectations. Moreover, our well-corroborated understanding of the physical processes of observation provides definite and, for that same understanding, impeccable grounds for assessing the reliability of observations. How one goes about using such diverse criteria in the progress of experience is well known from our daily lives. More sophisticated examples are provided by the history and current practice of scientific research. Beyond such bland generalities, philosophy has very little to say about the scientific procedures for collating observation data and the criteria for judging their worth. They pertain to the methodology of each field of inquiry and are decided upon by the practicing experts in the light of their current understanding of the matter at hand.

One philosophical generality should never be forgotten: if all our knowledge of physical objects is corrigible, it must be self-correcting, for there is no outside authority to which one could turn for help. Quine's famous dictum that "our statements about the external world face the tribunal of sense experience not individually, but only as a corporate body" (Quine 1961, p. 41) is apt to be misleading. For in the trial of empirical knowledge the defendants are at once the prosecution, the witnesses, and the jury, who must find the guilty among themselves with no more evidence than they can all jointly put together. This truism can be stated less aggressively, and perhaps more sensibly, as follows. In philosophy, things are said to be as we understand them to be, but we are well aware that they might not be that way. Such awareness, however, does not result from our transcending our understanding and glimpsing at things beyond it. It simply expresses our discontent with our own views and thoughts, which we feel to be incomplete, murky, or plainly inconsistent. But improvement can only be had by thinking harder, and we alone must see to that. (On the other hand, should one ever achieve perfect intellectual self-satisfaction, one would find no occasion for distinguishing between truth and appearance, between the way things are and the way one thinks and says they are—except, indeed, to contrast one's present knowledge with one's former ignorance and with the beliefs of others.)

1.3 On the manifest qualities of things

In personal observation, are we observationally aware of the object of our attention or merely of its effect on our minds? This is a question I think we ought to face before going any further. The answer will not change the way physics is done and may not be required for its proper elucidation, but a discussion of the question itself, its motivations and presuppositions, and a brief but clear statement of my own stance on the matter might prevent misunderstandings of what I shall be saying later.

To someone untainted by philosophy the question sounds silly, a typical example of the idle sport practiced by tenured professors on their captive audience. Isn't the juicy pineapple chunk I am now chewing the very same thing that feels cold and tastes sweet on my tongue, indeed the same one that looked yellow and smelled of pineapple a moment ago, when I carried it with a fork to my mouth, in front of my eyes and my nose? Normally, I should not have the slightest doubt about this, if 'is' and 'same' and the other ordinary English words employed are being used properly. It could happen, of course, that I fell into a swoon while still holding the pineapple in my hand and that another piece was put into my mouth while I lay unconscious. But such a situation can only be contemplated and diagnosed by contrasting it with the more common one, in which I eat what I pick and I do not pick anything without first seeing it. Of course, the food must act on my senses, or I would not be aware of it. But it is *the food* that I am aware of, as it stands alluringly on the plate, and later, as it yields to my teeth; not the changes caused by it in the cones and rods in my eyes and the papillae on my tongue, or in this or that lobe of my brain, or—as some are fain to say—in my soul.

Yet a venerable philosophical tradition maintains that, among the apparent properties of a body, only size and shape, position and speed can really belong to it and that the colors, sounds, tastes, odors, etc., which it sports are only the effects that its action on our bodies causes on our minds. The oldest known statement of this thesis was madc by Democritus of Abdera (ca. 375 B.C.):

> νόμῳ γλυκὺ, νόμῳ πικρόν, νόμῳ θερμόν, νόμῳ ψυχρόν, νόμῳ χροιή, ἐτεῇ δὲ ἄτομα καὶ κενόν.
>
> By custom, sweet; by custom, bitter. By custom, hot; by custom, cold. By custom, color. In truth: atoms and void.
>
> (Democritus, in Diels-Kranz, 68.B.9)[8]

Democritus' motivation is well known. He had learned from Parmenides that being cannot come from not-being or not-being from being, but he would not stomach Parmenides' denial of variety and change. So he made allowance for not-being (μηδέν) in the guise of the void (κενόν), to make room for his atoms—tiny unborn unchangeable indestructible Parmenidean beings—to differ from each other and to move. But the face which this reality presents to us he still regarded as a man-dependent appearance, to be accounted for by the changing configuration of the atoms that make up our souls.

Democritus' curt dismissal of the manifest qualities of bodies found little following among latter-day atomists in antiquity,[9] but was revived and collectively embraced by the founders of modern science. P. M. S. Hacker (1987, Chapter 1) has gathered a series of passages from Galileo, Descartes, Boyle, Newton, and Locke. Allow me to quote a few of them for the benefit of readers who do not have Hacker's book at hand.[10] The following is from *The Assayer* (1623):

> Che ne' corpi esterni, per eccitare in noi i sapori, gli odori e i suoni, si richiegga altro che grandezze, figure, moltitudini e movimenti tardi o veloci, io non lo credo; e stimo che, tolti via gli orecchi le lingue e i nasi, restino bene le figure i numeri e i moti, ma non già gli odori né i sapori né i suoni, li quali fuor dell'animal vivente non

> credo che sieno altro che nomi, come a punto altro che nome non è il solletico e la titillazione, rimosse l'ascelle e la pelle intorno al naso.
>
> I do not believe that anything is required in external bodies besides their size, shape, multitude, and motions, fast or slow, in order to excite in us tastes, odors, and sounds; and I think that if ears, tongues, and noses are removed, the shapes and numbers and motions will remain, but not the tastes nor the odors nor the sounds. Apart from the living animal, the latter—I believe—are nothing but names, just as tickling and titillation are mere names if the armpit and the skin lining the nose are removed.
>
> (Galileo, *Il Saggiatore*, §48 [EN, VI, 350])[11]

Robert Boyle made the same claim in *The Origin of Forms and Qualities according to the Corpuscular Philosophy* (1666):

> We have been from our infancy apt to imagine, that these sensible qualities are real beings in the objects they denominate, and have the faculty or power to work such and such things [. . .] whereas [. . .] there is in the body to which the sensible qualities are attributed, nothing of real and physical, but the size, shape and motion or rest of its component particles, together with that texture of the whole, which results from their being so contrived as they are; nor is it necessary they should have in them anything more, like to the ideas they occasion in us.
>
> (Boyle, WW, II, 466)

He admits, however, that the ordinary names of colors, tastes, etc., may be used "metonymically" (Boyle, WW, II, 7) to designate those textural features by virtue of which a body can effect the homonymous sensible qualities in our minds.

> I do not deny but that bodies may be said, in a very favorable sense, to have those qualities we call sensible, though there were no animals in the world; for a body in that case may differ from those, which now are quite devoid of quality, in its having such a disposition of its constituent corpuscles that in case it were duly applied to the sensory of an animal, it would produce such a sensible quality, which a body of another texture would not: as though, if there were no animals, there would be no such thing as pain, yet a pin may, upon the account of its figure, be fitted to cause pain in case it were moved against a man's finger.
>
> (Boyle, WW, II, 467)

In this, Boyle was generally followed by later writers, notably Locke, who liberally granted us the right to go on saying that grass is green and ice is cold because of the standing disposition of such things to make us see the color green or feel a sensation of cold. Boyle and his followers did not explain why pins are never said to be painful, let alone pained, in that metonymical sense in which Henri Rousseau's paintings are colorful and colored. Presumably they counted this among the vagaries of standard English.

The philosopher-scientists of the 17th century had no truck with Parmenides, and their position concerning manifest qualities was chosen on epistemic grounds. They had this vision that the Book of Nature is written in the language of mathematics (Galileo, *Il Saggiatore*, §6 [EN, VI, 232]), but mathematics as they knew it could deal only with the size, shape, position and change of position of bodies. Modern differential geometry has taught us to conceive of mass and force and temperature and heat flow as "geometric objects" on a differentiable manifold, but such notions were still a long way off at that time. So if the new mathematical science of nature was to tell us how things really are, things had first to be stripped of their manifest qualities. Descartes expressed it with his usual lucidity: Material things can be clearly and distinctly conceived, and thus their nature known with certainty, only "insofar as they are the subject-matter of pure mathematics" ("quatenus sunt puræ Mathescos objectum").[12]

But, of course, if the visual redness of a red cube is just a mental effect of its presence in our line of vision, its visual shape or position will not be otherwise. One may concede, perhaps, that visual shapes and positions resemble—whereas perceived colors do not resemble—their homonymous counterparts in the body we see. But they cannot themselves belong to it, unless the colors do. For such visual shapes and positions are identical with, or are compounded from, the shapes and positions of visual displays of color.

As analogous considerations are extended to all the manifest qualities of things, bodies fade into inferred entities, whose real presence—endowed with such-and-such properties and relations—is postulated in order to account for the perceived features of the mental kaleidoscope of sense appearances, which are all that, properly speaking, we see and hear and taste and smell. In his recent defense of such a view, Frank Jackson assimilates the philosopher's postulation of bodies as the source of mental "sense-data" to the physicist's postulation of molecules to account for the ostensible behavior of gases.[13] There is, however, a great difference between them. If you postulate the existence of small bodies as components of a larger one you observe, you are making a hypothesis about its parts, which must exist if it does. But if you maintain that all you are observationally aware of are mental objects, and postulate bodies as causes for them, you are introducing an entirely new category of being for which, by your own claim, you have no proper evidence.

I shall now describe an imaginary—though practically unimaginable—condition in which an observer might plausibly understand himself as being observationally aware only of his own mental states. The difference between that condition and our own will make clear, I hope, why we cannot persist in speaking, for any significant length of time, as if our senses made us aware only of so-called sense appearances. Consider a purely contemplative observer who sees static scenes, one after the other. He would have little or no inducement to analyze the scenes into parts or to associate parts of different scenes unless such parts were equal; and even if they happened to be so, he would have no reason for distinguishing the object of observation from its momentarily perceived aspects. Suppose now that the scenes observed change gradually and flow into each other, as in a motion picture. The observer could then perhaps discern patterns in the flow and come to view parts of successive scenes as diverse aspects of the same object. Such an object, however, would be no more than the series of its presentations, or rather the law of that series. A Humean analysis would unmask such laws, exposing them as mere habits. We can add sound and even smells to the motion picture without essentially changing the situation. We humans differ, of course, from a purely contemplative observer in that we have an interest, often a vital one, in the objects we perceive and are sometimes able to change them. But even if we let our fictitious observer resemble us in this, if we allow, say, some of the movie sequences, which are all he is aware of, to be pleasant or painful, and if we let him will and occasionally achieve the removal of pain, the renewal of pleasure, he still would not be one of us. He lacks the complex array of muscular, postural, thermal, tactile, or—to name them all by a single Greek word—*haptic* experiences in which we perceive ourselves as bodies incessantly interacting with other bodies, dangerously exposed to them, and also, through that very interaction, capable of manipulating them and observing them. The pencil I hold in my hand and press between my fingers, the chair I sit in, the table I write on, are grasped in observation as true bodies because through the pressure I exert on them, the movements I make against them, the thermal gradients they generate on my skin, I sense their bodily presence on a par with my own. Dr. Johnson refuted Berkeley by kicking a stone. The Greeks fought Pyrrho's scepticism by letting a dog loose on him. Professors smile with condescension on such wordless arguments, but there is a wisdom in them. Macbeth would clutch the dagger that he saw before him or else dismiss it as "a dagger of the mind."

Awareness of our interaction with the bodies surrounding us is the key to our construal of personal observation as a physical process, with our body as the receiver. It is convenient to recall how this construal introduces a measure of order and consistency into the diverse and often baffling appearance of things. If the observer becomes aware of the physical objects about him by their action on his body, his observational awareness must depend not only on the objects themselves but also on the condition of his body and all other circumstances influencing the observation process. Thus, our grasp of the physical basis of vision enables us to understand why a Gothic steeple should look different through the fog and under a blazing sun, why a pencil should show a kink when partially submerged in a glass of water, why the police van catching up with my car from behind should turn up in the mirror in front of me, why a supernova should now flare up in the sky in the direction where it faded out forever several million years ago. On similar grounds we can account also for our seeing visibles (and hearing audibles, etc.) that are not judged to be an aspect of anything, such as the red, semitransparent disks that we see wherever we direct our eyes after we have been looking intently for a while at a strong source of light, or the colorless little worms that we see wriggling about in the air if we stare at a bright cloudy sky. Since visual (acoustic, etc.) awareness closely depends on the state of the body, it is to be expected that it will often be stirred by changes in that state which are not a part of any process of observation (just as, say, a short-circuited loudspeaker will emit a noise which is not a part of any music being played). It is fortunate, indeed, that such occurrences, though frequent, rarely become obtrusive. But it is a perversion of philosophy to choose such marginal events as the prototype of all our sense experience and then to wonder how it may come to pass that by far the greater part of it is so neatly ordered as a display of physical objects. In fact, outside this order in which we normally perceive things in their manifold aspects, it is hard to conceive that there could even exist an awareness of objectless *sensibilia*.

Philosophers sometimes run into difficulties with the manifest qualities of things because, obfuscated by half-unconscious *theologoumena*, they unwittingly set standards of determinacy for things which the latter in effect do not meet. They persist in understanding things as *res*—in the scholastic medieval sense—or things-in-themselves, when all one ever meets and has to do with are *pragmata*, or things-in-our-environment. What one ordinarily means by a *thing* is, of course, colored or transparent, quiet or noisy, tasty or insipid, pretty or ugly, and though modern physics has thrived by methodically neglecting such features, its mathematical constructs are designed to represent and to assist us in understanding and handling the very things that sport them.[14]

1.4 Our understanding of the process of observation

In personal observation the observer apprehends his own body in physical interaction with the objects observed. Observational awareness never lacks this feature, at least where haptic perceptions are at play. Throughout our lives this is practically always the case, so it is no wonder that, in ordinary usage, the statement that a person *x* observes a thing or event *y* implies the statement that *y* causes *x* to be in a state in which she succeeds in observing it. Indeed this usage extends to all modes of observation, visual, auditive, etc., even where no reference is made to haptic awareness. I do not take this linguistic practice to mean that, say, purely visual observations—if such exist—must involve a claim to being caused by their objects (except perhaps when they are on the verge of being painful due to excess of light, in which case vision becomes proprioceptive like touch and kinesthesia).[15] But visual observations are made by us, men and women of flesh and blood, who must sit or stand or walk or run or turn or stoop or stretch or, at the very least, strain our eyes to see. Haptic awareness is thus pervasive and discloses, in one way or another, that we are committed to the physical world. Our everyday handling—holding, pressing, pulling, pushing, twisting—of all sorts of bodies and our continual exposure to bumping and falling, heat and cold, wind and water, light and noise, furnish the prototypes of our original notions of physical existence and physical action. It is therefore most unlikely that we shall ever find occasion of rejecting our grasp of ourselves as bodies interacting with other bodies. Philosophical attempts at replacing this ingredient of our self-understanding have hitherto been little else than exercises in the abuse of language. As John Dewey wrote at the beginning of his *Logic*: "It is obvious without argument that when men inquire they employ their eyes and ears, their hands and their brain."[16]

Yet while men have never seriously hesitated in their grasp of observation as a physical process, their general understanding of such processes has undergone great changes. For example, Aristotle conceived of a manner of physical action that was designed to account for perception and observation. By virtue of it, the constitutive "form" of the observed object could be transmitted "without matter," through an appropriate intervening medium, to the "sensitive faculty" of the observer.[17] This doctrine was taught at school to the founders of modern science, who later rejected it and replaced it with a different conception of physical action which in part revived pre-Aristotelian notions. Towards the end of the 17th century the new conception had taken such hold of the best minds in Europe that, for example, John Locke "found it impossible to conceive that body should operate on *what it does not touch* [. . .], or when it does touch, operate any other way than by motion."[18] Whence, when he comes to consider "how bodies produce ideas in us," he declares that it "is manifestly by impulse, the only way which we can conceive bodies to operate in."[19] This early modern idea of physical action was considerably modified by successive generations of natural philosophers, first by the 18th century theorists of instantaneous action at a distance, then in the 19th century by the creators of field theory. One capital ingredient of it survives, however, to this day: for us, as for Descartes, Huygens, etc., all physical action boils down to a transfer of momentum—or, as we would rather put it now, of four-momentum.[20]

The modern philosophy of nature has presided over great advances in the physiology of perception. It has also been associated, from its inception, with the modern development of means and methods of impersonal observation, which not only have tremendously expanded the scope of our knowledge but should also help us, through our growing familiarity with them, to achieve a better grasp of the nature of personal observation. On the other hand, the modern idea of physical action has burdened us—also from its inception—with the so-called mind-body problem. For, as the 17th century occasionalists were quick to see, transfer of momentum will neither account for nor be explained by a change of mind. All the attention devoted to the problem since Descartes has not brought us any nearer to understanding how a man's decision can initiate a definite outward flow of energy and momentum across his skin, or how an inward energy-momentum flow across it can modify his state of awareness. And we still do not know how to coordinate our particular states of awareness of observed objects with any well-defined, particular effects of the action of such objects on our bodies. It is unlikely that this rift between the two sides of observation can be closed without some radical, incalculable innovations in our understanding of physical action. But since our current understanding lies at the heart of so much valuable knowledge, there is little inducement to change it.

Even if our present understanding of the observation process is thus limited and beset with difficulties, we are deeply committed to it, and we cannot well imagine how some of its implications could be denied. Thus it seems clear that, no matter how we conceive physical action, in every observation the observed object interacts with a receiver. Such interaction is critical to the acquisition of knowledge by observation, for the observer cannot ascertain any more features of the observed object than become discernible to him through their recorded effect on the receiver. Indeed, a state of the receiver can furnish information about a feature of the object observed only to the extent—and within the range of ambiguity and imprecision—that the said feature is, under the circumstances, a necessary condition for the attainment of that state. The receiver's "power of resolution," its capacity to separate—or its tendency to blur—the imprint of different attributes and states of the object, is a measure of its cognitive value. From this point of view, impersonal observation, carried out by means of an increasingly diverse and efficient panoply of precision instruments, enjoys a distinct advantage over personal observation.

1.5. Personal versus impersonal observation

Observation processes have their peculiarities, without which they would not serve their purpose, but they are not generically different from other physical processes which are not observational. Observational interaction instantiates the same types and obeys the same laws as ordinary physical interaction. Indeed, the development of impersonal observation in the modern age could only get under way on the understanding that such was the case. Observation devices exploit known properties of well-typified natural processes for the sake of collecting information. Inference from the state of the receiver to the state of the observed object must rest on our knowledge of those properties, and can therefore hold good only if observation processes are not, physically speaking, a class apart.

Nonetheless, observation processes do differ from their nonobservational analogues in that they are ordered to an end: they are always embeddable in a quest for information. It is a requisite of this teleological order that, among the many factors that contribute to a physical process of observation, some should stand out as the objects of observation and their observed features, while others constitute the receiver and its data-recording states.

In impersonal observation, the receiver is usually artificial and is singled out by its human manufacturer. It is expressly designed to register the interesting effects of the intended object of observation, which has been previously singled out by some human research project. Since the object-receiver interaction is nevertheless immersed in nature's flux, great ingenuity must usually be devoted to filtering out the "noise" that hinders the clean flow of information from the object to the receiver. The status of these several items is indeed notional and depends on the project which the observation is meant to serve. Thus, by timing the eclipses of one of Jupiter's moons—intermittently hidden behind the planet—you can ascertain its period and thence, by Kepler's Third Law, its average distance from the planet, provided that you know the speed of light and use it to correct the anomaly of the observed period due to the Doppler effect consequent upon the relative motion of Jupiter and the Earth. If, on the other hand, you note the said anomaly but do not know the speed of light, you can, like Ole Rømer, use the timing of the eclipses to measure it—provided that you know the relative velocity of Jupiter and the Earth.[21]

The significance of the receiver's states is a matter of interpretation, depending, of course, on the circumstances of the observation—a thermometer reading will not tell us much about a child's fever if, on coming out of the child's mouth, the thermometer has fallen into a bowl of hot soup—but also, decisively, on the observer's understanding of the experimental situation. On the frontier of research, such understanding is apt to be flimsy. Thus, for example, the negative result of Michelson's famous attempt to measure the relative motion of the Earth and the ether was understood to indicate (*a*) that the ether is dragged by the Earth's atmosphere, the laboratory walls, the protective box in which Michelson's apparatus was enclosed, etc. (this was Michelson's own conclusion in 1881); (*b*) that the motion of the apparatus across the ether modifies the molecular forces that hold its parts together, shortening one of its perpendicular beams while merely narrowing the other (this was independently suggested by Fitzgerald and Lorentz); and (*c*) that we live in a Minkowski spacetime in which light pulses in vacuo follow null worldlines, so that the speed of light measured in an inertial laboratory in which time is defined by Einstein's method is the same in every direction. Or, to mention another, more recent example of an observation with positive results: the isotropic background noise that has been recorded in ultrasensitive microwave radio receivers around the world since 1965 and is generally regarded as the effect of thermal radiation of approximately 3 K, is understood as the manifestation of a very hot early global state of the universe. However, this cosmological reading of the phenomenon would have to be dismissed if it were determined that in another galaxy the noise is absent or is significantly anisotropic (or if it had turned out that outside the Earth's atmosphere its intensity does not peak at the frequency prescribed by Planck's law of thermal radiation).[22]

While the informative aim of an impersonal observation accrues to its underlying physical processes by human initiative, such a goal is, so to speak, endogenous to personal observation. Here the receiver has not been segregated from the mainstream of nature for fact-gathering purposes by an external agency but has grown of itself into a distinct, fairly stable physical system, suitably disposed to pick out specific effects of its interaction with specific objects. The information-bearing receiver states are not presented on a dial to the observer's interpretative acumen but translate spontaneously into observational awareness. The objects of personal observation do not have to be inferred from the states they induce in the receiver, for they are simply and straightforwardly perceived. In fact, it is rather from his direct awareness of them that the observer eventually learns—by inference—what receiver states are instrumental to their observation. Thus we have come to know that—though we are still quite incapable of explaining how—the recorded difference of less than 1/3,000 s between a sound's arrival in our left and in our right ear enables us to distinguish the direction from which the sound came; that our visual awareness of the volume of nearby bodies rests on the slight difference in the optical input from such bodies into each one of our eyes; that our sense of balance and orientation in the gravitational field in which we live depends on the flow of liquid along the sensitive walls of the semicircular canals in the internal ear. In inquiries leading to these and other results about the material conditions of perception, the physical objects of our perceptual awareness are the grounds, not the goals, of inference. Indeed the very notions of physical object, physical state, physical process—sophisticated though they have grown through the exertions of modern scientific thought—are rooted in the manner in which men and women, physically interacting with their surroundings, naïvely articulate their awareness of that interaction.

We normally have a more or less definite grasp of the objects of our personal observations and of their relations of place and time, and in some cases also of their causal relations with our bodies. This grasp is the source from which the theory and practice of impersonal observation ultimately draw their sustenance and motivation. Thus, personal observation may justly claim metaphysical priority over impersonal observation. But that does not bestow on it an epistemic privilege with respect to the latter. For personal observations and the "natural," unreflecting grasp of things that goes with them are both fallible and corrigible and are being continually rectified and qualified, not only by mutual comparison but also in the light of impersonal observations. Thus, we habitually compare the readings of outdoor thermometers or of wristwatches with the estimates of air temperature or of time elapsed based on our feelings—a practice which not only serves to control and to correct such estimates, but can also contribute to improving their accuracy. Personal observation is not only not superior to impersonal observation as a source of knowledge about physical objects, but, in both scope and precision, it is on the whole markedly inferior. The confusion that still prevails in some philosophical circles on this fairly obvious matter is due perhaps to a vicious craving for certainty. Of course, such craving will never be satisfied by impersonal observation, with its intricate scaffolding of theories. But neither can it be quenched by contracting one's knowledge claims to the bare subsistence level of commonsense judgments and naked eye observations.

Human perception must indeed always intervene at some stage of the harvest of impersonal observation data for use in science. Should not this obliterate the superior precision and reliability of those data? After all, a system for the transmission of information cannot perform better than its weakest link. However, the human sensors are not equally deficient at every task. They are rather bad for discriminating weights or temperatures or light intensities, and they are utterly useless for detecting small changes in atmospheric pressure; but they are pretty good for apprehending neatly printed digits and may be trusted to note a coincidence between a pointer and a thin black line on a white dial. Observation devices are designed to translate the often imperceptible effects of the observed object on the receiver into such easily perceivable receiver states. That persons should thus learn through their senses the outcome of impersonal observations has led some philosophers to think that a faithful description in plain everyday language of the apposite sense experiences can give the full "cognitive meaning" of the statements, couched in esoteric, "theoretical" terms, in which scientists normally report their findings. Of course, in real life things stand just the other way around: digital and pointer readings get their distinctive interpretation from the theory of the respective instruments, and without it they all look quite insignificant and very much the same.

1.6 On the relation between observed objects and receiver states

No difference can be observed in an object that is not recorded as a difference in the receiver. This principle is central to our current understanding of observation;[23] and it does not seem possible to deny it, no matter how we revise or refine that understanding. Indeed, the principle is so deeply ingrained in our language that we would never be said to *observe* a change we know to occur in the object, but which our bodies and the instruments at our disposal do not reflect.

It follows that in any personal observation receiver states must mediate between the observed features of the object and the observer's perception of them. We are far from understanding the relation between those states, of which we are mostly unaware, and our awareness of the objective situations they disclose. That there is no simple correspondence between the information-bearing states of our sense organs and any relevant states of the mind can be readily gathered from the examples of stereophonic and stereoscopic perception mentioned in Section 1.5. Only by sinking the cognitively significant receiver states deeper and deeper into the unexplored recesses of the brain can one hope to map them one-to-one onto the contents of our sense awareness. As neurology advances, such terræ incognitæ become increasingly unavailable, and one sees ever more clearly that a mind-brain isomorphism, if at all possible, can be established only on the basis of a thoroughly innovative, physically unorthodox description of the brain. On the other hand, the relation between the said receiver states and the matching features of the object can be handled by the standard methods of physics. In this respect there is no essential difference between personal and impersonal observation. And indeed practically all progress in the physiology of perception, since Kepler first conceived the eye on the analogy of the camera obscura, has been achieved by treating the organs of sense as impersonal receivers.

Object-receiver relations in personal and impersonal observations take varied forms and their study pertains to diverse fields of science. But they all share at least one common trait. *A receiver state conveys information about the presence of a certain feature in an object only if—or insofar as—this feature is judged to be a necessary condition of that state.*[24]

Consider impersonal observation. Although there may be no difficulty in classifying and recognizing observationally significant receiver states, a definite receiver state often does not unambiguously point to an equally definite feature in the object. That state may normally arise due to several conditions, some of which may not even involve the intended object of observation. (Precision measurements can be severely impaired by thermal variations in the instruments employed.) But even where such perturbing factors are negligible, the distinguishable states of the receiver may not suffice to discriminate between significantly different properties of the object. A gray shadow on a medical X-ray picture can reflect all sorts of conditions in the patient's body. To judge what is actually disclosed by it, an observer must rely on his experience of similar X-ray pictures and on his general knowledge of medicine. A coupled pair of spots in a telescope photograph of a piece of sky is usually taken as evidence that in the direction of those spots there are two, possibly associated, astronomical light sources; but the spots might exceptionally be caused by a single source, if the beam of light it sends towards us is split, on the way to our telescope, by a gravitational lens. To decide that the latter is indeed the case, a scientist must examine the circumstances in the light of gravitational theory.

There are, indeed, plenty of cases in which the record of an impersonal observation tells a person exactly what she wants to know about an object, although she has no inkling of how the observation works and of what precisely is recorded by the receiver. Thus, if, blindly following the instructions in a manual, I connect the terminals of a voltmeter to the knobs on the upper face of my car's battery, the position reached by the needle on the voltmeter's dial will let me know without further ado whether the battery is strong enough to start the car promptly on a cold morning. I require no theory, almost no experience, and very little judgment to draw the appropriate inference from the actual reading. Most of us ordinarily employ instruments of observation to learn about our surroundings in such a thoughtless way. But we can do so only because a vast repertoire of object-receiver correlations has been firmly established by scientific and technological research. Such research is all but thoughtless. It does not simply proceed by trying out any old instrument on a class of objects and setting up by straight-rule induction[25] a correspondence between the alternative states of the former and the interesting differences among the latter. The impersonal receivers in current use in all walks of life have for the most part been painstakingly developed in the light of scientific theories which entail certain necessary connections between diverse features of interest in our environment and directly observable receiver states. In Chapter 5 we shall consider what type of necessity scientific theorizing discovers—or should we say induces?—in nature. But it should already be clear that impersonal observation is impossible without it. A particular receiver state can disclose a particular state of affairs only if the latter is, under the circumstances, a necessary condition of the former. To judge it so, one must grasp them both as instances of general types which stand to one another in suitable relations of entailment. Such typifications are not ready-made but are the product of scientific thought. We may indeed unreflectingly profit from the impersonal observations with well-established significance which are taking place all about us. But we could not without reflection and theory-guided invention have brought them under way.[26]

So-called accidental discoveries might seem to be exceptions to this rule but ultimately tend to confirm it. Thus, for example, the first observation of radioactivity was recorded in a photographic plate stored with a preparation of uranium salts from 27 February to 1 March 1896 inside a drawer in Henri Becquerel's laboratory. The plate was exposed notwithstanding the absence of light in the drawer; but it took Becquerel's alertness and preparedness—he himself had mounted the uranium salts on the plate to study their phosphorescence under sunlight and had stored them in the drawer while waiting for propitious weather—to grasp as an observation record what another one would have discarded as a spoiled plate.

Physically, personal observation is no different from impersonal observation. A person cannot become aware, by observation, of a change in an object unless the latter effects a change in her body. A state of a human body cannot convey information about a feature of its surroundings unless this feature is, in the circumstances, a necessary condition of that state.[27] However, not every state of the body is a source of observational awareness; nor do those that are disclose every one of their necessary conditions. Observational awareness is selective: the observer's attention, guided by his interests and preconceptions, falls at any given time only on a small part of the current range of his consciousness. Observational awareness is self-transcending: it is no mere epiphany of organic states but the grasp of an object against the background of a world. Hence, while in impersonal observation the facts of the matter must be inferred from a suitable description of the receiver states in the light of scientific theories and a general assessment of the circumstances (or by means of the "inference tickets" supplied by the user's manual that comes with the instrument), in personal observation the actual presence of such-and-such an object is not a conclusion to be drawn deductively or inductively from the momentary state of one's body, for we are, so to speak, preprogrammed to jump to it straight away. (See Fodor 1984.) The observer's grasp of the object can be rectified to comply with earlier or further experiences, with scientific theories, or even with philosophical criticism. But it cannot be suppressed from observational awareness without destroying the latter's observational character. Thanks to this grasp of the environment in which his body is placed, the human observer develops an understanding of observation as a physical process and devises increasingly sophisticated theories about object-receiver links. Such theories are not required to get personal observation going—indeed, they would not even be possible if observational awareness did not precede them—but they are certainly apt to modify our grasp of what we observe personally.[28]

2. Concepts

In this chapter we begin our exploration of creative understanding in physics. In Section 2.1 I take a new look at the familiar view of scientific explanation as inference. I contend that such explanations require a re-thinking of the facts, in order to bring them under the scientific theories that explain them. In Section 2.2 I illustrate this with some examples from Newton. Section 2.3 raises a question we must face if the facts of observation are grasped and regrasped under changing concepts: "Can the facts remain the same as the framework of description varies?" A negative answer to this question would warrant the so-called incommensurability of scientific theories, proclaimed in the 1960s by T. S. Kuhn. Section 2.4 criticizes two classical ways of forestalling such incommensurability, favored, respectively, by Kant and by Carnap. Section 2.5 studies the internal connection between two incompatible theories, one of which arises through criticism of the other. Section 2.6 discusses the theory of meaning introduced (and subsequently abandoned) by Hilary Putnam to rescue the stability of reference under radical conceptual change. Section 2.7 elucidates the notion of a conceptual scheme, implicit in the problem of incommensurability, and proposes a new approach to it which should go a long way to solving that problem. The mathematical appendix in Section 2.8 sketches the notion of structure that will pervade Chapter 3 but that is already employed in the present chapter for "speaking of quantities" in Section 2.6.4.

2.1 Explaining and conceiving

To explain the facts of observation, their occurrence and their recurrence, has been said to be the "distinctive" and "one of the foremost" and even the sole aim of empirical science (Nagel 1961, p. 15; Hempel 1965, p. 245; Popper 1972, p. 191).[1] According to a philosophical tradition that issues from John Stuart Mill's *System of Logic* but can be traced to earlier sources, a scientific explanation takes the form of an inference whose conclusion describes the fact or facts to be explained (the explanandum), while its premises (the explanans, i.e., 'that which explains') consist of the statement of a law of nature and the description of some other facts. This idea of explanation as inference, carefully articulated within logical empiricism (Hempel and Oppenheim 1948; Braithwaite 1953; Hempel 1965), was relentlessly criticized from different standpoints in the sixties, when that philosophical movement, which like Bauhaus architecture and Comintern politics had posed as definitive, turned out to be even more ephemeral than such worldly fashions (Scriven 1958, 1962; Toulmin 1961; Feyerabend 1962; Bromberger 1966; Harré 1970, etc.). I do not intend to repeat here those criticisms or the replies they elicited but rather to concentrate on one feature of deductive explanation which, to my mind, contains the clue to its significance and yet has rarely been in the limelight of philosophical debate.

A deductive explanation, better known as a deductive-nomological or DN explanation (the term 'nomological' being built from νόμος, the Greek word for law), infers the statement of an observed fact *F* from the joint statement of a law or laws *L* and of factual conditions *C*.[2] For the explanation to work, each of the statements *L*, *C*, and *F* must meet certain requirements that need not concern us.[3] *F* can be inferred from *L* and *C* if and only if the conditional $(L \supset (C \supset F))$ is a logical truth. However, the conditional $(C \supset F)$ ought not to be one, or the law *L* would be superfluous. Therefore, the fact under consideration must be described by *F* in terms that also occur in *L*. Typically, the law *L* will link the terms descriptive of the fact to be explained *F* with those descriptive of the factual conditions *C*. We may express this by saying that in DN explanation the explanandum and the law in the explanans must be conceptually homogeneous. This principle is readily illustrated by the following example, long a favorite in philosophy classes. I can infer—and thereby supposedly explain—the fact that

(F^*) This thing here is black

from the general law

(L^*) All ravens are black

and the known condition

(C^*) This thing here is a raven

But the inference from L^* and C^* to the observed fact will not go through if I grasp this thing here as being warm or winged or noisy but do not grasp it as black—as may well be the case if it happens to be a raven that I touch or hear but do not see.[4]

Although the foregoing example meets the stated conditions for DN explanation and clearly illustrates the requirement of conceptual homogeneity, it will probably not be recognized outside philosophical circles as an instance of scientific explanation. Explanation, in the ordinary meaning of the word, should be enlightening; yet the uniform blackness of ravens throws no light at all on the fact that this raven here is black. If we turn to a more likely example—e.g., if we derive the current generated by a particular alternator when it rotates with a given frequency from classical electrodynamical laws and suitable factual conditions—we shall see that it differs from the black raven case in at least the following two respects:

(i) The laws adduced for explanatory inference in real science normally involve concepts alien to prescientific discourse. In order to achieve conceptual homogeneity the facts of observation which are to be explained must somehow be grasped under those same concepts.

(ii) Such strictly scientific concepts are always part of a coherent and explicit—or, at any rate, progressively self-explicating—system of thought that links the explanandum through the laws in the explanans to a variety of other facts, derivable from the same or related laws.

These noteworthy features of standard scientific explanatory inferences are not independent of one another. If natural philosophers and scientists had remained content with the stock of notions of prescientific common sense, instead of developing novel intellectual systems, they would never have been able to bring together such seemingly disparate phenomena as falling apples, orbiting satellites, and receding galaxies, and to have each of them illuminate the others and bestow relative necessity upon them. By bringing their innovative thought to bear on the facts of observation they have succeeded in producing explanatory inferences that truly increase our understanding. One will admit to having understood why some particular thing or event is as it is if one gets to see that it could not be otherwise. Such physical necessities are relative, not absolute, inasmuch as they depend on what the rest of things and events is like. By grasping different facts under concepts bound together in a system, we achieve just this kind of understanding: unless each explanandum within the scope of the system follows from the relevant laws and a suitable description of the prevailing circumstances, we must rethink all other facts within that scope. When a collection of facts is thus incorporated into an intellectual system, each one of them is, so to speak, held in place by the rest. While we grasp it as we do, we cannot conceive it to be otherwise than we think it is, unless we reconceive the other facts in the collection as well. Inferential explanations in genuine science thus differ sharply from the classroom example proposed above. If albino ravens are found in Alaska, we should not feel compelled to rethink our zoology. And the logical necessity with which the conclusion "This thing here is black" (F^*) follows from the premises L^* and C^* does not create even a mirage of physical necessity with regard to the fact F^* itself. In the light of the classroom explanation, this thing here could just as well be green *and* a raven, and all other ravens remain unchanged. On the other hand, should we ever establish with reasonable certainty that a particular planet does not obey the accepted law of gravitation, we would have to revise our thinking about gravitational phenomena throughout the universe.

2.2 Examples from Newton

In order to see better how a systematic rethinking of facts is at work in scientific explanations, we shall now consider a few applications of Newton's Law of Gravity.

Take the motion of the Moon around the Earth. On a first approximation we ignore the presence of the Sun and other heavenly bodies, and we treat the Earth as fixed. The explanation of lunar motion by Newton's Law of Gravity rests then on the assumption that the Moon is freely falling towards the Earth in accordance with that law. Although this is nowadays a trite commonsense idea, it was far from being one in the 17th century. Indeed it must have seemed paradoxical, inasmuch as the falling Moon never reaches the ground. And yet, unless we conceive of the Moon in some such way it would be madness to try to infer a statement of its several positions from a law of gravity.

Newton's conception of the Moon as a falling body derived some plausibility from Galileo's analysis of the motion of projectiles near the surface of the Earth. According to a view current in Galileo's day, a heavy body such as a cannonball will naturally move downwards to the center of the Earth if it is not stopped. However, by force it can be made to move unnaturally upwards or sideways in any direction. But no body will move both naturally and against its nature at the same time. Hence, a cannonball, after being shot, will first be driven by the exploding gunpowder in the direction in which the cannon points, and only when the force of the explosion is spent will it fall—vertically—on its target.[5] This entails that the range of a cannon is greatest when it points horizontally, a prediction not confirmed by experience. Galileo dismissed the assumption that different motions cannot coexist in the same body and chose to think of a cannonball as falling freely from the moment it left the cannon's muzzle. According to Galileo's ideas about free fall, this entails that a flying cannonball suffers at all times the same downward acceleration, regardless of the material of which it is made. Galileo showed by a clever calculation that, on the stated assumptions, a projectile issuing from a horizontal cannon on top of a parapet describes a parabola. If Galileo's principle of inertia is extended to nonhorizontal motion, the result holds also for projectiles shot in any direction.[6]

Galileo's calculation presupposes that the acceleration of gravity is constant in both magnitude and direction. But, of course, if it points to the center of the Earth it can keep steady only within a small region—namely, that within which the Earth's surface may be regarded as approximately flat. Newton assumes, moreover, that its magnitude is the same only at equal distances from the center of the Earth and varies as the inverse square of that distance. It can then be shown that the projectile's trajectory is a conic section, generally an ellipse or a hyperbola. If we think of the Moon as such a projectile and let $\mathbf{v}(t)$ stand for its velocity at a given time t, while $\mathbf{r}(t)$ denotes its position at that time, referred to the Earth's center $\mathbf{0}$, we can readily calculate the acceleration $d\mathbf{v}(t)/dt$ if we are given, say, the values ρ and g of the radius of the Earth and the acceleration of gravity at the poles. For then it follows from the said assumption of Newton's that

$$\frac{d\mathbf{v}(t)}{dt} = \mathsf{g}\frac{\rho^2}{\mathbf{r}^2(t)}\frac{\mathbf{r}(t)}{|\mathbf{r}(t)|} \qquad (1)$$

The value calculated from this equation agrees passably well with the acceleration needed to account for the observed motion of the Moon.[7]

Newton's Law of Gravity is far bolder and more speculative than the modest and fairly straightforward extension of Galileo's Law of Free Fall that I have sketched here. Yet even within the narrow bounds in which I have deliberately kept our example we can readily see how the familiar data of observation must be rethought before explanatory inference can do its job. Not only must the Moon be conceived on the analogy of a cannonball, but its motion must be described under novel concepts of time, space, velocity, and acceleration, whose systematic interconnections provide the means for comparing and coordinating the lunar data among themselves and with the phenomena of falling bodies. Without these concepts, the falling Moon is no more than a suggestive metaphor; but thanks to them the analogy of projectiles takes a precise and pregnant meaning: ballistic trajectories, that of the Moon included, are solutions, under diverse conditions, of the same set of differential equations. (Whence, by judiciously choosing and effecting still other conditions, we have been able to put all those tiny man-made moons into the sky.)

Had Newton been content with extending earthbound Galilean gravity to the Moon in the manner proposed above he would have provided a good example of cautious generalization from proven facts, but his success would have been short-lived. For the trajectory that can be obtained from eqn. (1), though strikingly accurate as a first approximation, still differs noticeably from the one observed.[8] But Newton thought of mutual gravitation as a universal law of matter. He expressed this law in terms of his original concept of impressed force. By definition, "an impressed force is an action exerted upon a body in order to change its state, either of rest or of uniform motion in a straight line" (Newton, *Principia*, Def. IV). According to Newton's Third Axiom or Law of Motion, to every such action there is always opposed an equal reaction, so that "the mutual actions of two bodies upon each other are always equal, and directed to contrary parts." Newton's own statement of the Second Law of Motion indicates that he meant by impressed force what we now call impulse (with the dimension of mass times velocity); but his use of the concept in actual proofs warrants our understanding of Newtonian force as a cause of acceleration and its familiar representation by a vector proportional to the acceleration caused by it (the factor of proportionality being equal to the mass or quantity of matter of the accelerated body).[9] Building on these ideas, Newton attributed the accelerated fall of bodies towards the center of the Earth and the continual deviation of planets from rectilinear motion to an attractive force exerted by every material particle on all the others. By a liberal application of his professedly inductivist methodology he concluded that this force is directly proportional to the mass of both the attracting and the attracted particle, and inversely proportional to their distance squared.[10]

This Law of Universal Gravitation furnished Newton and his successors with an extremely supple and efficient instrument for calculating planetary motions. Yet at first blush it might seem to raise an unsurpassable difficulty. It is a truism often forgotten in empiricist discussions of empirical science that the data collected by observation cannot be explained by inference from general laws unless they have been rendered comparable. This requirement is met in Newtonian physics by referring all data on matter and motion to a common space and time. Now, Newton's "absolute, true, and mathematical" time and space cannot be observed but must be constructed from the relative times embodied in mechanical clocks and the relative spaces sustained by material frames of reference. But if every speck of matter is continually being pulled in every direction by all the rest, any reference frame or clock one may chance to choose is likely to be accelerated in a wholly unpredictable way. How can one expect to gather in a single coherent system of comparable kinematic data the results of observations referred to frames and clocks whose true state of motion is unknown? This difficulty, however, is satisfactorily resolved by invoking two principles inherent in Newton's basic kinematic and dynamic assumptions, that is, in his concepts of space and time and in his Laws of Motion. By the justly celebrated Newtonian Principle of Relativity (Corollary V to the Laws of Motion),

> The motions of bodies included in a given space are the same among themselves, whether that space is at rest, or moves uniformly forwards in a straight line without any circular motion.

This means that an inertially moving frame and a good mechanical clock affixed to it are adequate substitutes for absolute space and time (provided that one assumes, with Newton, that "every moment of time is diffused indivisibly throughout all spaces"—Hall and Hall 1978, p. 104). Yet in a world held together by universal gravitation one is hard put to find a body in true inertial motion. By the no less significant Newtonian Principle of Equivalence (Corollary VI to the Laws of Motion),

> If bodies, moved in any manner among themselves, are urged in the direction of parallel lines by equal accelerative forces, they will all continue to move among themselves, after the same manner as if they had not been urged by those forces.

Thanks to this principle, the Newtonian astronomer need not be worried by the farfetched but not impossible thought that the entire firmament of the fixed stars may be gravitating towards a remote and invisible but enormous concentration of matter. Indeed he may comfortably ignore the circumambulation of the Sun in the Galaxy and of the Galaxy in the Local Group—not discovered until the 19th and the 20th century, respectively—and carry out his investigation of planetary motions as if the center of gravity of the solar system were at rest. Moreover, he may neglect, on a first approximation, as we did above, the conspicuous acceleration of the Earth towards the Sun and treat the Earth-Moon system as if it were isolated. For during the short intervals required for ascertaining the velocity and acceleration of the Moon in the relative space of the Earth, the force exerted on the system from the Sun does not vary appreciably in magnitude or direction. (See Stein 1977, pp. 19ff.)

We thus see how radically committed to the Newtonian mode of thought is even the humblest explanation of an observed fact by Newton's Law of Gravity. The reason for this is plain enough. To be covered by the Law the fact has to be embedded in the structure of Newtonian kinematics: the times of observation must be instants—or very short intervals—of universal Newtonian time, the observed positions must be located in an admissible Newtonian relative space, measured distances must satisfy the applicable theorems of Euclidian geometry, velocities must behave as smooth vector-valued functions of a real variable. Moreover, Newtonian kinematics is inextricably intertwined with dynamics (see, e.g., Torretti 1983, Chapter 1). In contrast with such hackneyed "laws" as "All ravens are black," or "Water cleans," which demand little by way of intellectual commitment, the laws of mathematical physics, exemplified by Newton's Law of Gravity, are always deeply involved with some exacting theoretical system, apart from which they have no definite meaning. While empirical generalizations of the former kind can hardly be said to explain anything, a theory-laden law, such as Newton's, has great explanatory power thanks to the links induced—or should we rather say disclosed?—in a rich variety of facts by embedding them in the law's underlying theoretical structure.

The intellectual efficacy of such links can be further clarified by considering a case in which Newton's Law of Gravity does not provide a satisfactory explanation of facts presumably within its scope. Mercury's perihelion advances each year by somewhat less than 1 minute of arc. 90% of this advance—some 5,000″ per century—is due to the precession of the axis of the Earth, to which astronomical coordinates are referred. An additional 9% can be accounted for by the action of the other planets in agreement with Newton's Law (280″ per century can be ascribed to the action of Venus, 150″ to that of Jupiter, 100″ to the rest). But there remains a balance of approximately 43″ per century which, under the known circumstances, cannot be explained by Newton's Law. If the planet Vulcanus, invented for just this purpose, had been discovered or if the oblateness of the Sun (i.e., the ratio of its equatorial diameter to the distance between its poles) were significantly larger than it appears to be or if the dust surrounding the Sun were dense enough, the small secular anomaly of 43″ would be readily covered by Newton's Law. On the other hand, the anomaly agrees uncannily well with the prediction of Einstein's theory of gravitation, designed, as against Newton's, to fit the new Einsteinian conceptions of space and time. Since Einstein (1915) made this agreement known, the anomalous advance of Mercury's perihelion has come to be generally regarded as one of the classic instances in which the Newtonian explanation of planetary motion fails.[11] Now, the interesting thing to note is that, when one takes this stance, one cannot simply view Mercury—as one would a green raven—as a doubtless repeatable but on the whole unlikely exception to the otherwise well confirmed ordinary course of nature, but one must conclude that the other planets, whose observed behavior has hitherto agreed well with Newton's Law, follow it only approximately, within a margin compatible with current observational imprecision, due to their particular circumstances.[12] This inference from the failure of Newton's Law in a single case to its universal invalidity is forced on us by the very concepts of mass, space, velocity, etc., by which planets and their motions must be grasped in order to subject them in scientific discourse to that law. It is applicable to them only if they are bodies of gravitationally homogeneous mass moving in homogeneous Euclidian space. If they are such, either Newton's Law is true of them all or the recorded compliance of most planets with it is a mere coincidence. Hence, if one gives up all hope of accounting for a proven anomaly by Newton's Law and some hitherto undetected factual condition (as Leverrier and Adams explained the anomalous motion of Uranus by the Newtonian attraction of the then unknown planet Neptune), there are only two viable ways of dealing with it: either one tries a different law conceived in the same terms as Newton's—e.g., one in which the factor r^{-2} has been replaced by a less simple function of the distance between the interacting particles—or one builds upon a different conceptual foundation and comes up with a different understanding of gravitational phenomena.[13] But under no circumstances can one maintain, in the face of a single avowedly insoluble anomaly, that "Eight planets out of nine attract each other (and the Sun) directly as their masses and inversely as their distances squared"—as one may still endorse "Water cleans" even while trying in vain to wash out with water a stain in one's clothes. Systematic thought breeds necessity in such a way that, when the latter is found wanting, the system itself loses its hold on things.

2.3 Questions raised by conceptual innovation

Examples similar to those of Section 2.2 can be found in all fields of mathematical physics. They tend to show that—at any rate in this branch of science—explanatory inference is only a step or a facet of a process of thought whose decisive stage consists in producing concepts appropriate for grasping a variety of facts and linking them together in an intellectual system. Systematic linkage of many ostensibly diverse facts serves to corroborate the appropriateness of the concepts by which each of them is grasped, and is indeed the mainstay of inferential explanation. In Chapter 3 I shall have more to say about such intellectual systems. But let us first consider a question that one is bound to face sooner or later if it is true that facts of observation are grasped and regrasped under changing concepts.

Somewhat schematically that question can be introduced as follows: Science responds to puzzling facts, which it seeks to explain by reconceiving them. The facts are embedded in a conceptual system within which they follow from general laws, given their particular circumstances. In the course of this process, science discards the original description of the facts, which set the inquiry in motion, and proposes a new description, under which they are no longer puzzling. To what extent and by virtue of what device does the new description, required for the proposed explanation to work, refer to the same facts as the old description, which caused an explanation to be sought for them?

More pointedly, we may ask:

Q1. Can the facts remain the same as the framework of description varies?

Q2. Does the apparent need for a steady reference set some permanent limits to conceptual innovation?

Q3. Is such a steady reference really necessary, or may the researcher, without detriment to the rationality of his enterprise, sometimes forget along the way the facts he originally had in mind?

The breakdown of reference due to conceptual innovation is a recurrent theme in the historicist school of philosophy of science initiated in the fifties by Norwood Russell Hanson and Paul K. Feyerabend. Hanson bids us imagine the 16th century astronomer Tycho Brahe, a firm believer in the fixity of the Earth, and his Copernican assistant, Johannes Kepler, as they watch the dawn from the top of a hill. Hanson asks, *"Do Kepler and Tycho see the same thing in the east at dawn?"* (Hanson 1958, p. 5). He argues that, though "Tycho and Kepler are both aware of a brilliant yellow-white disc in a blue expanse over a green one," they cannot properly be said to witness the same fact. For "Tycho sees the sun beginning its journey from horizon to horizon. He sees that from some celestial vantage point the sun (carrying with it the moon and planets) could be watched circling our fixed earth. [. . .] But Kepler will see the horizon dipping, or turning away, from our fixed local star" (Hanson 1958, pp. 7, 23).

In a series of papers, Feyerabend (1958, 1960, 1962, 1965) repeatedly emphasized that terms inevitably change their meaning as they pass from the context of one scientific system into that of another. It is therefore very difficult to compare how well several such systems "fit the facts." A valid comparison is downright impossible when the systems under consideration concern the basic elements and properties of the universe. Each system will then "possess its own experience, and there will be no overlap between these experiences. [. . .] A crucial experiment is now impossible [. . .] because there is no universally accepted *statement* capable of expressing whatever emerges from observation" (Feyerabend 1965, p. 214).

Similar ideas were voiced, somewhat fuzzily, but with great rhetorical efficacy, by Thomas S. Kuhn (1962). To him a scientific revolution involves "a displacement of the conceptual network through which scientists view the world" (1962, p. 101). As a consequence of such a displacement, the new scientific tradition that issues from a scientific revolution "is not only incompatible but often actually *incommensurable* with what has gone before" (1962, p. 102; my italics). Therefore, one "may want to say that after a revolution scientists are responding to a different world" (1962, p. 110).

Kuhn's strong claims concerning the incommensurability of alternative modes of scientific thought and the substitution of one world for another in the course of a scientific revolution are clearly unjustified in those cases where enough remains of the prerevolutionary conceptual setup to allow a shared description of crucial facts. Thus, for example, deep though it was, Darwin's revolution in biology did not affect the distinction between living organisms and inanimate bodies nor such descriptions of the structure and behavior of the former as may be adduced for resolving the dispute between Darwin and his adversaries. Hence, someone who does not share Darwin's vision may dismiss the evidence gathered in *On the Origin of Species* as inconclusive, but not as unintelligible, as he might do if it were expressed in esoteric terms peculiar to the doctrine he rejects, and not in plain English.

It would seem, however, that reference to the selfsame facts will not survive conceptual renewal when this involves the very notions in terms of which the phenomena of motion and the states of physical systems are described. Yet even in this case, we can think of three conditions any one of which is sufficient to ensure the comparability of scientific claims in the face of such radical conceptual innovation. The continuity of scientific discourse can be preserved if:

C1. Some concepts are immune to change, and they provide a stable reference to decisive facts.

C2. The new concepts are arrived at through internal criticism of the old, by virtue of which the facts purportedly referred to by the earlier mode of thought are effectively dissolved.

C3. Reference to facts does not depend on the concepts by which they are grasped.

These three conditions are closely linked to the three questions regarding conceptual innovation I raised earlier in this section. Thus, an affirmative answer to question Q2 entails condition C1. Condition C2 would warrant an affirmative answer to the second part of question Q3. Finally, even if Q2 were to receive a negative answer, the fulfillment of C3 would justify an affirmative reply to Q1: if reference does not depend on concepts it may very well remain steady even when concepts change. In the next two sections I shall examine C1 and C2 in the tacit understanding that C3 does not hold. Then, in Section 2.6 I shall argue that, notwithstanding recent allegations to the contrary, C3 must be denied.

2.4 Are there limits to conceptual innovation in science?

2.4.1 Self-classifying sense impressions

Conceptual innovation will be confined within definite limits if all our experience of the world is compounded by association from simple, repeatable, self-classifying sense impressions. Scientific concepts would then represent different kinds of combinations or combinations of combinations, etc., of such simple impressions, and the vocabulary of science would fall into two parts:

(i) A basic vocabulary V_O, each term of which would designate one of the known classes of simple sense impressions or the simple relation of association between them

(ii) A derived vocabulary V_T, whose terms would signify the several ways of combining those sense impressions or their combinations or combinations of combinations, etc., to any order, which are felt to merit a label

The extension of each term of V_O would then be fixed once and for all by the natural self-classification of sense impressions. The list of such terms could only grow or decrease together with our capacity for receiving different sorts of impressions, and would therefore be stable, except during periods of major change in the genetic makeup of man. Innovation would be confined to terms in V_T and the concepts they express, which every scientist would indeed be free to fashion and refashion at his pleasure. Such terms, however, would be completely meaningless unless connected by definitions or other meaning-bestowing devices with terms of the basic vocabulary V_O. Any descriptions of particular facts involving terms in V_T would be replaceable, without prejudice to truth, by equivalent descriptions of the same fact that use only terms in V_O. The latter descriptions would anyway be shared by alternative systems of scientific explanation, in spite of any differences in their V_T vocabulary. Thus, it would not be impossible to compare them and to ascertain which one among them provides more appropriate premises for inferring the description of any given fact observed.

The scheme just sketched is, of course, chimerical. In real life, private sense impressions, far from providing the ultimate foundation of all experience, turn up—e.g., at the ophthalmologist's or while tasting wines in a winery—only in settings firmly anchored to public physical objects. Moreover, they are never simple, and they display in and of themselves no indication as to how they ought to be classified. Thus, for example, nothing in the sheer visual appearance of a rainbow could constrain us to see just the six colors we normally distinguish in it, instead of three (with Aristotle) or seven (with most nursery school teachers). Indeed, as the same example suggests, any classification of sense appearances is open to refinements and displacements due not to genetic mutations but to our changing interests and attention.

2.4.2 Kant's forms and categories

A very different treatment of the question of concept stability can be extracted from Kant's *Critique of Pure Reason*. Kant was firmly committed to a conception of man as a "finite" subject of knowledge, who can only learn about an object by the way the object "affects" him. But Kant saw clearly that a knowledge of objects could not result from the mere association of subjective affections. The core of his book is an inquiry concerning the "conditions of possibility" of our human experience of the physical world. He divides them into two classes. There are, in the first place, the "forms" of sense awareness, which make it possible that the manifold of sense appearances be ordered in certain relations.[14] These forms he identifies with time and space, which he regards as inherent conditions of our "receptivity" to sense impressions. In the second place, there are the "functions" by which our understanding combines and unifies the given manifold of sense in such a way that it is construed as a presentation of objects.[15] The "categories," or fundamental concepts of ontology, exactly correspond to the said "functions" of the understanding "insofar as the manifold of a given intuition is determined with respect to them" (Kant 1787, p. 144). Kant derives an allegedly complete list of the "categories" from the classification of "judgments" found in contemporary textbooks of logic, suitably enriched with two unfamiliar items, viz., "singular" and "infinite" judgments, to meet the desiderata of ontology. He claims that this classification reflects the several functions of the understanding. He argues, more convincingly, that temporal self-awareness presupposes awareness of enduring objects in space. He takes it for granted that the ordering of sense appearances in relations of space and time necessarily complies with the principles of Euclidian geometry and Newtonian chronology. He contends that every distinct content of sense awareness (every shade of color, tone of sound, etc.) must be grasped as belonging to some continuous scale of intensities that goes from the presence of that qualitatively peculiar content right down to its total suppression, passing through every conceivable intermediate degree. In a long chapter on "The Analogies of Experience" he offers proof that the objective time order of phenomena—as opposed to the merely subjective succession of appearances[16]—can only be established by grasping them under the categories of "subsistence and inherence (substance and attribute)," "causality and dependence (cause and effect)," and "community (reciprocity between agent and patient)," subject to the principles of conservation of the quantity of matter, causal determinism, and thoroughgoing instant interaction. The "forms" of time and space, the categories of the understanding, and the principles that govern the application of the latter to the manifold displayed in the former are, according to Kant, permanent features of human reason, such that "we cannot form the least conception" of a cognitive faculty which worked differently. We must, therefore, deem them necessary, although we cannot give any grounds "why we have just these and no other functions of judgment, or why space and time are the only forms of our possible intuition" (Kant 1787, p. 146).

Kant says that the "unity of consciousness" built by exercising the "functions" of the understanding on the manifold of sense "is that which alone constitutes the reference of representations to an object" (Kant 1787, p. 137). If this is granted, the Kantian system of categories and "forms" can certainly fix the reference of our factual descriptions in the face of conceptual innovation. For such innovation can then concern only empirical concepts that do no more than specify the categories, and it will be constrained by the principles of the understanding that preside over the articulation of sense appearances into an experience of physical objects. Reference to *the same thing* can be secured, in this view, regardless of any changes in its attributes or in the way we describe them and classify them, by following a constant "quantity of matter" in time as it moves in space. Thus, going back to the examples of Section 2.2, we can see at once that, if we refer to the several components of the solar system by giving the position at each instant of their respective masses, we fix thereby all the facts that any dynamic theory of planetary motion, consonant with Kant's philosophy, should seek to account for.

One could still object that the method of identifying bodies by timing (in Newtonian time) the positions (in Euclidian space) of their real-valued masses was inaugurated by Newton and his contemporaries, and therefore could not secure the stability of reference at the transition from pre-Newtonian to Newtonian physics. But a Kantian may well counter this complaint by recalling that "the highway of science" ("der Heeresweg der Wissenschaft"—Kant 1787, p. xii) has been entered upon by each branch of inquiry at a certain point in history; and that only from then on—i.e., in the case of physics, only from the 17th century (Kant 1787, p. xii)—can it boast an objective representation of phenomena, which both sets the task and provides a test for alternative scientific explanations of them.

A much more damaging criticism of the Kantian position results from considering the actual scope of conceptual novelty in the two major systems of physical thinking that have replaced Newton's in the 20th century, namely, Relativity and Quantum Theory. The former took issue with Newtonian physics on the selfsame notions of time, space, and mass that were for Kant the key to objective reference; the latter gave up the principle of causal determinism, without which, according to Kant, there could be no experience of objective succession.[17] As to the principle of thoroughgoing instant interaction, which he had considered indispensable for establishing the objective simultaneity in space, Relativity dismissed it from the outset. Thus, in the first major revolutions in mechanics after the publication of Kant's book, his system of "forms," categories, and principles could not stem the tide of conceptual innovation but was swept away by it.

We cannot go further into this matter here (though I shall have something to say on Einstein's criticism of Newtonian time in Section 2.5 and on his several concepts of mass in Section 2.6). I take it that the bare mention of those theories is sufficient to remind us that the Kantian restrictions on conceptual innovation have proved untenable. Some philosophers believe that Kant's "metaphysics of experience" was too strong and trespassed on matters pertaining to empirical science but that if only it is conveniently weakened, it can and must be upheld (Rosenberg 1980; Stevenson 1982). I shall touch again on this issue in Section 2.7. But here I wish to consider another way of stabilizing factual reference by restricting the admissible range of conceptual novelty, which was developed in the second third of the 20th century, mainly by Rudolf Carnap.

2.4.3 Carnap's observable predicates

In *Der logische Aufbau der Welt* (1928), Carnap proposed a method of "reducing" all objects—in the widest sense, including things, states and events, properties and relations—with the aid of modern logic, to a "basis" of homogeneous "ground elements" and fundamental relations between them.[18] Carnap admitted the possibility of adopting a physical basis with one of the following alternative sets of ground elements: (α) electrons and protons, (β) spacetime points, (γ) point-events on the worldlines of matter.[19] He chose, however, a solipsistic psychical basis, whose ground elements are the *Elementarerlebnisse,* or instantaneous cross sections of the total stream of a person's mental life.[20] Following Nelson Goodman (1966, p. 154), I call such ground elements *erlebs.* As fundamental relation Carnap chose *Ähnlichkeitserinnerung* (literally, 'recollection of resemblance'), i.e., the relation between two erlebs *x* and *y* that are known to be similar by comparing *y* to a memory of *x*. Carnap's preference for this basis was motivated by his conviction that knowledge of one's own stream of erlebs was presupposed by one's knowledge of anything else. Reduction to a solipsistic basis was therefore required in order to be faithful to the epistemic hierarchy of objects.[21] That every object of science can be reduced to erlebs he proved as follows:

> If any physical object were not reducible to sensory qualities and hence to psychical objects, that would mean that there are no perceptible criteria for it. Statements concerning it would then dangle in a void. At any rate, it would have no place in science.
>
> (Carnap 1961, p. 78)

The reductive program of Carnap's *Aufbau* has time and again enticed talented writers to devote their energy and ingenuity to its advancement (Goodman 1951; Moulines 1973). However, Carnap himself moved further and further away from it, dropping its main assumptions one by one. Yielding to Otto Neurath's criticism he substituted a "physicalistic" basis for the solipsistic erlebs (Carnap 1932). Later, he liberalized the requirements of reduction, to make allowance for scientific talk of dispositions that might never be actualized—such as the solubility in water of a substance that will never be removed from a waterless planet (Carnap 1936/37). Finally, he gave up the very idea of reduction and replaced it with a program of "partial interpretation" (Carnap 1956). Yet he did not relinquish his dream of science anchored forever to the rock bottom of the epistemic hierarchy. It is indeed ironic that he should have once sought to build his cathedral of knowledge on the drifting sands of *Erlebnis.* But when he opted for physicalism, the basis of spacetime points he originally countenanced was soon discarded, for it clearly did not enjoy epistemic primacy. Speaking the "formal idiom"—talk of words—which he now preferred to the more familiar but potentially misleading "material idiom"—talk of objects—he had used before,[22] he demanded that scientific discourse be reduced to *thing-language*— i.e., "that language which we use in every-day life in speaking about the perceptible things surrounding us" (Carnap 1936, p. 466)—and, more specifically, to the "observable predicates of the thing-language" (1936, p. 467). The key notion of an *observable predicate* Carnap explicated as follows:

> A predicate 'P' of a language L is called *observable* for an organism (e.g. a person) N, if, for suitable arguments, e.g. 'b', N is able under suitable circumstances to come to a decision with the help of a few observations about a full sentence, say 'P(b)', i.e. to a confirmation of either 'P(b)' or '~P(b)' of such a high degree that he will either accept or reject 'P(b)'.
>
> (Carnap 1936, pp. 454ff.)

He noted that "there is no sharp line between observable and non-observable predicates because a person will be more or less able to decide a certain sentence quickly." However, "for the sake of simplicity," he chose to draw a sharp distinction—"in a field of continuous degrees of observability"—between observable and non-observable predicates (Carnap 1936, p. 455).[23]

Twenty years later, Carnap (1956) still held to this sharp distinction but acknowledged that some nonobservable terms of science may be irreducible. He took for granted that scientific discourse is expressed in a formal language—or in a readily formalizable segment of a natural language—which he called "the language of science," *L*. This language falls neatly into two parts, the "observational language" L_O, and the "theoretical language" L_T. L_O is an interpreted first-order language,[24] whose variables range over "concrete, observable entities (e.g. observable events, things, or thing-moments)" and whose predicates—the "observational vocabulary" V_O—designate "observable properties of events or things (e.g. 'blue', 'hot', 'large', etc.) or observable relations between them (e.g. '*x* is warmer than *y*', '*x* is contiguous to *y*', etc.)" (1956, p. 41). L_T is a predicate calculus that *may* include logical and causal modal operators, besides the usual quantifiers and truth-functional connectives. The domain D over which the variables of L_T may range in a given interpretation is subject only to the following conditions: (i) D includes a distinguished countable subdomain, and (ii) D is closed under the operations of n-tuple formation (for every positive integer n) and class formation (1956, p. 43). Let T denote a finite set of sentences of L_T. The set of the logical consequences of T is a "theory," also designated by T, for which T, the finite set, provides the postulates.[25] The predicates of L_T which occur in T (in either meaning) form the "theoretical vocabulary" V_T of the theory. Carnap believes that any meaningful scientific use of "the language of science" L involves a choice of such a theory T.[26] According to him, any such use also requires the stipulation of a set C of "correspondence rules," which license the drawing of conclusions in L_O from premises in T (usually conjoined with premises in L_O). T owes its cognitive meaning to the consequences in L_O which thus accrue to it through C. Its empirical support depends on their truth. The correspondence rules C take the form of additional postulates or of rules of inference. In either case, their formulation in L must include predicates from both V_T and V_O. But not every predicate in V_T must occur in C. It is enough that the correspondence rules link *some* of the theoretical predicates to observational predicates. The interpretation of theoretical discourse by its observable consequences need only be partial.

Carnap's theory of scientific "theories" was not designed to cope with the problem of scientific change—to which, indeed, its author was notoriously insensitive. But it may be readily adapted for that purpose. Restrict the innovative action of thought to the theoretical vocabulary of science and the postulates and correspondence rules in which it is embedded. Then, the observational vocabulary, changing, if at all, at a much more leisurely pace, undisturbed by the vicissitudes—and the achievements—of theory, provides the stability of reference required for a comparison between the observable consequences of any two rival theories. Since Carnap and his followers issued no disclaimers when they were criticized by Feyerabend and Hanson for just this type of restriction, one may conclude that they approved it. And yet one cannot but feel astonished at the sheer extravagance of supposing that the language in which scientific observations are reported should be out of bounds for scientific thought.

I shall not review the arguments against a distinction between observational and theoretical terms in science. If the reader is not acquainted with them, I suggest reading Hilary Putnam's "What Theories Are Not" (1962) and Mary Hesse's "Is There an Independent Observation Language?" (1970). After two decades during which the distinction was unfashionable (but see, however, Shimony 1977, which is the printed version of a lecture dating from 1969), trendy writers are now saying that there was something to it. Ian

Hacking (1983, p. 175) pokes fun at the notion that all terms are "theory-laden"—and quite rightly, I dare say, for surely we do not wish to claim that when Monsieur Jourdain shouted, *"Nicole, apportez-moi mes pantoufles, et me donnez mon bonnet de nuit,"* his words were loaded with anything we would call a theory. In a more conservative vein, W. H. Newton-Smith, while dismissing "the alleged O/T dichotomy," sponsors

> a rough and ready differentiation between the more observational and the more theoretical [. . .] determined by the following principles:
>
> 1. The more observational a term is, the easier it is to decide with confidence whether or not it applies.
> 2. The more observational a term is, the less will be the reliance on instruments in determining its application.
> 3. The more observational a term is, the easier it is to grasp its meaning without having to grasp a scientific theory.
>
> (Newton-Smith 1981, pp. 26–27)

As a rough and ready characterization, the above is neat enough. But are the three stated criteria mutually consistent? To avoid disputes about the meaning of 'meaning' I leave the third one aside and concentrate on the first two.[27] Many familiar words score well on both counts. But confidence in the use of a term is not necessarily stronger because one does not rely on instruments in determining its application. Having just weighed 263 ± 0.2 grams of corn flour in a precision balance I am certainly more confident that it weighs 263 grams (to the nearest gram) than that it is corn flour. Roughly speaking, the terms that we use most confidently have to do with the practice of life—including, of course, the life of science—and their applicability can be tested in action. I rest assured that it is an axe I hold in my hands if I can chop a log with it. But our confidence need not decrease because the practical test involves the use of instruments. I am certain that power is back after a blackout as soon as I hear music on my tuner; I do not have to test my belief by plugging my fingers into an outlet. There is, however, one common circumstance that could perhaps suggest that our confidence in the application of terms is greater, the less we rely on instruments for deciding it. Many ordinary terms are employed very confidently just because the standards for applying them are quite loose. Such terms are not usually the sort whose application is controlled by means of instruments. But neither are they a source of "cognitive meaning" for the theoretical language of science.[28]

A major obstacle to the ordering of scientific terms from the more observational to the more theoretical—and, a fortiori, to their partition into two classes—is that the same term often functions, in different contexts, at either end of the proposed scales. For example, the term 'free fall', as instantiated by a falling stone, is highly observational by all three of Newton-Smith's criteria. But the same term is applied to the motion of the Moon on the strength of the Newtonian or the Einsteinian theory of gravity and the many instrument-assisted observations of planets, pendula, the Moon itself, etc., which corroborate those theories. In this use, therefore, the term is very theoretical by Newton-Smith's criteria 2 and 3, even though, after all the successful experimenting with artificial satellites in the last twenty years, it surely is very observational by criterion 1. The astronomical and the terrestrial uses of the term are not just homonymous, nor are the former only a metaphorical extension of the latter. (As a metaphor, it would be a rather poor one.) When Newton conceived the Moon as a freely falling body, he at the same time implied that cannonballs were but slow moons. His bold thought changed both the denotation and the connotation of 'free fall'. We are now quite certain that he was substantially right, that the term applies in its new sense both to its new and to its old extension, because we know from practice that a marginal increment in energy can convert an earthbound missile into a heavenly body.

As the preceding example shows, there is a legitimate distinction between the more familiar and the more technical uses of language, between everyday words and terms of art (see Wittgenstein, BB, p. 81; PU, §18). But words move back and forth from one category to the other, and—what is more important for our philosophical discussion—scientific usage, once established, claims—and gradually achieves—a controlling role over ordinary language. 'Distance', 'force', 'heat', 'light' are everyday words that modern physics has reclaimed and made precise; and we would not dream of using them, except metaphorically, in a way incompatible with their technical meaning. 'Gas' and 'electricity', first introduced as terms of art, have become kitchen words, but physics is still the acknowledged keeper of their primary meaning. Who is to be the master is less clear in the case of a term like 'energy', originally invented for strictly technical use within a system of thought that is no longer accepted. Modern physics employs it in her own way—different from Aristotle's—but she has not been able to inhibit or regulate its use in journalism and pseudoscience. Physical terms of art naturally belong to theories and are generally applied with the aid of instruments. They are therefore "less observational" by Newton-Smith's criteria 2 and 3. Whether they are so also by criterion 1 will depend mainly on the trustworthiness of the relevant theory. Misologists, to whom anything that smacks of intellect is suspicious, believe that familiar words, like 'soap' and 'clean', can be employed more confidently than terms of art, like 'entropy' or 'inductance'.

Yet if engineers were often wrong in their applications of the latter, the costs would be unbearable.

Clearly, it is very difficult—perhaps impossible—to isolate a family of meaning-invariant English words that regularly satisfy Carnap's criterion for observable predicates or Newton-Smith's criteria for "more observational" terms. No wonder that Carnap and his followers seldom give any examples of the "correspondence rules" by which scientific discourse is supposedly anchored to the observational vocabulary, and when they produce one it leaves much to be desired. No such example is to be found in Carnap's paper of 1956; but in his "Foundations of Logic and Mathematics" (1939) he had proposed the following:

> Let us imagine a calculus of physics constructed [. . .] on the basis of primitive specific signs like 'electromagnetic field', 'gravitational field', 'electron', 'proton', etc. The system of definitions will then lead to elementary terms, e.g. to 'Fe', defined as a class of regions in which the configuration of particles fulfils certain conditions, and 'Na-yellow' as a class of space-time regions in which the temporal distribution of the electromagnetic field fulfils certain conditions. Then semantical rules are laid down stating that 'Fe' designates iron and 'Na-yellow' designates a specified yellow color. [. . .] In this way the connection between the calculus and the realm of nature, to which it is to be applied, is made for terms of the calculus which are far remote from the primitive terms.
>
> (Neurath, Carnap, and Morris 1971, vol. I, pp. 207–8)

Now, if this example is to be taken seriously, one must point out at once that never in the history of chemical nomenclature has 'Fe' designated just any old piece of metal which an ironmonger might accept as iron. Far from being "anchored to the solid ground of observable facts" through the stable reference of the ordinary, "observable" predicate 'iron' as in the stipulation put forward by Carnap, the chemical and physical theories to which the term 'Fe' belongs have set new standards for the application of 'iron', and for the classification, evaluation, and improved production and utilization of what goes by that name. Through its association with 'Fe', 'iron' ceases to be a common everyday predicate, decidable "after a few observations," and becomes a scientific term, whose accurate application relies on appropriate laboratory procedures. Needless to say, if 'iron' had not been thus linked to 'Fe' it would not apply to the gaseous element of atomic number 26 which can be spectroscopically detected in stars. As to the other part of Carnap's example, involving the term 'Na-yellow', I must confess that I do not know what Carnap means here by "a specified yellow color." In ordinary English one may say that a can contains paint of a certain color, e.g., canary yellow, although the paint has never seen the light. But the "temporal distribution of the electromagnetic field" inside a closed can of paint cannot meet the conditions for 'Na-yellow'. One is therefore tempted to believe that, notwithstanding his avowed conversion to physicalism, Carnap understands here by "a specified yellow color" the chromatic quality sensed by a healthy human observer who sees a surface of that color under so-called normal illumination.[29] Now, a scientist who wishes to learn how to recognize Na-yellow radiation at a glance must develop the habit of associating the term 'Na-yellow' with the—presumably stable—color he sees when his eyes receive radiation of that frequency (as measured with the appropriate instruments). One may certainly say that a habit of this sort establishes a semantic rule. The scientist's subjective chromatic experience and any name he may have for it in a private language acquire thereby an objective significance. But the transfer of cognitive meaning effected by such a semantic rule follows a direction exactly opposite to the one indicated by Carnap, viz., *from* the frequency measured in the laboratory *to* the class of erlebs by which the scientist will henceforth diagnose it. That 'Na-yellow' does not simply designate a specific yellow color, in the sense explained, can be inferred from the fact that "the class of space-time regions" that satisfy the requirements for Na-yellow includes some in which the radiation energy is too weak to be recorded by a human eye and some in which it is so strong that it will burn away any organism.

But maybe Carnap's example was only a didactic ploy, not intended for critical scrutiny.[30] To judge the efficacy of correspondence rules one ought then to look for a presentation of a genuine physical theory that actually introduces some of its terms by means of such rules. Carnap's own helpless attempt at formalizing the foundations of Relativity does not reach the point of furnishing an interpretation of the axioms (Carnap 1958, pp. 197ff.). But Hans Reichenbach's *Axiomatik der relativistischen Raum-Zeit-Lehre* (1924) does include what appear to be the equivalent of Carnap's correspondence rules, under the name of coordinative definitions (*Zuordnungsdefinitionen*). The following are typical examples:

> *Definition 9.* Light rays are **straight lines**.
> *Definition 18.* A **natural clock** is a closed periodic system.
> *Definition 19.* A **rigid rod** is a solid rod that is isolated from all external forces.[31]

I have set the definienda in boldface. Each is a term of art of the theory. The light rays mentioned in Definition 9 should of course traverse a perfect vacuum. Thus, this concept is plainly not observational by logical empiricist criteria. Neither is the concept of physical isolation or closure that occurs in the definiens of Definitions 18 and 19.[32]

2.5 Conceptual criticism as a catalyzer of scientific change

Thomas S. Kuhn has noted that "in periods of acknowledged crisis [. . .] scientists have turned to philosophical analysis as a device for unlocking the riddles in their field" (Kuhn 1962, p. 88). In a paper on thought experiments first published in 1964 he explains how a *Gedankenexperiment* proposed by Galileo "helped to teach [. . .] conceptual reform":

> The concepts that Aristotle applied to the study of motion were, in some part, self-contradictory, and the contradiction was not entirely eliminated during the Middle Ages. Galileo's thought experiment brought the difficulty to the fore by confronting readers with the paradox implicit in their mode of thought. As a result, it helped them to modify their conceptual apparatus.
>
> (Kuhn 1977, p. 251)

Kuhn, however, does not sufficiently stress that when a new mode of thought issues from conceptual reform, the problems raised by its purported incommensurability with what went on before are automatically—and trivially—solved. For there can be no question of choosing between two modes of thought if the very existence of the one issues from a recognition of the conceptual failings of the other. If the old is disqualified by the same exercise in criticism that ultimately leads to the new, a comparison between them is not even called for.

The First Day of Galileo's *Dialogo sopra i due massimi sistemi del mondo* contains several fine examples of conceptual criticism, aimed at dislodging the Aristotelian cosmology. The thought experiment to which Kuhn refers is one of them, and here I shall touch on another. Aristotle's system of the world rests on his doctrine about the natural local motion of the elements. Being simple, elements must move simply, unless they are compelled by an external agent to move otherwise. Aristotle recognized two kinds of simple local motion, corresponding to the two classes of lines from which all trajectories are compounded, viz., the straight and the circular. Since the four known elements, earth, water, air, and fire, move naturally in straight lines to and from a particular point, Aristotle concludes that there must exist a fifth element which naturally moves in circles about that same point (*De Caelo*, I, ii–iii; see in particular, 268^b11, 269^aff., 270^b27ff.). This element is the material out of which the heavens are made, and the said point is therefore the center of the world. This is Aristotle's reason for separating celestial from terrestrial physics, and as Galileo's spokesman Salviati points out, it is indeed "the cornerstone, basis, and foundation of the entire structure of the Aristotelian universe" (Galileo, EN, VII, 42). But even granting the premises, Aristotle's conclusion does not follow, for, as Galileo's Sagredo is quick to note,

> if straight motion is simple with the simplicity of the straight line, and if simple motion is natural, then it remains so when made in any direction whatever; to wit, upward, downward, backward, forward, to the right, to the left; and if any other way can be imagined, provided only that it is straight, it will be suitable for some simple natural body.
>
> (Galileo, EN, VII, 40)

Similarly, any circular motion is simple, no matter what the center about which it turns. "In the physical universe there can be a thousand circular motions, and consequently a thousand centers," defining "a thousand motions upward and downward" (Galileo, EN, VII, 40). Salviati goes even further:

> Straight motion being by nature infinite (because a straight line is infinite and indeterminate), it is impossible that anything should have by nature the principle of moving in a straight line; or, in other words, toward a place where it is impossible to arrive, there being no finite end. For nature, as Aristotle well says himself, never undertakes that which cannot be done.
>
> (Galileo, EN, VII, 43)

Therefore, "the most that can be said for straight motion is that it is assigned by nature to its bodies (and their parts) whenever these are to be found outside their proper places, arranged badly, and are therefore in need of being restored to their natural state by the shortest path" (Galileo, EN, VII, 56); but in a well-arranged world only circular motion, about multiple centers, is the proper natural local motion of natural bodies. Although the Copernican physics that Galileo was reaching for would eventually be built upon the primacy of straight, not circular, motion, the Aristotelian cosmology and its underlying physics could not survive the conceptual criticism of Galileo. For, as he lets Salviati say, "whenever defects are seen in the foundations, it is reasonable to doubt everything else that is built upon them" (Galileo, EN, VII, 42). No wonder that, pace Feyerabend, Aristotelianism ceased to be, for Galileo's ablest readers, a viable intellectual option.

The most famous and perhaps also the clearest example of conceptual criticism issuing in a scientific revolution is Einstein's discussion of the classical concept of time in §1 of "Zur Elektrodynamik bewegter Körper" (Einstein 1905b). To understand him properly we must bear in mind that the kinematics of Newton's *Principia*, purportedly based on the transcendent notions of absolute space and time, gave way in the late 19th century—at least in the more enlightened circles—to the revised critical version of Newtonian kinematics proposed by Carl Neumann (1870) and perfected by James Thomson (1884) and Ludwig Lange (1885). Neumann and his followers developed the concept of an inertial frame of reference, which is Einstein's starting point. In fact, Lange's definition of an inertial frame—which, by the way, is very close to Thomson's—is much more appropriate to Einstein's needs than the one that he himself, somewhat carelessly, gives[33] and was therefore appositely prefixed by Max von Laue to his masterly exposition of Special Relativity.[34]

Lange defines an "inertial system" as a frame of reference in whose relative space three given free particles projected from a point in non-collinear directions move along straight lines. Following Neumann, Lange defines an "inertial time scale," i.e., a time coordinate function adapted to an inertial frame, by the following stipulation: A given free particle moving in the frame's space traverses equal distances in equal times, measured by the scale in question. Let F be an inertial frame endowed with an inertial time scale t. Relatively to F and t the Principle of Inertia can be stated as an empirically testable law of nature: Any free particle that is not involved in the definition of F or t travels with constant speed in a straight line.

What apparently nobody realized until Einstein made it obvious is that the Neumann-Lange definition of an inertial time does not determine a unique partition of the universe into classes of simultaneous events. If F and t are as above and x, y, and z are Cartesian coordinate functions for the relative space of F, then any real-valued function linear in t, x, y, and z,

$$t' = a_0 t + a_1 x + a_2 y + a_3 z + a_4$$

is also an inertial time scale adapted to F. (The transformation $t \mapsto t'$ rotates each hyperplane $t =$ const. about its intersection with the axis $x = y = z = 0$.)

Einstein overcame this ambiguity with his famous definition of time by means of radar signals emitted from a source at rest in the chosen inertial frame:

> If at a point A of space there is a clock, an observer at A can time the events in the immediate neighborhood of A by finding the positions of the hands that are simultaneous with these events. If there is at the space point B another clock—and we wish to add, "a clock with exactly the same constitution as the one at A"—it is possible for an observer at B to time the events in the immediate neighborhood of B. But without further stipulations it is not possible to compare, with respect to time, an event at A with an event at B. We have so far defined only an "A time" and a "B time" but no common "time" for A and B. The latter time can now be defined by stipulating *by definition* that the "time" required by light to travel from A to B equals the "time" it requires to travel from B to A. Let a ray of light start at the "A time" t_A from A towards B, let it at the "B time" t_B be reflected at B in the direction of A and arrive back at A at the "A time" t'_A. By definition the two clocks synchronize if
>
> $$t_B - t_A = t'_A - t_B$$
>
> (Einstein 1905b, pp. 893–94)

Einstein's stipulation determines a time coordinate function unique up to the choice of origin and unit, the *Einstein time* of the frame. Let t be the Einstein time of an inertial frame F. Relatively to F and t the Principle of the Constancy of the Velocity of Light can be stated as an empirically testable law of nature: Any optical signal that is not involved in the definition of t travels in vacuo with the same constant speed in a straight line, regardless of the state of motion of its source. Einstein's Principle of Relativity says that the laws of physics take the same form when referred to any kinematic system consisting of Einstein time and Cartesian space coordinates adapted to an inertial frame. The joint assertion of the Principle of Relativity and the Principle of the Constancy of the Velocity of Light entails that any two such kinematic coordinate systems are related to each other by a Poincaré transformation.[35]

All the revolutionary implications of the Special Theory of Relativity follow from this result. Let me recall one only. If t and t' are time coordinate functions defined by Einstein's method, employing the same time unit, for two inertial frames F and F' that move past each other with speed v, then the partition of nature into classes of simultaneous events determined by t is different from and incompatible with the one determined by t'. Specifically, for any event E and any arbitrary positive real number T there always exist events (or, at any rate, possible event locations) E_1 and E_2 such that $t(E) = t(E_1)$

$= t(E_2)$, but $t'(E) - t'(E_1) = t'(E_2) - t'(E) = T$. In other words, for any event E, there are events simultaneous with E by t that, by t', precede or follow E by as much time as one chooses.[36] From the inception of Special Relativity in 1905 this feature of the theory has been regarded as a radical departure from the classical conception of time. Since time enters into the definition of the basic kinematical concepts of velocity and acceleration and the latter is tied by Newton's Second Law of Motion to the key dynamic concepts of force and mass, the breach between Special Relativity and Newtonian mechanics could well be such as to make them truly incommensurable. Indeed, the conceptual differences between both theories apparently run so deep that one may even come to doubt that they are genuine alternatives. For obviously, if two pieces of scientific discourse refer by incommensurable concepts to incommensurable matters, neither of them can be offered as a substitute for the other.[37]

However, Einstein's point at the beginning of "Zur Elektrodynamik bewegter Körper," §1, is not that the classical conception of physical time is *wrong* or *inconvenient* and therefore ought to be replaced by his, but rather that classical kinematics *does not have* a definite notion of time sufficient to determine the relations of simultaneity and succession between distant events. Einstein does not give there any reason for *modifying* a given time concept but proceeds to *establish* one where none so definite and far-reaching was yet available. He takes for granted the standard notion of an inertial frame of reference endowed with Cartesian space coordinates. He notes that "if we wish to describe the motion of a particle, we give the value of its coordinates as functions of the time." But, he goes on to say, "such a mathematical description has a physical meaning only if one is quite clear as to what is here to be understood by time" (Einstein 1905b, p. 892). Implying that such clarity was missing in the extant literature of mathematical physics, Einstein then proceeds without further comment to fill this gap.

He considers three procedures for dating and timing events.[38] The first one we continually use. It consists in assigning to any given event the time shown by a clock when that event happened.

> Such a definition suffices in fact for the purpose of defining a time for the place where the clock is located; but it is not sufficient where it is a question of temporally connecting series of events which occur in different places or—what amounts to the same—of assigning temporal values to events which occur at places distant from the clock.
>
> (Einstein 1905b, p. 893)

The second procedure is the naïve extension of the first one to remote events.

It consists in assigning to them the time shown in the observer's clock when he sees them happen. This is the method I would normally use to record the time at which an aeroplane fell into the lagoon I see from my window, or a telephone call was made to me from overseas. But, as Einstein says, "we know by experience"[39] that this procedure "has the disadvantage that it is not independent of the standpoint of the observer with the clock" (Einstein 1905b, p. 893).[40] Therefore Einstein came up with the third procedure, whose description I have earlier quoted. He assumed *as a matter of physical fact* that this procedure can be consistently applied and that it does not depend on the spatial location of the base point A (nor presumably on the time t_A at which the synchronization is performed).

On these assumptions, Einstein has little difficulty in showing that, when the time variable that occurs in the equations of classical mathematical physics is understood in the manner just proposed by him, Maxwellian electrodynamics satisfies the Principle of Relativity, i.e., the Maxwell equations hold in their standard form in every inertial frame of reference if they hold in one. The experimentally recorded insensitivity of the speed of light to a substitution of frames can therefore be accounted for in a most natural way. No difference in the speed of light is *measured* when it is referred to different inertial frames because, when time is understood in Einstein's sense, the speed of a given light signal in vacuo happens to *be* the same in all such frames.

By exposing the lack of definiteness of the Newtonian time concept Einstein undermined the entire stock of notions built upon or intertwined with it. However, neither he nor his fellow physicists yielded to the cheap temptation of dismissing Newtonian science as one big connected piece of nonsense. On the contrary, they were at pains to show how, in the light of the new mode of thought, the Newtonian system possessed both meaning and truth, within appropriate limits. The equations of relativistic kinematics and mechanics collapse into Newtonian equations in the limit $(v/c)^2 \to 0$ (where c denotes the speed of light in vacuo and v the greatest speed achieved by the material objects under consideration). Consequently, *according to the new theory*, the Newtonian equations *hold good* wherever $(v/c)^2$ is negligible. But, of course, as P. M. Churchland (1979, p. 85) aptly notes, what is thereby vindicated is not the Newtonian theory as conceived by its founders but only a simulacrum of it, intellectually parasitic on Special Relativity, from which it obtains its meaning. Thus, with hindsight, we can confer definite denotations to the Newtonian concept of time and the other Newtonian concepts that depend on it; namely, the same as the homonymous relativistic concepts would have when $v \ll c$. By this move, Special Relativity inherits all the empirical evidence that was once supposed to corroborate the Newtonian theory, while the latter attains finality within its henceforth definite domain of validity.[41]

Einstein's critique of Newtonian chronometry is exceptional, for both its deadly efficacy and the pervasiveness of the concept under fire. It is no wonder, therefore, that the transition from classical to relativistic mechanics has been so often adduced as an example in discussions of the incommensurability thesis. From what we have just seen, it follows that in this particular case the thesis is adequate, but innocuous. There can be no factual basis for comparing two theories of kinematics when one possesses and the other lacks a definite criterion for timing events. But this difference alone is sufficient to give the former a crushing advantage over the latter.

In other transitions in the history of scientific thought the effects of conceptual criticism, though important, have been less decisive. Einstein (1905a) argued that the possibility of deriving the so-called Rayleigh-Jeans law of blackbody radiation from classical electrodynamics and statistical mechanics proved the inconsistency of classical physics.[42] But the successive quantum theories introduced in the 20th century to account for the emission and absorption of radiation and related phenomena have not been hitherto more satisfying from a purely intellectual point of view than the classical theories they dislodged. It is not on account of their greater conceptual perfection that the quantum theories have so far prevailed. Indeed, it is unlikely that anybody has ever thought of justifying a preference for the quantum approach to microphysics only, or mainly, because the classical theory of radiation is inconsistent.

Issuing from the very mode of thought it dissects and dissolves, conceptual criticism keeps a steady grip on whatever facts the former had got hold of while at the same time improving the way they are understood. Thus, any chasms that might arguably open between successive stages of intellectual history can be effectively bridged. Shall they, however, remain unbridged in major scientific revolutions in which conceptual criticism is secondary and indecisive? As I noted in Section 2.3, conceptual chasms need not be feared if our reference to objects and objective situations is altogether independent of the concepts by which we grasp them. To counter the claims of incommensurabilism with such a conception of reference was the aim of the theory of meaning we shall now examine.

2.6 Reference without sense

The realization that scientific thought is not referred to its proper objects through a set of easily decidable, theory-neutral, "observable" predicates was one of the main motives for the new theory of meaning developed by Hilary Putnam (1973, 1975).[43] Putnam thought he could kill the incommensurability thesis of Feyerabend and Kuhn by severing the traditional link between the reference or denotation of scientific terms and their connotation or sense. For then the former can remain stable even as the latter undergoes upheaval. Since Putnam no longer believes in reference without sense (see Section 2.6.5), I have some qualms about criticizing him on this issue. However, other philosophers still back Putnam's former theory of meaning,[44] and now they may even draw sustenance from Putnam's revival of some of his early arguments and examples in Chapter 2 of *Representation and Reality* (1988).[45] Rather than argue with them, I have addressed my polemic to their source, although it involves fighting the straw man that Putnam left behind as he moved to a more insightful philosophical position.

2.6.1 *Denoting and connoting*

Traditionally, a general term is said to *denote* any and every object of which it is true, and it is said to *connote* the conditions that an object meets if the term is true of it.[46] These characterizations suggest that by fixing the denotation of a term its connotation will be automatically taken care of; for it comprises the features shared by the denotata as a matter of fact. But fixing the denotation is not without difficulty, as the following examples should make clear.

Consider a binary predicate, such as 'x is heavier than y', or 'x sits on y', or 'x laughs at y'. Each of these terms is true of an ordered pair $\langle\alpha,\beta\rangle$ if a true statement is obtained by substituting a name of α for x and a name of β for y. Therefore, the denotation of a given binary predicate consists of the ordered pairs of which that predicate is true. But, what is an ordered pair? One is tempted to say that an ordered pair is a pair of things taken in a certain order, and leave it at that; for one can hardly come up with clearer words to give a more perspicuous answer. But 20th century logicians, intent on explicating everything in terms of sets—i.e., of amorphous collections identified only by their members, regardless of any order or other relations between them—produced other, presumably less naïve, definitions of an ordered pair. To my mind, the one that best renders our naïve intuitions is the following:

$$\langle\alpha,\beta\rangle =_{\mathrm{Df}} \{\{\alpha,1\},\{\beta,2\}\} \tag{1}$$

(In other words, the ordered pair with first element α and second element β is a set of two sets, each containing two elements, viz., the object α and the number 1, and the object β and the number 2, respectively.)

But definition (1) will not satisfy someone who feels that the positive integers are not sufficiently perspicuous. Logicians have therefore adopted the ingenious definition of an ordered pair proposed by Kuratowski (1921):

$$\langle\alpha,\beta\rangle =_{\mathrm{Df}} \{\{\alpha\},\{\alpha,\beta\}\} \tag{2}$$

Yet evidently the following would do just as well:

$$\langle\alpha,\beta\rangle =_{\mathrm{Df}} \{\{\alpha,\beta\},\{\beta\}\} \tag{3}$$

Or, recalling that the empty set Ø is included in every set (see Section 2.8.1), and is therefore universally available, one can use it as a marker and define an ordered pair with Wiener:

$$\langle\alpha,\beta\rangle =_{\mathrm{Df}} \{\{\alpha\},\{\beta,\varnothing\}\} \tag{4}$$

Surely, the reader can conceive of still other alternatives.[47] Now, when one sets up the formal semantics of an artificial language one may indeed fix by decree the denotata of binary predicates by arbitrarily choosing any plausible definition of an ordered pair and excluding all others. But this solution is not open to the student of living languages. Thus, for example, there is no serious reason for maintaining that the English expression '*x* owns *y*' denotes each and every set $\{\{\alpha\},\{\alpha,\beta\}\}$ such that α owns β rather than, say, each and every set $\{\{\alpha\},\{\beta,\varnothing\}\}$ such that α owns β.

If the world is indeterministic, the very fact that all discourse is temporally and spatially localized severely restricts the determinateness of reference. General terms used today in ordinary conversation denote—unless otherwise indicated—objects that exist now or have existed in a not too distant past or are expected to exist in a more or less imminent future, on or near the Earth. 'Cow', spoken by a farmer, refers to the domestic animal whose milk we drink, not to its wild forebears, let alone to creatures biochemically indistinguishable from our cows living in a remote galaxy. But scientific discourse—at any rate in physics and chemistry—purports to be universal in scope. When we say 'electron' or 'sodium chloride', presumably we denote every electron or every molecule of sodium chloride that was, is, or will be. In a deterministic universe, the set of all electrons, past, present, and future, can be regarded at any time and place as a perfectly definite aggregate, identified by its members, to which the term 'electron' refers. But in an indeterministic world, in which electrons are being continually created and annihilated at random, no such a *set of all electrons* can—here and now—be identified by membership. It appears, therefore, that in a world like ours, any attempt to conceive the reference of general terms of universal scope as a relation in the sense of set theory, i.e., as a set of ordered pairs, is doomed not just for linguistic but for physical reasons. For suppose that the reference of a term *T* is understood as a binary relation whose domain—i.e., the set of the first elements of each pair—consists of all the utterances or inscriptions of *T* and whose codomain—i.e., the set of the second elements of each pair—consists of all the objects denoted by *T*. Evidently, if the identity of a set is fixed exclusively by the identity of its members, such a relation will be defined at a given time and place only if the elements of both its domain and its codomain are determinate then and there. Now, although the domain of the said relation can often meet this requirement (viz., if the term *T* belongs to a dead language or is no longer in use), the codomain normally cannot meet it if *T* is a general scientific term of universal scope and the universe is indeterministic.

The set of denotata of a general term does not give rise to the stated difficulty if its identity is given not by its actual membership but by the necessary and sufficient conditions for belonging to it; in other words, if the denotation of the term depends on its connotation. This is, of course, the traditional view, according to which the referents of a general term are picked out by the concept expressed by that term. Putnam developed his ideas on meaning in overt opposition to this view.

2.6.2 *Putnam's attack on intensions*

Putnam holds against the traditional view the fact that nobody has yet explained what it is "to grasp an intension" (Putnam, PP, vol. II, pp. 199, 263). Now, this expression, often employed by Carnap to speak of the mental operation of conceiving the connotation of a general term, is no doubt unfortunate. For it suggests that the intension of such a term, i.e., the condition or conditions that an object must satisfy in order to be denoted by it, stands in some supernatural place—such as the proverbial "museum in which the exhibits are meanings and the words are labels" (Quine 1969, p. 27)—ready to be seized with the mind's hand.[48] But if one ignores this anyway silly suggestion, it is not easy to see the force of Putnam's objection, for there are plenty of situations in life that nobody has ever explained and yet undeniably occur. To mention just one example: I cannot explain what it is for me to perceive the pencil I hold between my fingers, and indeed I cannot even imagine what it would be for me to explain it; but my inability in both these respects does not detract from the truth of the statement that there are pencils and that I am now writing with one. We might, perhaps, list the conditions—physical or otherwise—without which my perception of the pencil could not take place; but no such list could teach, say, a bodiless spirit what it is to see and to feel and to hold a pencil. The task of elucidating what it is "to grasp an intension" is still more hopeless. For surely one cannot, without circularity, explain in general the understanding of conditions by listing the conditions of understanding.

Putnam assumes for the sake of the argument that grasping the intension of a given term *T* consists in a specific, as yet unanalyzed, perhaps unanalyzable, state of mind and proceeds to show, by means of counterexamples, that in order to refer successfully to the objects denoted by *T* it is neither necessary nor sufficient to be in such a state. It is not *necessary*, for Putnam himself succeeds in referring exhaustively and exclusively to beech trees with the term 'beech' although he is not able to distinguish them from elms or to explain the difference between these two kinds of trees. Now, this example does not show that Putnam is in the same state of mind when he uses the term 'beech' as when he uses the term 'elm'; for surely in the former case he is aware of saying or writing 'beech', not 'elm', and these words do not sound or look alike. But, of course, the denotation of a word cannot be determined by its acoustic or visual shape. Indeed, I can imagine a planet exactly like ours—which I call, after Putnam, Twin Earth—in which a language is spoken, exactly like English, except that in it 'beech' denotes elms and 'elm' denotes beeches. A professor of philosophy in Twin Harvard can then be in exactly the same mental state when he refers to elm trees as our Professor Putnam is when he refers to beech trees.

The example points to a fact about language which, according to Putnam, nobody seems to have noticed before him:

> There is *division of linguistic labor.* We would hardly use such words as 'elm' and 'aluminum' if no one possessed a way of recognizing elm trees and aluminum metal; but not everyone to whom the distinction is important has to be able to make the distinction.
>
> (Putnam, PP, vol. II, p. 227)

But the example does not prove that the expert guardians of the social meaning of a given term can fix its denotation without conceiving its connotation. Suppose that Professor Putnam moves to a large estate in the south of England that he has received as a birthday present from a grateful and wealthy student. Suppose that he asks the agricultural advisor he has brought over from America to mark with a 𝔅 all the beech trees in the estate, so that he may in his daily rides learn to recognize this kind of tree. As he imparts this instruction Professor Putnam succeeds in referring to just the trees he wants marked, although he is himself unable to single them out, because other speakers of English can, if need be, do it for him. One of them is the advisor, who, however, can accurately fulfil his employer's wishes not because he is acquainted with each and every beech tree in the estate but, I dare say, because he has mastered what the term 'beech' connotes.

Maybe Putnam would have granted this much even in his early days.[49] He maintained, however, that no purported mastery of connotations is *sufficient* to ensure successful reference to the objects denoted by a general term. To prove it, Putnam introduced Twin Earth and invited the reader to imagine that the oceans, lakes, brooks, plants, animals, of that planet are filled with a liquid that possesses all the phenomenological properties of water as they were known ca. 1750 but which is not H_2O but some other chemical compound we may call *XYZ*. Then, according to Putnam, our word 'water' has never referred to the liquid on Twin Earth; for 'water', even as it was used in the early 1600s, e.g., in the King James Bible, denoted the chemical compound in our seas and rivers, not one that only outwardly resembles it. Nor has 'water' in Twin English ever referred to water, but only to *XYZ*, even before Twin Cavendish established its chemical composition. Therefore, assuming that science and literature have evolved in both planets more or less in the same way, we must conclude that, when Twin Shakespeare made Twin Antony say, pointing at the swiftly changing clouds,

> That which is now a horse, even with a thought
> The rack dislimns, and makes it indistinct
> As water is in water,

he could not refer to the same fluid as our Bard in the familiar homophonic lines; although both poets may have had very much the same in mind as they wrote these passages. Thus, two terms can connote exactly the same conditions to their users and yet denote different things.

Perhaps it is no accident that Putnam had to resort to this farfetched story to drive his point home. If we take him at his word, we must believe that *XYZ* will quench a man's thirst, in spite of the fact that, as J. B. S. Haldane once remarked, even the pope is 70% water.[50] Contemporary chemistry would be unable to account for such an extraordinary phenomenon. But, having consented to play Putnam's game of science fiction, we may just as well imagine that Twin Earthian scientists have in this respect been more fortunate than ours. As they do not parse matter in the manner of Lavoisier and his successors, their conception of the substance in their sea has continued to fit water also. They will not be surprised, therefore, if they find that our liquid mixes well with their blood. Imagine, moreover, that finally they have been able to explain, from their non-Lavoiserian standpoint, the hitherto baffling phenomena of electrolysis, etc. that lend support to our chemistry and its untenable distinction between H_2O and *XYZ*. In this version of the story it may still be true that 'water' in current scientific English refers to the terrestrial and not to the Twin Earthian fluid, but in scientific Twin English 'water' refers to both. It is up to the reader to decide what the two Shakespeares meant by this word in the last line of the above quotation. Maybe, with poetical insight, each referred to H_2O by one occurrence of 'water' and to *XYZ* by the other. Thus, if we let science fiction embrace not only the subject matter of science but also the manner in which scientific thought deals with it, it can teach us a lesson directly opposed to the one that Putnam tried to derive from it. Give a mild Kuhnian twist to Putnam's tale and the denotation of 'water' will turn out to depend on accepted scientific theory.[51]

2.6.3 The meaning of natural kind terms

But we do not have to indulge in fiction to perceive the inanity of Putnam's attempt to stabilize the reference of scientific terms by divorcing it from scientific thought. It is enough that we take a look at the theory of meaning proposed by Putnam himself. He developed it for two classes of general terms only, namely, for natural kind terms, such as 'elm', 'molybdenum', 'neutrino', and for physical magnitude terms, such as 'mass', 'electric charge', 'heat'. According to him the meaning of a natural kind term is fully specified by four components: (i) a *syntactic marker* indicating the grammatical classification—and behavior—of the term; (ii) a *semantic marker* indicating the ontological type or category to which the entities denoted by it belong; (iii) a *stereotype* or list of features that are normally shared by those entities and which a language user is required to know before he can be said to have mastered the term (e.g., *to be striped* is part of the English stereotype of 'tiger'—although albino tigers are not striped); and (iv) the term's *extension*. Thus, the meaning of 'water' may be given by the following ordered quadruple: (i) mass noun, concrete; (ii) natural kind,[52] liquid; (iii) colorless, transparent, tasteless, thirst-quenching, etc.; (iv) H_2O (give or take impurities). Putnam stresses that when the meaning of 'water' is explained in this way "this does *not* mean that knowledge of the fact that water is H_2O is being imputed to the individual speaker or even to the society. It means that (*we* say) the extension of the term 'water' as *they* (the speakers in question) use it is *in fact* H_2O" (Putnam, PP, vol. II, p. 269). But the extension of a natural kind term must be somehow determined, if not *by* the language users, at any rate *for* them. According to Putnam this can be achieved only if their use of the term is connected "by a certain kind of causal chain to a situation" in which the term is instantiated (Putnam, PP, vol. II, p. 200; cf. p. 176). The extension of the term is thereby fixed; it comprises anything that is *the same* as the said instance, in the manner of sameness indicated by the semantic marker of the term.[53]

> Our theory can be summarized as saying that words like 'water' have an unnoticed indexical component: 'water' is stuff that bears a certain similarity relation to the water *around here.* Water at another time and in another place or even in another possible world has to bear the relation $same_L$ [i.e., *same liquid as*] to *our* 'water' *in order to be water.*
>
> (Putnam, PP, vol. II, p. 234)

In this theory of meaning, despite Putnam's protestations,[54] the extension of a natural kind term still depends on its intension in two respects. In the first place, the particular instance of the term—e.g., the water around here—to which its denotation is anchored must be conceived under the category—stuff, liquid, chemical compound, or whatever—designated by the term's semantic marker. Users of the term must therefore understand that it refers, e.g., to a liquid body and not, say, to its surface or its glitter, etc. In the second place, the conditions for being *the same* as the said instance with regard to that category—e.g., for being the same liquid or the same chemical compound as the water around here, or the same sort of tree as the elms over Abbie Cabot's roof—must be fixed, lest the reference of the term be unsteady. I do not contend that very many natural kind terms in common use meet this last requirement but only that their denotation is well defined just to the extent that they do meet it. Now, the conditions for being the same as a given object with respect to a certain appropriate category constitute precisely what one would normally call the connotation or intension of a natural kind term. And surely it is not too bold to claim that both the categories distinguished by the semantic markers of our language and the conditions of sameness with regard to them fall under the jurisdiction of philosophical and scientific thought and are liable to change with it.

2.6.4 *Speaking of quantities*

Putnam treats physical magnitude terms more cursorily than natural kind terms. This is a pity, for magnitudes are central to modern science, and their manner of being—and of being signified—is of great interest to philosophy. Putnam's approach to the matter is, if I may say so, one of studied laxity. He labels a section of one of his semantical papers "The Meaning of Physical Magnitude Terms" (Putnam, PP, vol. II, pp. 198–207), and he tells us—in a different paper—that it contains "a causal theory of reference in connection with magnitude terms" (Putnam, PP, vol. II, p. 176n.). But, except for the remark that the users of such terms know that they refer to "putative physical *quantities*—capable of more and less, and capable of location" (Putnam, PP, vol. II, p. 199), we do not find in the said section any considerations about meaning that would not apply equally well to natural kind terms. Yet surely speaking of quantities implies some commitments that are not presupposed when one talks about natural kinds.

It is not clear to me whether Putnam considers physical magnitudes a species of physical quantities or whether he regards 'quantity' and 'magnitude' as synonymous. Aristotle, who first discussed the category of quantity, said that a 'quantum' (πόσον) is a 'multitude' (πλῆθος) if it can be counted, a 'magnitude' (μέγεθος) if it can be measured. In the same passage, Aristotle defined a quantum as "that which is divisible into constituents each of which is by nature a *one* and a *this*." He also noted that multitudes can be divided into discrete parts, whereas magnitudes are potentially divisible into constituents that are mutually connected or continuous (συνεχῆ) (Aristotle, *Metaphysica*, Δ, 13, 1020^{a}7–11). Thus, for Aristotle, a 'magnitude' was a quantity of a certain sort, viz., continuous quantity, the only sort that can be measured.

Of course, modern physics has scored great successes in measuring properties like temperature or angular momentum that are not divisible into constituents each of which is a *one* and a *this.*[55] Therefore, we no longer subscribe to Aristotle's definitions but allow the category of quantity to encompass any attributes of things, processes, or events that "can be reasonably represented numerically" (Krantz et al. 1971, p. xvii). In an interesting essay on "The Mathematical Classification of Physical Quantities," J. C. Maxwell attributes to Hamilton the "most important distinction" between "Scalar quantities, which are completely represented by one numerical quantity, and Vectors, which require three numerical quantities to define them" (Maxwell 1890, vol. II, p. 259).[56] For example, Newtonian mass is a scalar quantity and Newtonian force a vector quantity. And there is, of course, no reason why we should not also speak of tensor quantities, like stress. As to scalar quantities, some—like probability—are real-valued (i.e., representable by real numbers); while others—like the amplitude of a quantum-mechanical probability wave—are complex-valued. Does Putnam's term 'magnitude' refer to all these different sorts of quantities? An answer to this question cannot be extracted from his text. I shall assume, however, that Putnam's 'magnitudes' are real-valued scalar quantities. This will greatly simplify our discussion without restricting its scope, for the main conclusion we shall reach regarding such quantities holds a fortiori for the rest.

An attribute of physical objects is a magnitude in the stated sense if, but only if, its particular instances can be assigned numbers in such a way that some of the distinctive structural features of these numbers come to represent physical relations between those instances. For example, our ordinary concept of mass requires that the mass of a body *A* should be assigned a number equal to *n* times that ascribed to the mass of another body *B* whenever *A* balances a body formed by putting together *n* copies of *B*. (The copies in question need only be equal *in mass*, i.e., they must balance one another.) The investigation of the exact import and requirements of such numerical representations of physical attributes was initiated by Helmholtz (1887) and carried forward by Hölder (1901) and N. R. Campbell (1920). It reached a high level of sophistication in the treatise *Foundations of Measurement*, by Krantz, Luce, Suppes, and Tversky (1971). Further advances are recorded in Louis Narens' *Abstract Measurement Theory* (1985). In the light of this research tradition, the numerical representation of an attribute of physical objects is best conceived as a mapping of the set of its instances into the field **R** of real numbers, so contrived that some of the structural features of **R** faithfully represent—by way of the homonymous mappings induced on Cartesian products and power sets—relations characteristic of that set. The existence of such a mapping obviously entails that the particular instances of the attribute provide a concrete realization of a species of structure—in Bourbaki's sense—whose distinguished components meet specific conditions. (The terminology employed and the general mathematical ideas involved in the two foregoing sentences and in the rest of this section are explained in the appendix on mathematical structures, which is Section 2.8. Readers who are bored by mathematical definitions may proceed at once to the conclusion in the last paragraph of the present subsection, on page 65, provided that they are willing to accept it on faith.)

To give an idea of the nature and strength of such conditions I shall report on the two simplest types of extensive magnitude characterized in *Foundations of Measurement.* Roughly speaking, extensive magnitudes—like mass—are numerically representable attributes whose instances can be compared as to size and can be added to one another to make larger instances. The numerical representation of such magnitudes makes use of two structural features of the real number field, namely, its additive group structure and the ordering of the real numbers by the binary relation '≥'. As a matter of fact, what is used is not the full additive group structure of $\mathbf{R}$ but the semigroup structure that remains when addition is restricted to the set $\mathbf{R}^+$ of the positive reals, while their inverses and the neutral element 0 are forgotten. It is thus apparent that in order to conceive a physical attribute as an extensive magnitude one must distinguish at least two structural features in the extension Q of that attribute, viz., (i) a reflexive, connected and transitive relation P such that $\langle Q,P\rangle$ is a weak order, and (ii) an associative binary operation $\oplus$ such that $\langle Q,\oplus\rangle$ is a semigroup. ($\oplus$ is therefore a mapping of Q^2 into Q. We write $a \oplus b$ for the value of $\oplus$ at $\langle a,b\rangle$. $\oplus$ is associative, so that for any $a,b,c \in Q$, $(a \oplus b) \oplus c = a \oplus (b \oplus c)$.) A faithful numerical representation of $\langle Q,P,\oplus\rangle$ will then be given by a mapping μ: $Q \rightarrow \mathbf{R}^+$ such that for all $a,b \in Q$,

M1. aPb if and only if $\mu(a) \geq \mu(b)$; and

M2. $\mu(a \oplus b) = \mu(a) + \mu(b)$.

Note that if μ satisfies M1 and M2, and $\alpha \in \mathbf{R}^+$, $\alpha\mu$: $a \mapsto \alpha\mu(a)$ is a mapping of Q into $\mathbf{R}^+$ which also satisfies M1 and M2 and thus provides another faithful numerical representation of the quantity Q. μ and $\alpha\mu$ are said to be different *scales* for measuring the same quantity.

A mapping μ meeting conditions M1 and M2 exists if and only if $\langle Q,P,\oplus\rangle$ is what Krantz et al. call a closed positive extensive structure, i.e., if and only if its components satisfy the axioms C1–C5 given below.[57] For greater perspicuity, I write '$a \sim b$' for 'aPb and bPa'; and 'na' for '$a_1 \oplus \ldots \oplus a_n$', where $a_i \sim a$ for every value of the index i $(1 \leq i \leq n)$.

$\langle Q,P,\oplus\rangle$ is a *closed positive extensive structure* if and only if

C1. $\langle Q,P\rangle$ is a weak order, and for all $a,b,c \in Q$,

C2. $(a \oplus b)Pa$, but it is not the case that $aP(a \oplus b)$;

C3. $(a \oplus b) \oplus c \sim a \oplus (b \oplus c)$;

C4. The following three conditions hold whenever any one of them holds: aPb, $(a \oplus c)P(b \oplus c)$, and $(c \oplus a)P(c \oplus b)$;

C5. If aPb but it is not the case that bPa, then for any $x, y \in Q$, there is a positive integer n such that $(na \oplus x)P(nb \oplus y)$.

Axioms C1–C5 are consistent if the theory of real numbers is consistent, for the structure $\langle \mathbf{R}^+, \geq, + \rangle$ evidently would satisfy them.

One may well doubt that the instances of any extensive physical magnitude can in fact be weakly ordered by size. If they could, the relation 'x is neither greater nor smaller than y' (here symbolized by '~') would be transitive. Evidently, this requirement is not satisfied by observed magnitudes, for due to the limited power of resolution of our instruments and organs, any three instances a, b, and c of an extensive magnitude can be such that the size of a is indistinguishable from that of b, and the size of b from that of c, and yet a is perceptibly larger than c. However, it is ordinarily assumed that the actual instances of an extensive magnitude that our observations do but partially and imperfectly record are weakly ordered by size (just as, mutatis mutandis, the actual instances of an intensive physical magnitude, such as temperature, are understood to be weakly ordered by intensity).[58] On this understanding, if P designates the ordering relation, every extensive magnitude that is studied in physics satisfies axiom C1. Moreover, the instances of such a magnitude can, under certain conditions, be joined in a standard way to make larger instances. Evidently, if $\oplus$ designates the operation of joining two instances of a given magnitude in the appropriate way, axiom C2 is fulfilled by any instances of it, a and b, which can be thus joined. The reader will be persuaded that C4 is satisfied if c is an instance that can be joined to a and to b, and that C3 holds good if c can be joined to $a \oplus b$. Now, two instances of an extensive magnitude cannot be joined to make a bigger one if any of them happens to be the product of joining the other with a third one. But this limitation is easily overcome by treating instances of equal size as interchangeable copies. Then, even if $a = b \oplus c$, $a \oplus b$ is defined if there is an instance $b' \sim b$, and a can be joined to b'. The solution fails, though, if we run out of copies. A similar—and equally insuperable—limitation arises if one of the instances is too large to be joined with the other by the appropriate standard method. To cope with this difficulty, one can include in the structure of any extensive magnitude Q for which it arises a distinguished subset D of Q^2, on which alone the operation of joining is defined, and treat $\oplus$ as a mapping of D into Q. A faithful numerical representation of such a structure $\langle Q,P,D,\oplus \rangle$ is given by a map $\mu: Q \rightarrow \mathbf{R}^+$ that meets condition M1, but satisfies the following requirement instead of M2:

M2*. For every $a, b \in D$, $\mu(a \oplus b) = \mu(a) + \mu(b)$.

Krantz et al. have established sufficient conditions for the existence of a mapping μ satisfying M1 and M2*. These conditions involve, of course, the replacement of C2, C3, and C4 by other axioms that make allowance for the fact that the operation ⊕ is now defined only on a part of Q^2. They also require a drastic revision of axiom C5, which we shall now discuss.

Axiom C5 is a version of the so-called Archimedean postulate. It says in effect that if the set Q of instances of an extensive magnitude is the base set of a closed positive extensive structure, no two instances of that magnitude are incommensurable. No matter how small is x and how large is y, if a is but slightly larger than b, there is an integer n such that x joined to n copies of a is larger than y joined to n copies of b. But if ⊕ is not everywhere defined on Q^2, it may not be possible to form na and nb for a sufficiently large integer n. Yet the faithful representation of an extensive magnitude in the semigroup $\langle \mathbf{R}^+,+\rangle$ certainly demands that the structure of that magnitude should satisfy some form of postulate. The version chosen by Krantz et al. to overcome the present difficulty can be simply stated if we introduce two new terms. Let a *standard sequence in Q* designate either an infinite sequence $a_1, a_2, \ldots$ or an m-tuple $\langle a_1, \ldots, a_m\rangle$ of elements of Q, such that for each positive integer n less than the number of terms in the standard sequence, $a_{n+1} = a_n \oplus a_1$. Let us say that a standard sequence in Q is *strictly bounded* if there is a b in Q such that for every term a_k in the sequence, bPa_k is true but a_kPb is not. The Archimedean postulate for $\langle Q,P,D,\oplus\rangle$ says then that *every strictly bounded standard sequence in Q is finite* (i.e. is an m-tuple, for some positive integer m). In effect, Krantz et al. (1971, p. 87) prove that there is a mapping $\mu: Q \to \mathbf{R}^+$ satisfying M1 and M2* if $\langle Q,P,D,\oplus\rangle$ is what I shall call an *ordinary extensive structure*, that is to say, if

E1. $\langle Q,P\rangle$ is a weak order, and for all $a,b,c \in Q$,

E2. If $\langle a,b\rangle \in D$, then $(a \oplus b)Pa$, but it is not the case that $aP(a \oplus b)$;

E3. If $\langle a,b\rangle \in D$ and $\langle a \oplus b,c\rangle \in D$, then $\langle b,c\rangle \in D$, and $\langle a,b \oplus c\rangle \in D$, and $((a \oplus b) \oplus c)P(a \oplus (b \oplus c))$;

E4. If $\langle a,c\rangle \in D$ and aPb, then $\langle c,b\rangle \in D$ and $(a \oplus c)P(c \oplus b)$;

E5. If aPb but it is not the case that bPa, then there is some $x \in Q$ such that $\langle b,x\rangle \in D$ and $aP(b \oplus x)$;

E6. Every strictly bounded standard sequence in Q is finite.

Krantz and his associates characterize further structure for periodic extensive magnitudes (e.g., angle), extensive magnitudes with an essential maximum,[59] and many types of nonextensive magnitudes. Additional structures are defined by Narens (1985). But the two examples examined here sufficiently illustrate the complexity and far-reaching scope of some of the structural conditions involved in quantitative thinking. To use a physical magnitude term correctly one certainly does not have to know the Krantz-Luce-Suppes-Tversky axioms for the corresponding structure, any more than one needs to know the Hilbert axioms for the Euclidian plane in order to pave the kitchen floor with square tiles. But just as a commitment to thus pave a floor presupposes that the Hilbert axioms apply to it within the admissible margin of imprecision, so a meaningful reference to a physical attribute as an extensive magnitude of a certain type implies that its instances stand to one another as elements of the pertinent structure. Indeed even the humblest thought of an extensive quantity—e.g., the thought that regulates the pouring of oil into a food processor to make mayonnaise—conceives its instances as orderable by size and as liable to increase by the addition of further instances. Such notions, plus an idea of the standard method or methods for joining its instances, belong to what, after Putnam, we may call the stereotype of a given extensive magnitude term. But, of course, in a Putnamist theory of meaning, the truly critical factor in the semantics of a general term is not its stereotype but its extension. The extension of a natural kind term is the set of its instances. Insofar as this set is open-ended, it can only be determined by a list of conditions (an intension). The same can be said of the instances of a magnitude, but with an added proviso: a collection of particulars can pose as the set of instances of a certain magnitude only to the extent that it is a realization of the corresponding abstract structure. While the extension of a natural kind term is normally conceived as an unstructured set, a set can be the extension of a given physical magnitude only if it is the base set of a suitable structure. Of course, no collection of particulars actually met in science is seen as *the* set of (all conceivable) instances of a magnitude. But in order to be grasped as *a* set of such instances it must be embedded in the full structure.[60] A fortiori, the same holds also for nonscalar physical quantities.

It would be interesting to inquire into the epistemological implications of this remarkable difference between the thought of quantities, prevalent in modern science, and the thought of natural kinds, favored by Aristotle. One surmises that Aristotle's neglect of quantitative concepts and the tightly knit structures they connote is the main reason why, notwithstanding his conviction that natural necessity is the distinguishing mark of the subject matter of natural science, he did so poorly at unravelling any particular examples of it. But here we must deal with the meaning of physical magnitude terms, and specifically with the likelihood that they preserve their reference when they are inherited from a scientific theory by its revolutionary successor. Putnam contends that a physical magnitude term securely enjoys a stable reference when its uses have the proper causal connection with a situation—an "introducing event"—in which the magnitude referred to by that term was actually singled out "as the physical magnitude *responsible* for certain effects in a certain way" (Putnam, PP, vol. II, p. 200). Even if the introducing event has been forgotten, the intention of referring to the same magnitude that was referred to by the same term in the past links our current use of it to those earlier uses. Indeed, the very presence of the term in our vocabulary is a causal product of earlier events and ultimately of the introducing event.

> If anyone knows that 'electricity' is the name of a physical quantity, and his use of the word is connected by the sort of causal chain I described before to an introducing event in which the causal description given was, in fact, a causal description of electricity, then we have a clear basis for saying that he uses the word to refer to electricity. Even if the causal description failed to describe electricity, if there is good reason to treat it as a mis-description *of electricity* (rather than as a description of nothing at all)—for example, if electricity was described as the physical magnitude with such-and-such properties which is responsible for such-and-such effects, where in fact electricity is responsible for the effects in question, and the speaker intended to refer to the magnitude responsible for those effects, but mistakenly added the incorrect information 'electricity has such-and-such properties' because he mistakenly thought that the magnitude responsible for those effects had those further properties—we still have a basis for saying that both the original speaker and the person to whom he teaches the word use the word to refer to electricity.
>
> (Putnam, PP, vol. II, p. 201)

I take it that when Putnam speaks of physical quantities being *responsible* for observed effects he does not mean to anthropomorphize them, but only to describe them as efficient causes. Now, causality has been conceived in philosophy as a relation between two things (or rather, between a person or, generally, an animal and a thing, as when Aristotle said that the sculptor was the cause of the statue) or between two events (as in 'the fire was caused by lightning'). Perhaps common sense would also acknowledge causal connections between things and events (as in 'a single fission bomb caused the death of thousands of Japanese children'). But I confess I had not heard of causal relations between magnitudes and phenomena. I must therefore wait for a new analysis of causality before I can accept the causal efficacy of magnitudes *as such*, not just of the things or events that sport them. Indeed, even if such an analysis were forthcoming, I doubt it could succeed in bestowing causal powers on magnitudes like distance, or time, or vertical acceleration. Putnam wisely chooses not to talk about them but takes as an example a physical magnitude term that was long used as the name for a putative physical substance. Some of this obsolete connotation must still attach to the current stereotype of 'electricity' if, as Putnam says, it includes the idea of "a magnitude which can move or flow." Such thinglike quantities are met perhaps at "introducing events," but for a straight physical magnitude term which is not thus categorially ambiguous it is very hard to imagine wherein such an event might consist. For my part, I do not think that if I had stood "next to Ben Franklin as he performed his famous experiment" and had heard him say that 'electricity' denotes that "which collects in clouds" until it suddenly "flows from the cloud to the earth in the form of a lightning bolt," I could have even guessed that he was introducing a name for a *magnitude,* not a fluid.

Be this as it may, the important lesson that can be drawn from our discussion of extensive magnitudes is that he who conceives a particular as an instance of a quantity must place it wittingly or unwittingly in the relational network of a structure constituted by it and its fellow instances. If a quantity is an attribute that can be represented numerically, its conditions of identity must anyway include the structural features portrayed by its numerical representations. The reference of a physical magnitude term cannot be impervious to changes in the structure that keeps its extension together.

2.6.5 *'Mass' in classical and relativistic dynamics*

The foregoing considerations lend support to the incommensurabilist claim that the physical referent of the term 'mass' in relativity physics is by no means identical with that of the homonymous Newtonian term (Kuhn 1962, p. 101). Feyerabend (1962) stated this claim very simply, as follows:

> In classical, pre-relativistic physics the concept of mass [. . .] was absolute in the sense that the mass of a system was not influenced (except perhaps causally) by its motion in the coordinate system chosen. Within relativity, however, mass has become a relational concept whose specification is incomplete without indication of the coordinate system to which spatiotemporal descriptions are all to be referred.
>
> (Feyerabend 1981, vol. I, p. 81)

The coordinate-dependent quantity to which Feyerabend refers here is presumably what is known as *relativistic mass* ('apparent mass' for Dixon 1978, p. 114), i.e., the scalar factor by which one must multiply the velocity of a particle to obtain its relativistic momentum. Since the relativistic mass of a particle *P* is the same with respect to all coordinate systems relative to which *P* has the same speed, its numerical representation may be regarded as defined on the set of ordered pairs $\langle P,v \rangle$, where v stands for the particle's speed and ranges over the closed-open interval $[0,c)$. This alone constitutes an unbridgeable difference between relativistic mass and the quantity called 'mass' in classical mechanics.[61]

However, besides relativistic mass, Relativity assigns to every material particle a scalar quantity known as *proper mass* (also 'rest mass'), which, like classical mass, is a function of state, i.e., is independent of the particle's motion. This bifurcation of the meaning of 'mass' makes it at first blush even harder to equate the term's reference in relativistic and in classical dynamics (see Field 1973).[62] But, on a closer examination, it will be apparent that the proper mass has a better right than the relativistic mass to take over the role of the classical concept of mass. Indeed, a very striking analogy can be drawn between the laws of motion of classical and relativistic dynamics when they are stated in an idiom that makes them comparable. In the standard four-dimensional formulation of Special Relativity developed by Minkowski (1908, 1909), the following equation holds:

$$\mathrm{F} = \dot{\mathrm{p}} = m\ddot{\mathrm{r}} \tag{5}$$

Here r is a four-vector representing the worldpoint or spacetime location of a particle (referred to an arbitrarily chosen origin in Minkowski spacetime); m, p, and F denote, respectively, the particle's proper mass, its four-momentum and the four-force acting on it; and a dot over a variable signifies differentiation with respect to the particle's proper time. Now, not only is eqn. (5) typographically almost indistinguishable from Newton's Second Law of Motion in its standard three-dimensional formulation (I have refrained, however, from using boldface for the four-vectors), but it is homologous with the corresponding law of classical mechanics in the neo-Newtonian four-dimensional formulation (as explained, e.g., by Friedman 1983). To obtain a statement of this law we do not have to rewrite equation (5); it is enough to reinterpret it by letting m, p, and F stand, respectively, for the particle's classical mass, its neo-Newtonian four-momentum, and the neo-Newtonian four-force on it, and by taking the dot over a variable to mean differentiation with respect to universal time. Even the latter seemingly drastic reinterpretation loses much of its sting if we recall that in a neo-

Newtonian theory universal time (i.e., the time coordinate into which all Einstein times collapse in the limit $c \to \infty$) agrees with each particle's proper time along its particular worldline.

There is, however, one very substantial difference between classical mass and the proper mass of Relativity: the proper mass of a particle changes in inelastic collisions (collisions in which the aggregate kinetic energy of the colliding particles is not conserved). Now, this is not one of those differences between two theories that concern only our beliefs about some entity but not the meaning of the term denoting it. For inelastic collisions are of course inevitable in any process of putting masses together to make larger masses. In fact, according to Relativity, when several particles are brought together, the aggregate's proper mass includes not only the mass equivalent of the kinetic energy lost as their relative velocities vanish but also the mass equivalent of any work done against forces that tend to keep the particles apart (*minus*, of course, the work that would have to be done, to separate the particles, against their binding forces). In sharp contrast with this, the total classical mass of a system of classical particles is simply the sum of the masses of each, no matter what their nature and our manner of joining them. Since the standard physical methods of adding the instances of a physical extensive quantity play a key role in the constitution of its structure, the relativistic term 'proper mass' cannot share the same reference with the classical term 'mass' even if they occur in typographically identical and conceptually kindred equations. This may not be altogether obvious as we draw worldlines on paper, labelling each with a little number equal to its proper mass, while at the same time forgetting that those lines stand for material bodies which exert forces on each other and harbor an internal structure which will be set in commotion whenever the lines touch. But as soon as one is reminded of it, it becomes clear that if proper mass has an extensive structure, it must be a very peculiar one, quite different from the straightforward ordinary extensive structure of classical mass.[63]

It is amusing to fancy what an "introducing event" for 'mass' would look like according to the causal theory of reference. It evidently will not do to have an imaginary Newton tell his assistant, after toiling in vain to move a large stone, "Look, it's the stone's big *mass* that won't let it budge," for if the stone were placed on top of a wheelbarrow he would readily carry it away, barrow and all. However, as writers of fiction we are entitled to move the scene to the age of railways. Our Newton could then point out to his hearers that in order to come to a full stop a loaded train needs, ceteris paribus, a much longer piece of track than an empty train. "*Mass,*" he might then proclaim, "is the physical quantity responsible for this effect." The trouble is that if the fancy world he lives in happens to be one in which there is an upper bound to the speed of signals, the quantity thus designated—like the track length used for estimating it—would be frame-dependent. In other words, our mock-Newton would have introduced a magnitude akin to relativistic mass, not to classical and proper mass. To have him introduce by ostension the quantity *m* in equation (5) we must place him with his students in a more recherché situation. For instance, they may travel together in a squadron of freely falling spaceships of different sizes. He can then declare *mass* to be that which causes each ship to abide by its geodesic worldline, so that the larger the mass, the more fuel you must burn to make it deviate from free fall. Obviously, to catch their master's meaning, the students must do more than just look and hear: they must think hard and recall a good deal of the differential geometry and the geometric theory of gravity they learned before taking off. There is no such thing as thoughtless ostension, but in cases like this one the intellectual sophistication involved in ostensively introducing a term is apt to be considerable. As a matter of fact, in real life 'mass' was not introduced by any of the methods suggested here. It was in the course of theological discussions concerning the Holy Eucharist that Newton's term for classical mass, viz., 'quantitas materiæ', came to signify the quantity that remains unchanged when the volume of a body changes by condensation and rarefaction.[64]

2.6.6 Putnam's progress

After 1976, Putnam by and large disassociated himself from the views on reference we have examined in this section. He announced his new stance in his presidential address to the American Philosophical Association on December 29, 1976 (Putnam 1978), and gave a careful statement of its grounds and implications exactly one year later, in his presidential address to the Association for Symbolic Logic (Putnam 1980). Here he distinguished three main positions on reference and truth: "the extreme Platonist position, which posits non-natural mental powers of directly 'grasping' forms," "the verificationist position which replaces the classical notion of truth with the notion of verification or proof, at least when it comes to describing how the language is understood," and "the moderate realist position which seeks to preserve the centrality of the classical notions of truth and reference without postulating non-natural mental powers" (Putnam 1980, p. 464; also in Putnam, PP, vol. III, pp. 1–2). He then proceeded to show, on the strength of the Löwenheim-Skolem Theorem and related results of the theory of models for first-order formal languages, that his earlier "moderate realism"—i.e., realism without "non-natural" grasp of intensions—must be given up. This is not the place to go through Putnam's model-theoretic argument.[65] The gist of it is that the "moderate realist," who tries to carry on with scientific discourse, à la Putnam 1975, by means of general terms which denote without connoting, is incapable of singling out the "intended" interpretation of any scientific theory he may put forward (in a first-order language), from among the infinitely many distinct and even incompatible interpretations that satisfy it.

> Nor do 'causal theories of reference', etc., help. Basically, trying to get out of this predicament by *these* means is hoping that the *world* will pick one definite extension for each of our terms even if *we* cannot. But the world does not pick models or interpret languages. *We* interpret our languages or nothing does.
>
> (Putnam 1980, p. 482; PP, vol. III, p. 24)

> If 'refers' can be defined in terms of some causal predicate or predicates in the metalanguage of our theory, then, since each model of the object language extends in an obvious way to a corresponding model of the metalanguage, it will turn out that, *in each model M, reference*$_M$ is definable in terms of *causes*$_M$ but unless the word 'causes' [. . .] is already glued to one definite relation with metaphysical glue, this does not fix a determinate extension for 'refers' at all.
>
> (Putnam 1980, p. 477; PP, vol. III, p. 18)

As Putnam will have nothing to do with preternatural Platonistic insights into ready-made mind-independent intensions, he seeks a way out of his predicament in the example of mathematical intuitionism and constructivism.

> 'Objects' in constructive mathematics are *given through descriptions.* Those descriptions do not have to be mysteriously attached to those objects by some nonnatural process (or by metaphysical glue). Rather the possibility of *proving* that a certain construction (the 'sense', so to speak, of the description of the model) has certain constructive properties is what is asserted and *all* that is asserted by saying the model 'exists'. In short, *reference is given through sense, and sense is given through verification procedures and not through truth conditions.* The 'gap' between our theory and the 'objects' simply disappears—or, rather, it never appears in the first place.
>
> (Putnam 1980, p. 479; PP, vol. III, p. 21)

Putnam extends this approach to the entire philosophy of science. He disparages his former stance as "the perspective of metaphysical realism," for which "there is exactly one true and complete description of 'the way the world is'," and "truth involves some sort of correspondence relation between words or thought-signs and external things and sets of things." To this "externalist perspective" and its preferred "God's Eye point of view" he opposes "the internalist perspective," thus called "because it is characteristic of this view to hold that *what objects does the world consist of?* is a question that it only makes sense to ask *within* a theory or description" (Putnam 1981, p. 49).

> In an internalist view also, signs do not intrinsically correspond to objects, independently of how those signs are employed and by whom. But a sign that is actually employed in a particular way by a particular community of users can correspond to particular objects *within the conceptual scheme of those users.* 'Objects' do not exist independently of conceptual schemes. *We* cut up the world into objects when we introduce one or another scheme of description. Since the objects *and* the signs are alike *internal* to the scheme of description, it is possible to say what matches what.
>
> (Putnam 1981, p. 52)

> What is wrong with the notion of objects existing "independently" of conceptual schemes is that there are no standards for the use of even the logical notions apart from conceptual choices. [. . .] We can and should insist that some facts are there to be discovered and not legislated by us. But this is something to be said when one has adopted a way of speaking, a language, a "conceptual scheme." To talk of "facts" without specifying the language to be used is to talk of nothing; the word "fact" no more has its use fixed by the world itself than does the word "exist" or the word "object."
>
> (Putnam 1988, p. 114)

Those of us who woke early from dogmatic slumber can hardly be surprised by Putnam's "internalism," but we certainly welcome it. Its relevance to the Feyerabend-Kuhn incommensurability thesis and related problems of scientific change is clear enough.[66] Its implications, however, can only be gauged if the notion of a conceptual scheme is made more precise.

2.7 Conceptual schemes

While the phrase 'conceptual scheme' is a fairly new addition to the philosopher's stock-in-trade—there is still no entry for it in the index to Paul Edwards' *Encyclopedia of Philosophy* (1967)—the thoughts it is meant to express or suggest can be traced back to Kant's "Copernican revolution" in philosophy. Foremost among them is the distinction between the sensuous content or "matter" of empirical knowledge and its rational ordering or "form," i.e., the universal outline or *scheme* that is being progressively filled by the manifold "given" through observation and experiment.[67] This distinction leads almost at once to the idea that "the same" content might be structured—by extraterrestrials? by our children's children?—according to a different scheme. Kant was emphatic that "we cannot render in any way conceivable or comprehensible to ourselves" alternative "forms" of sense awareness and discursive thought, "even if they should be possible," and that "even assuming that we could do so, they would still not belong to experience, the only kind of knowledge in which objects are given to us" (Kant 1781, p. 230). But the thought of liberating one's fellow men from the bonds of their inherited mode of thinking and experiencing and of achieving immortality by conjuring up a newly shaped world has proved irresistibly attractive and it still rings titillatingly in many a comment on the revolutionary scope of the Theory of Relativity or the shattering implications of Quantum Mechanics. One further connotation of 'conceptual scheme' can be read out of the above quotation from Kant. Besides being formal or *schematic* and admitting alternatives (within or beyond our reach), a conceptual scheme is *systematic*—it weaves experience into one coherent whole, leaving no loose ends or poorly fitted links a different scheme could seize upon in an attempt to supplant it. This opinion of the wholesomeness of our "forms" of perceiving and understanding is not surprising in a philosopher who, for all his critical radicalism, continued to view them—in the manner of 18th century creationism—as springing from the God-given nature of human reason (see Kant 1781, p. 669); but from our present evolutionary perspective it looks indeed most implausible. Perhaps that is why nowadays the completeness of "our conceptual scheme" and its capacity to digest anything we might come across are not often argued for. And yet, if it were openly acknowledged that our variegated experience is being put together in accordance with diverse, imperfectly coherent, partly asystematic, unfinished, not altogether stable patterns of understanding, much recent philosophizing about conceptual schemes—pro or contra—would lose its edge.[68]

The two writers who have probably done most to disseminate the English expression 'conceptual scheme' on either side of the Atlantic are W. O. Quine and Sir Peter Strawson. The Kantian themes of form versus content and of all-embracing order, as well as the post-Kantian theme of alternative, disposable schemes, resound in the following passage of Quine's "On What There Is," published in 1948:

> We adopt, at least insofar as we are reasonable, the simplest conceptual scheme into which the disordered fragments of raw experience can be fitted and arranged. Our ontology is determined once we have fixed upon the over-all conceptual scheme which is to accommodate science in the broadest sense; and the considerations which determine a reasonable construction of any part of that conceptual scheme, for example, the biological or the physical part, are not different in kind from the considerations which determine a reasonable construction of the whole.
>
> (Quine 1961, pp. 16–17)

Here Quine appears too sanguine about our chances of finding a vantage point outside all conceptual schemes from which to judge them and make an intelligent choice among them. Later in life, he withdrew this suggestion, noting that the study and revision of a given conceptual scheme cannot be undertaken "without having some conceptual scheme, whether the same or another[. . .], in which to work" (Quine 1960, p. 276). The principles of simplicity and coherence that preside over conceptual reform are now said to be internal to the prevailing scheme.

In his profound and influential book *Individuals: An Essay in Descriptive Metaphysics* (1959), Strawson voices a far more moderate opinion than Quine's about the scope available for conceptual change:

> Certainly concepts do change, and not only, though mainly, on the specialist periphery; and even specialist changes react on ordinary thinking. [But] there is a massive central core of human thinking which has no history—or none recorded in histories of thought; there are categories and concepts which, in their most fundamental character, change not at all. Obviously these are not the specialties of the most refined thinking. They are the commonplaces of the least refined thinking; and are yet the indispensable core of the conceptual equipment of the most sophisticated human beings. It is with these, their interconnexions, and the structure that they form, that a descriptive metaphysics will be primarily concerned.
>
> (Strawson 1959, p. 10)

Throughout the book, Strawson refers to this unchanging "core of human thinking" by the phrase "our conceptual scheme."

Neither Quine nor Strawson nor—as far as I know—any of their followers have attempted to give a full, systematic description of such a scheme, in the manner of Kant's "metaphysic of experience," or of the many systems of categories that 19th century philosophers developed in Kant's wake. The cautiousness of our contemporaries stems, no doubt, in part from a deep seated aversion to heavy-handedness in philosophy; but it may also be due to some however dim realization that no such system is to be had. Be that as it may, what seems to be agreed on by all writers who countenance this general approach is that our current adult way of singling out, identifying, and reidentifying particular objects of diverse sorts pertains to our conceptual scheme. This fits well with what Putnam says in the passage quoted near the end of Section 2.6. Not all conceptual schemers, however, will accept Putnam's claim that, because we "cut up the world into objects when we introduce one or another scheme of description," it follows that objects "do not exist independently of conceptual schemes." A more subtle view has been firmly and clearly put forward by David Wiggins:

> [Philosophy] must hold a nice balance [. . .] between the extent to which the concepts that we bring to bear to distinguish, articulate and individuate things in nature are something invented by us and the extent to which these concepts are something we discover and permit nature itself to intimate to us and inform and regulate for us. Conceptualism properly conceived must not entail that before we got for ourselves these concepts, their extensions did not exist autonomously, i.e. independently of whether or not the concepts were destined to be fashioned and their compliants to be discovered. What conceptualism entails is only that, although horse, leaves, sun and stars are not inventions or artifacts, still, in order to single out these things, we have to deploy upon experience a conceptual scheme which has itself been fashioned or formed in such a way as to make it *possible* to single them out.
>
> (Wiggins 1980, p. 139)

Wiggins' balancing act is not without difficulty, but it is preferable—at any rate, for and within our human self-understanding—to Putnam's facile plunge into the quagmire of relativism. For, as Wiggins aptly emphasizes, to single out something, one must single *it* out. Reference must reach out for the referent, not bring it about, or it will have failed its purpose. Positing objects and pointing at them—by word or gesture—are both necessary, equally respectable, but altogether different mental functions. A philosophy that is unable to distinguish between them lacks, so to speak, sufficient power of resolution. The myth of the self-differentiating object ("the object which announces itself as the very object it is to the mind"—Wiggins 1980, p. 139) is in effect a remnant of primeval animism and we ought to disown it. But it is then all the more inevitable that we own the self-differentiating *subject* as the living principle of all experience and, indeed, of all truth. Now, self-differentiation presupposes that *from* which that which I call myself differentiates itself. It also imposes constraints on any conceptual scheme that might conceivably be my own. Let us take a somewhat closer look at this.

The method of "transcendental deduction," especially as practiced by Fichte, was mainly directed at spelling out such constraints. It was assumed that self-awareness in general (*Selbstbewußtsein überhaupt*) is possible, and it was purportedly shown that its conditions of possibility include all the main features of our physical and cultural world, such as space, time, causation, other minds, even private property. Such derivations normally suffered, in one way or another, from the same vice, viz., that a specific content was surreptitiously imported, at one or more stages of the argument, from our human experience into the generic conditions presupposed by the initial assumption. By artfully scaffolding several such steps at each of which a little more was taken for granted than had been validly proved by the preceding ones, the transcendental philosopher managed to go all the way from the abstract demand that a wholly unspecified form of self-awareness must be possible right down to the familiar structures of everyday life. The fallacies of transcendental deduction can be avoided by restricting the inquiry to our human self-awareness and the constraints on conceptual schemes that are manifestly inherent in it. This, if I am right, was Strawson's intent in *Individuals.* Like all genuine philosophy, an inquiry along these lines involves a considerable risk of error. But it is not condemned, like the argument from *Selbstbewußtsein überhaupt,* to a choice between sophistry and vacuity. Of course, it will not disclose the "only possible" conceptual scheme that "consciousness in general" can wield. But its findings may nevertheless be sufficient to curb relativism without yielding to superstition.

We are aware of ourselves as persons surrounded by things and other persons with whom we interact physically. These conditions of our human self-awareness do not entail that we may not wake up one day and find ourselves as disembodied spirits amidst a choir of angels—whatever that may mean. But even granting, for the sake of the argument, the unlikely premise that our own identity would not perish in such a sweeping change, it is apparent that the identity of our world cannot survive it. Now, a "form of experience" to which no transition can be made from ours without loss of our self-identity or of the identity of "the real" from which we self-differentiate ourselves is plainly irrelevant to the philosophy of human knowledge. Thus, self-differentiation, as we know it from our self-awareness, restricts the variety of conceptual schemes that, in any interesting sense, deserve our consideration.

It may well be that the idea of a conceptual scheme which is inaccessible from ours, and yet is exercised on the same contents, is not absurd in God's eye. But a God's eye viewpoint is clearly one that we cannot share. Besides, although we can make good sense of the abstract thought of a scheme of categories or basic modes of conception and predication apart from any empirical contents, nobody has ever been able to attach a definite meaning to the notion of a pure content of experience apart from a categorial framework; yet this notion is presupposed by the idea of mutually inaccessible or, as the saying goes, incommensurable schemes of thought that have the same field of application in God's eye.

In point of fact, none of the historical examples adduced by the incommensurabilists—neither the dislodgement of epicycle astronomy by Kepler or phlogiston chemistry by Lavoisier or ether electrodynamics by Einstein, nor even the displacement of Aristotelian by mathematical physics—has brought about an immediate, total remaking of the European mind. In particular, these turnabouts in science do not appear to have affected the enduring core of thought mentioned by Strawson in the passage quoted above. This comprises the conceptual means employed in identifying and reidentifying, localizing and dating—i.e., generally speaking, in objectifying—the diverse features of our everyday life. The stability of such a conspicuous part of our intellectual resources has no doubt provided some of the motivation for the philosophical teachings of Kant and logical positivism about unsurpassable limits to conceptual innovation, discussed in Section 2.4. Since deep and far-reaching conceptual changes are known to have happened in science, a critical point in all such doctrines is the relationship they establish between the inert core of everyday thinking and what Strawson superciliously describes as "the specialist periphery."

On one interpretation of Kant, the connection he saw here was simple enough: the basic concepts and principles of classical mathematical physics were only an improved, streamlined version of the scheme by which men had mentally organized their environment from time immemorial. This view, however, is not only questionable in itself (what precedent is there in everyday thought for the classical idea of the evolution of a physical system governed by differential equations?); it becomes utterly preposterous when one tries to extend it to the connection between the "permanent core" and 20th century physics.

The philosophy of logical positivism was developed in full consciousness of the volatility of fundamental physics, and its mature version keeps clear of such difficulties. No system of scientific thought has an intrinsic, privileged connection with human reason (as Newtonianism supposedly had according to Kant). The ordinary and the "specialist" concepts are mutually related by arbitrary "correspondence rules." Scientific theories may be invented and elaborated in blissful ignorance of actual experience. However, their application to it rests on the suitable choice of a partial interpretation of their esoteric terms in plain words. That such application might involve—and require—a new way of conceiving experience is out of the question. The aim of scientific theorizing is prediction, not understanding. Things are well understood as they are commonly understood.

Scientific change would certainly be less disquieting if all it had to offer were new, more successful ways of telling where the pointers will stop in certain dials given that they stand within such-and-such intervals in other dials. A dial, however, owes its specific difference to the instrument bearing it, which in turn, apart from the scientific ideas that presided over its construction, is just a meaningless "black box." Hence scientific thought must supply the interpretation of ordinary laboratory talk—explaining, for instance, what this or that pointer reading says—and not the other way around. But thanks to modern science and technology we now live among boxes which, for all their gleaming screens, are closed and "black" to the common understanding. With them the "specialist periphery" has penetrated our daily lives and undermined the "core of human thinking." Not that we need to understand their workings in order to use them for our ends. But they stand on every side bearing witness to the inadequacy of our prescientific conceptual scheme for properly grasping what we can see and handle.

It is a commonplace of contemporary culture that the differences between prescientific and scientific categories preclude a satisfactory description in ordinary terms of many of the commonest artifacts of science. We ought to realize, moreover, that there would have been no room for such differences to arise or for the insufficiency of the traditional scheme of thought to become manifest, had that scheme been the definite, comprehensive, coherent, and inherently stable system that Kant and other like-minded "descriptive metaphysicians" made it out to be. If our categorial framework were complete and closed in itself, it would leave no gaps through which to catch a glimpse of anything beyond it, and our understanding could grow only by the further specification of existing ideas, not by the invention of new ones. Fortunately, however, our so-called conceptual scheme is not the ready-made blueprint for a True Intellectual System of the Universe but rather a motley of patterns of understanding, to which others can still be added by varying, extending, or simply forgetting those already available.

The flexibility of our understanding is well demonstrated by the variety of criteria we bring to bear on the identification of the ordinary objects of our attention. Some philosophers maintain that we grasp all such objects as substances or as attributes—properties or relations—of substances. But the very difficulty they have in reaching a coherent and satisfactory account of substances should make us wary of oversimplification in this matter.[69] Suppose you are asked to hum a tune you have just heard played on the piano. Physics has taught us to understand a tune as a complex pattern of air waves; but this understanding is quite irrelevant to your task. Either the category of substance must be stretched to encompass musical notes and melodies, or you have here to do with an object which you grasp neither as a substance nor as an attribute. A flash of lightning, a gust of wind, a fire, a river, can, with some effort of the imagination, be brought under the substance-and-attribute scheme. But it is not under any such description that they are thought of when they ordinarily attract our attention. A waterfall is falling water, but it is individuated by the falling, not by the water. A philosopher might argue that it is a feature—an "accident"—of the place where it is located; but in ordinary experience it is the fall that makes the place, ordering the entire landscape around the stupendous downpouring of ever-renewed water. It is permissible and perhaps even plausible to hold with Aristotle that all happenings like those just mentioned are either motions, or alterations, or shrinkings and growths, or generations and destructions, of substances. But this is a principle of revisionary, not descriptive, metaphysics; a guide for the formation of scientific hypotheses, not a truth about the actual structure of prephilosophical human experience.

The inadequacy of the substance-and-attribute scheme for capturing our ordinary ways of thinking is especially obvious if one seeks to apply it to social and cultural realities. Naturalistically minded ontologists have labored in vain to reduce to it the familiar concepts and concept clusters under which those realities are effortlessly grasped, e.g., 'art', 'language', 'money', 'law'. What substances change as a language gradually loses the subjunctive or when a loan to a developing country is classified as "nonperforming"?

But even within its accustomed range of application the classical category of substance is too heterogeneous to play the unifying role usually assigned to it. Aristotle declared that "substance is thought to belong most obviously to bodies" and went on to explain that we therefore say that "animals and plants and their parts are substances," but also "the natural bodies, such as fire and water and earth and the like, and such things as are either part of them or composed from them, [. . .] like heaven [οὐρανός] and its parts, stars and moon and sun" (*Metaphysica*, Z, 2, 1028^{b}8ff.). Water and "earth," however, are not usually grasped as vast, disconnected, sprawling bodies, but rather as stuffs out of which bodies are made. (The same, presumably, must be said of fire, as Aristotle understood it; although one should be permitted to doubt that flames were ever perceived as substantive bodies except by philosophers and very small children.) The classical prototypes of substance are thus quite disparate: on the one hand, the "simple bodies" or stuffs; the living organisms, on the other. The tension between these two extremes was often resolved by allowing only one of them and discarding the other. While the godforsaken majority tended to equate primordial being or substance with stuff—a preference already implicit in the use of the word *substantia,* which latinizes the Greek ὑποκείμενον, not οὐσία—a bold and imaginative thinker like Leibniz chose the opposite option: according to him, it is from my self-awareness and my awareness of other beings like myself that I obtain my idea of substance.[70] Of course, Aristotle embraced stuffs and plants and animals and even his incorporeal astrokinetic intelligences in the category of substance, and much of his lasting influence may be due to this broad-mindedness. But for him it is "a man or a plant or some other thing such as these that we pronounce to be substances above anything else" (ἃ δὴ μάλιστα λέγομεν οὐσίας εἶναι—*Metaphysica,* Z, 7, 1032^{a}19). And in his doctrine of the simple bodies—earth, water, air, fire, and ether—each endowed with its own characteristic internal principle of motion and rest which is like a simple appetite, he somchow conceives their nature on the analogy of an organism's life. Aristotle laid thereby the foundations of an admirably coherent and comprehensive system of the world. But he also exposed himself to criticism such as that reported above (Section 2.5) and drove the study of motion into a blind alley.

Galileo and his successors, who in the 17th century established a new science of motion, turned away from substances and their natures and sought instead the laws of phenomena. They renounced Aristotle's encyclopedic ambitions and devoted their attention to special types of events displayed in well-circumscribed situations. They subjected them to a new way of thinking that is, to this day, the intellectual hub of physics. It consists in developing for any given such type of occurrence a *physical theory,* which in effect sustains the unity of the type and accounts for its internal diversity. In Chapter 3, we shall examine the more significant features of this mode of thought in the light of some recent work in the philosophy of science. Before proceeding to it, let me summarize our findings about the question raised in Section 2.3 regarding the continuity of scientific thought and the comparability of scientific claims in the face of radical conceptual innovation. I stated there three conditions, each one of which would be sufficient to warrant such comparability and continuity, viz.:

C1. Some concepts are immune to change, and they provide a stable reference to decisive facts.

C2. The new concepts are arrived at through internal criticism of the old, by virtue of which the facts purportedly referred to by the earlier mode of thought are effectively dissolved.

C3. Reference to facts does not depend on the concepts by which they are grasped.

In Section 2.6 I argued at length that reference to facts is not independent of the concepts by which they are grasped. In Section 2.4 I contended that there are no known grounds for believing that any concepts relevant to science are immune to scientific change. In Section 2.5 I tried to show how new concepts arrived at through internal criticism of the old preserve the continuity of scientific thought, but I noted that it is unlikely that all new scientific concepts are formed in this manner. Thus, the results of our inquest are quite contrary to C1 and C3, and only partly supportive of C2.

In view of such findings, it is fair to conclude that the incommensurabilists would be right *if* our understanding were a tightly knit system in which a change in any fundamental concept must bring about semantic displacements in all the rest. In that case, revolutions in basic science would shake and shatter the very roots of reference, and successive, conceptually diverse scientific theories could not, strictly speaking, be about the *same* objective situations or compete for confirmation by the *same* empirical evidence. However, our human reason is not the rigid, single-purpose, all-of-a-piece engine of war fancied by some philosophers but is a many-faced, makeshift bundle of intellective endeavors. Indeed, Kant himself, who set so much store by the purportedly systematic nature of our categorial framework, was aware that our scientific understanding cannot cope with our moral life and our experience of art, and proclaimed the separateness of these distinct spheres of reason.[71]

Here, indeed, we are not concerned with the all-too-obvious multiplicity of our modes of thought in diverse fields and walks of life, but with intellectual variety in science and specifically in physics. The modern understanding of natural phenomena by means of physical theories methodically segregates the domain of each new theory from the broad background of experience, as articulated by common sense and earlier science. To have a limited scope—even when the limits are not exactly known—is therefore a characteristic of all physical theories (and an important reason for their effectiveness). No physical theory lays claim to a global understanding of reality.

The physicist who substitutes one theory for another goes on living in the same neighborhood, working at the same institute, driving every day the same old road between them, to and fro, while reflecting on the politics of his nation or the moods of his teenage children or the failings of his car in the light of the same social, moral, or low-level mechanical concepts as before. Kuhn's dictum that "after a revolution scientists are responding to a different world" is either a piece of empty rhetoric or evidence of a misunderstanding of the nature and scope of physical theories. The founders of mathematical physics did of course abandon the Aristotelian system, which was indeed a worldview, not a physical theory. But that does not mean that they left the world in which they and their forebears had lived until then. For it was not an Aristotelian world, and nobody, not even Aristotle himself, had ever managed to see everything in it as depicted by the Aristotelian view. Nor do theory dislodgments such as physics has repeatedly known since the 17th century imply in and by themselves a change of worldview, for neither the dislodged nor the dislodging theories have ever concerned the *world* or entailed a definite conception of it.[72]

Kuhn's dictum is indeed true—and trivial—if by "a different world" he just meant the peculiar domain of the revolutionary theory. But even on this more temperate interpretation the phrase is apt to be misleading. The domain of a revolutionary physical theory is normally conceived so as to include that of the earlier theory which it is meant to surpass. The earlier theory's success within its own domain is then accounted for by the novel theory, which thereby draws the limits of the old one. Thus, new theories in mathematical physics typically do not dislodge their predecessors except to lodge them permanently in the appropriate epistemic niche (Rohrlich and Hardin 1983). The innovative physicist must indeed be able to refer to the old domain in order to rethink it as a proper part of the new one. But this is hardly surprising, unless we picture scientists, as in the caricature Kuhn has drawn of them, on the analogy of religious zealots, who may "convert" from one mode of thinking to another but cannot retain two together in their one-track minds. In the real world, of course, the creators of new physical theories have been trained in the extant ones and—as noted in Section 2.5—it is often by critically reflecting on them that they find a way of going beyond them. All special fields of inquiry must, moreover, be accessible from the same general background of human life. Thus, for example, a relativist and an ether theorist could both work with interferometers made to the same specifications by the same manufacturer or read about Michelson and Morley's experiment in the same issue of *The Philosopher's Magazine.* We do not hold the shortsighted opinion that physical theories have no further aim than that of enmeshing select parts or features of the common background in a web of calculations. But that is not to deny that the background lies there, ill-understood, confusing even, like a murky ocean joining the shiny islands of theory.

Summing up: The "reality" from which we differentiate ourselves as thinking agents is articulated by the concepts that shape our thought and regulate our action. They provide the intellectual means by which particular objects are distinguished from one another, identified and reidentified. Radical conceptual change would therefore wreak havoc on objectivity, as the incommensurabilists maintain. Their strictures, however, cannot impair the continuous, coherent development of mathematical physics, because this is not a succession of comprehensive, mutually incompatible views of that "reality" (a succession of *Weltanschauungen*) but a plurality of interrelated attempts at conceiving definite features or aspects or parts of it by means of intellectual systems of limited scope ("physical theories," in the peculiar sense to be explored in Chapter 3).

In spite of the often deep conceptual differences between such systems, they are protected against incommensurability by the following factors:

(i) They all belong to a connected historical tradition, in which new formations normally grow from the older ones by critical reflection about them and with explicit reference to them.

(ii) They all draw many of their concepts from the same fairly coherent body of mathematical thought.

(iii) They pick out their respective domains from the one background variously known as "reality," "human experience," or, more pompously, "the world."

(iv) There is an inevitable fuzziness in the way each domain is inserted or "embedded" in the background—this, in turn, favors the gross prima facie identification of some objects referred to by diverse theories, even if the latter conceive them very differently.

2.8 Appendix: Mathematical structures

Mathematical structures came up in our discussion of physical quantities in Section 2.6.4 and will be central to our consideration of physical theories in Chapter 3. I give here a repertory of terms and symbols I use for speaking of such structures. It can serve as a refresher to readers who have some acquaintance with the subject and also as a means for controlling my terminology.

[Following the advice of one of the book's referees, I have interspersed the abstract exposition with examples, which I hope will make it accessible and useful also to readers who know very little about modern mathematics. The examples can be readily spotted because, like this paragraph, they are enclosed in brackets.]

2.8.1 Sets

I take the standpoint of a more or less naïve set theory (see Section 2.8.7). The expression '$\{a, b, c\}$' denotes the set whose members or elements are a, b, and c. If '$S(x)$' stands for a sentence in which all occurrences of a noun and all pronouns proxying for it have been replaced by the variable x, the expression '$\{x \mid S(x)\}$' denotes the set of all objects x such that $S(x)$, provided that there is such a set.

[In standard mathematics it is assumed that, given any set, the existence of certain other sets related to it is assured. Two such conditional existential assumptions are mentioned at the beginning of Section 2.8.3. It is also assumed that there is a set corresponding to the expression '$\{x \mid x$ is a natural number$\}$'. On the other hand, consider the expression '$\{x \mid x$ is a set and x is not a member of $x\}$'. Evidently, there cannot be a set which this expression denotes, for if there were such a set, it would both be and not be a member of itself.]

If a is an element of a set A, we say that a is in A, that a belongs to A, that a is contained in A, or that $a \in A$. A set's identity is completely determined by the elements it contains. If every element of a set A is also an element of the set B, we say that A is included in B, that A is a part of B, that A is a subset of B, or that $A \subset B$. Note that according to this definition, if A is any set, $A \subset A$. If $A \subset B$, the set of all elements of B that do not belong to A is called the *complement* of A in B and is denoted by $B \setminus A$.

The *intersection* $A \cap B$ of sets A and B is the set of all elements that belong to both A and B. The *union* $A \cup B$ of sets A and B is the set of all elements that belong to A, or to B, or to both. If $\mathcal{S}$ is a set of sets, the intersection of $\mathcal{S}$, denoted by $\bigcap\mathcal{S}$, is the set of all elements that belong to every set in $\mathcal{S}$; the union of $\mathcal{S}$, denoted by $\bigcup\mathcal{S}$, is the set of all elements that belong to at least one set in $\mathcal{S}$.

If A is a set, the set of all its subsets, $\{X \mid X \subset A\}$, is called the *power set* of A, and is denoted by $\mathcal{P}(A)$. We write $\mathcal{P}^2(A)$ for $\mathcal{P}(\mathcal{P}(A))$ and generally $\mathcal{P}^{n+1}(A)$ for $\mathcal{P}(\mathcal{P}^n(A))$ $(n \geq 2)$. If A and B are sets, the set formed by all ordered pairs[73] the first term of which belongs to A and the second term of which belongs to B—in other words, the set $\{\langle a,b\rangle \mid a \in A \,\&\, b \in B\}$—is called the *Cartesian product* of A by B and is denoted by $A \times B$. We write A^2 for $A \times A$, and generally A^{n+1} for $A^n \times A$ $(n \geq 2)$. A^n may be called, for brevity, the nth Cartesian product of A.

The *null* set or *empty* set $\emptyset$ is the set $\{x \mid x \neq x\}$; in other words, $\emptyset$ is the set of all objects x such that x is not identical with itself. Evidently, no object at all belongs to $\emptyset$. Thus, if A is a set, the following statement is true: Any object that happens to be an element $\emptyset$ is also an element of A. Therefore, $\emptyset \subset A$. Hence, according to the (standard) definitions adopted here, $\emptyset$ is a subset of every set. A set is said to be *non-empty* if it is not identical with $\emptyset$. Two sets A and B are said to be *disjoint* if $A \cap B = \emptyset$.

2.8.2 *Mappings*

A mapping f from a set A to a set B assigns to each element a of A one and only one element $f(a)$ of B. We often refer to such a mapping as 'the mapping $f\colon A \to B$ which maps A into B by $a \mapsto f(a)$'. We say that f sends a to $f(a)$. A is the *domain* of f, B its *codomain*. Every a in A is an *argument* of f; $f(a)$ is the *value* of f at a. If C is a subset of A, the mapping $f|_C\colon C \to B$, which maps C into B by $x \mapsto f(x)$, is called the *restriction* of f to C. If A is a subset of D, and $g\colon D \to B$ is a mapping such that $g(a) = f(a)$ for any $a \in A$, g is said to agree with f on A and to be an *extension* of f. Note that f is then the restriction of g to A. Given two mappings, $f\colon A \to B$ and $g\colon B \to C$, the *composition* of f by g is the mapping $g \circ f$ that maps A into C by $a \mapsto g(f(a))$.

The set $\{x \mid x = f(y)$ for some $y \in A\}$ is a subset of the codomain of f known as the *range* of f. If the range of f is equal to the codomain of f, one says that f is a *surjective* mapping or a *surjection*, and that it maps its domain *onto* its codomain. By the *fiber* of f over b I mean the set $\{x \mid f(x) = b\}$. If every non-empty fiber of f is a singleton (i.e., if it contains a single element of A), f is said to be an *injective* mapping or an *injection*. A mapping that is both surjective and injective is said to be a *bijective* mapping or a *bijection*. If f is bijective, there is an *inverse* mapping $f^{-1}\colon B \to A$ such that for each $a \in A$, $f^{-1}(f(a)) =$

a. f^{-1} is plainly a bijection. Two sets are *equinumerous* if and only if there is a bijection from one to the other.

An important instance of surjection is furnished by the projection mappings attached to any Cartesian product. If $A = B_1 \times \ldots \times B_n$ is the Cartesian product of component sets $B_1, \ldots, B_n$, then A is mapped surjectively onto its kth component ($1 \leq k \leq n$) by the *projection* π_k: $\langle x_1, \ldots, x_k, \ldots, x_n \rangle \mapsto x_k$.

The *graph* of f is the set $G_f = \{\langle a, f(a) \rangle \mid a \in A\}$. Bourbaki identifies the mapping $f: A \to B$ by $a \mapsto f(a)$ with the ordered triple $\langle G_f, A, B \rangle$ (Bourbaki 1970, E.II.13). To the working mathematician this identification must seem frightfully artificial. However, by providing for the definition of a mapping in set-theoretical terms, it paves the way for the fairly simple universal characterization of mathematical structures given below.

[For example, suppose that Allison is loved by Peter and Jack, and Anne is loved by Edward and Joseph, while Mary is not loved by anybody. Let G be the set of the three girls, and B the set of the four boys. Let us also denote by A the subset of G formed by the girls whose names begin with the letter A and by J the subset of B formed by the boys whose names begin with the letter J. There is a mapping $f: B \to G$, which assigns to each boy in B the girl he loves. The fiber of f over Anne is the set {Edward, Joseph}. The restriction of f to J is an injection. There is also a mapping $\mathsf{g}: B \to A$, which also assigns to each boy in B the girl he loves. g is however a different mapping, because, although it has the same domain as f, it has a different codomain. Note that g is a surjection. Its restriction to J is a bijection. The inverse of this bijection assigns to each girl in its domain the boy in J who loves her. If we equate the ordered pair $\langle a,b \rangle$ with the set $\{\{a\},\{a,b\}\}$ (see Section 2.6.1), the graph of g turns out to be the set

{{{Peter},{Peter,Allison}},{{Jack},{Jack,Allison}},
{{Edward},{Edward,Anne}},{{Joseph},{Joseph,Anne}}}.

The reader is invited to write down the *set* g, in accordance with Bourbaki's definition of a mapping as a set. (Hint: View ordered triples as ordered pairs whose first term is an ordered pair, viz., $\langle a,b,c \rangle = \langle \langle a,b \rangle, c \rangle$.)]

2.8.3 Echelon sets over a collection of sets

In set theory the following two assumptions are normally taken for granted:

(i) If a set A is given, the power set $\mathcal{P}(A)$ is also given.

(ii) If sets A and B are given, the Cartesian product $A \times B$ is also given.

Thus, if a set of sets $\mathfrak{S}$ is given, then, by repeated application of conditions (i) and (ii) we can specify an endless array of sets that are supposedly given together with $\mathfrak{S}$. We refer to it as the array of *echelon sets* over $\mathfrak{S}$. A set A is an *echelon set* over the set of sets $\mathfrak{S}$ if and only if A meets one of the following conditions:

(α) A is one of the sets in $\mathfrak{S}$;

(β) $A = \mathcal{P}(B)$, and B is an echelon set over $\mathfrak{S}$; or

(γ) $A = B \times C$, and both B and C are echelon sets over $\mathfrak{S}$.

If the set $\mathfrak{S}$ is just the singleton $\{A\}$, we speak of echelon sets over A.

Suppose now that, after the Polish fashion, we write '$\mathcal{P}A$' for '$\mathcal{P}(A)$' and '$\times AB$' for '$A \times B$'. Then, if every set in the set $\mathfrak{S}$ is assigned a unique and exclusive name, each echelon set over $\mathfrak{S}$ can be described without ambiguity by a unique expression formed with the names of one or more sets in $\mathfrak{S}$ and one or more occurrences of the symbols $\mathcal{P}$ and $\times$. Let A be an echelon set over the set of sets $\mathfrak{S} = \{S_1, \ldots, S_n\}$ and let $Q(A)$ denote the unique expression that describes A in this way. The expression formed by substituting in $Q(A)$ the integer i for every occurrence of the name of the set S_i $(1 \leq i \leq n)$ may be called *the scheme for the echelon construction of A*. Two echelon sets over different sets of sets are said to be homologous if the schemes for their echelon construction are identical. Obviously, if the sets $\mathfrak{S}$ and $\mathfrak{S}'$ are equinumerous, then to each echelon set over $\mathfrak{S}$ there is one and only one homologous echelon set over $\mathfrak{S}'$.[74]

Consider now two equinumerous sets of non-empty sets, $\mathfrak{S} = \{S_1, \ldots, S_n\}$ and $\mathfrak{S}' = \{S'_1, \ldots, S'_n\}$. Let $f{:}\bigcup\mathfrak{S} \to \bigcup\mathfrak{S}'$ be such that, for every k $(1 \leq k \leq n)$, $f(S_k) = S'_k$. f determines a mapping from each echelon set A over $\mathfrak{S}$ into the homologous echelon set A' over $\mathfrak{S}'$. I call it the *homonymous mapping induced by f in A*, and when there is no danger of confusion I also denote it by f. The homonymous mapping induced by f in any echelon set over $\mathfrak{S}$ is readily defined in terms of the three conditions (α), (β), and (γ) involved in the definition of echelon sets.

(α) Let A be one of the sets in $\mathfrak{S}$. The homonymous mapping induced by f in A is the restriction of f to A.

(β) Let A and A' be homologous echelon sets over $\mathfrak{S}$ and $\mathfrak{S}'$, respectively. The homonymous mapping induced by f in $\mathcal{P}(A)$ sends each set $X \subset A$ to the set $f(X) = \{y \mid y = f(x) \text{ for some } x \in X\}$.

(γ) Let A and A', B and B', be two pairs of homologous echelon sets over $\mathfrak{S}$ and $\mathfrak{S}'$. The homonymous mapping induced by f in $A \times B$ sends each pair $\langle a,b \rangle$ in $A \times B$ to $\langle f(a), f(b) \rangle$ in $A' \times B'$.

Note, in particular, that if $V = ((U_1 \times U_2) \times \ldots \times U_n)$, where $U_1, \ldots, U_n$ are any echelon sets over $\mathcal{S}$, and f_i denotes the homonymous mapping induced by f in U_i $(1 \le i \le n)$, the homonymous mapping induced by f in V sends each ordered n-tuple $\langle u_1, \ldots, u_n \rangle$ in V to the n-tuple $\langle f_1(u_1), \ldots, f_n(u_n) \rangle$, which clearly belongs to the echelon set homologous to V over the codomain of f.

[Consider again the mapping f from the set B of boys to the set G of girls defined at the end of Section 2.8.2. Figure out the values of the homonymous mapping induced by f in the power set $\mathcal{P}(B)$. For instance, $f(J)$ = {Allison, Anne}, whereas f({Jack, Peter}) = {Allison}.]

2.8.4 *Structures*

I shall characterize a mathematical structure as a finite list or ordered n-tuple of components meeting requirements of a certain kind. The characterization could be extended to the case of a countably infinite list of components, but I shall not attempt it here.

By an *(m,n)-list of structural components* I mean an ordered n-tuple $\langle S_1, \ldots, S_m, Q_1, \ldots, Q_{n-m} \rangle$ such that (i) the first m terms $S_1, \ldots, S_m$ $(m < n)$ are distinct non-empty sets, and (ii) each one of the remaining $n - m$ terms $Q_1, \ldots, Q_{n-m}$ is a distinguished element of some echelon set over $\{S_1, \ldots, S_m\}$. The terms of class (i) are known as the *base sets,* and I shall often refer to those of class (ii) as the *distinguished components* of the (m,n)-list.

Two (m,n)-lists of structural components will be said to be *similar* if the kth term of the one is homologous to the kth term of the other $(1 \le k \le n)$. Let $\mathcal{L} = \langle S_1, \ldots, S_m, Q_1, \ldots, Q_{n-m} \rangle$ and $\mathcal{L}' = \langle S'_1, \ldots, S'_m, Q'_1, \ldots, Q'_{n-m} \rangle$ be two similar (m,n)–lists of structural components and put $\mathcal{S} = \{S_1, \ldots, S_m\}$ and $\mathcal{S}' = \{S'_1, \ldots, S'_m\}$. Let $f\colon \bigcup\mathcal{S} \to \bigcup\mathcal{S}'$ be a bijection such that for every k $(1 \le k \le m)$, $f(S_k) = S'_k$, and let f_r denote the homonymous mapping induced by f on the echelon set over $\mathcal{S}$ containing Q_r $(1 \le r \le n-m)$. If, for every such index r, $f_r(Q_r) = Q'_r$, I shall say that f *transports* $\mathcal{L}$ to $\mathcal{L}'$ and that $\mathcal{L}'$ is related to $\mathcal{L}$ by the *transport mapping* f. A set of conditions C jointly satisfied by the terms of an (m,n)–list of structural components $\mathcal{L}$ is *transportable* if the terms of any other such list related to $\mathcal{L}$ by any transport mapping f jointly satisfy the conditions obtained by replacing in C every occurrence of the name of each term τ of $\mathcal{L}$ by the name of $f(\tau)$.

A *species of structure* is an (m,n)–list of structural components whose terms jointly satisfy a transportable set of conditions. If $m = 1$, one usually designates any particular instance of the species of structure by the name of its sole base set.[75]

[In what follows, I shall characterize, by way of illustration, some of the more important species of structure in mathematics. However, before looking into their definitions, some readers might wish to try a hand at describing as an instance of a species of structure the situation involving two groups of boys and girls introduced at the end of Section 2.8.2. That situation involved a set B of 4 objects, a set G of 3 objects, and a mapping $f: B \to G$. One need not refer to the set of natural numbers in order to specify that G has just 3 objects. It is enough to demand that there are objects x, y, and z belonging to G such that $x \neq y \neq z \neq x$, and that any object w belonging to G is identical with either x, y, or z. This condition is clearly transportable. In a similar fashion one can say that B has just 4 objects. However, in order to define f by means of transportable conditions we need some further structure. I propose the following: Prescribe linear orders for B and G (as defined by L1–L4 below). Let $<_B$ and $<_G$ symbolize, respectively, the linear order on B and on G. I define f as a mapping which sends the first two elements of $\langle B,<_B\rangle$ to the first element of $\langle G,<_G\rangle$, and the last two elements of $\langle B,<_B\rangle$ to the second element of $\langle G,<_G\rangle$. The situation described at the end of Section 2.8.2 will be conceived as an instance of the species of structure $\langle B, G,<_B,<_G,f\rangle$ meeting the said conditions if the linear order on the set of girls is understood to agree with the alphabetical order of their names, and the linear order on the set of boys is understood to agree with the reversed alphabetical order of their reversed names.]

Let us now characterize some species of structure that are mentioned elsewhere in this book. A *group* is a quadruple $\langle G,e,f,g\rangle$, where G is an arbitrary non-empty set, e is an element of G, f is a mapping of G into G, and g is a mapping of G^2 into G such that for any elements a, b, and c of G, the following characteristic conditions are fulfilled:

G1. $g(a,g(b,c)) = g(g(a,b),c)$ (g is associative).

G2. $g(a,e) = g(e,a) = a$.

G3. $g(a,f(a)) = g(f(a),a) = e$.

g is called the group product; f, the inversion mapping; e, the neutral element.[76] It can be shown that e is unique: no further element of G can satisfy the conditions imposed by G2 and G3 on e. The group is said to be *abelian* if the product is commutative, i.e., if $g(a,b) = g(b,a)$ for all a and b in G.

A pair $\langle G,g\rangle$ meeting condition G1 is said to be a *semigroup*. A triple $\langle G,e,g\rangle$ satisfying G1 and G2 is called a *monoid*.

[For example, let G be the set of letter strings defined as follows: (i) if 'α' denotes a letter of the English alphabet, α is a letter string; (ii) if α and β are letter strings, their concatenation $\alpha\beta$ (i.e., α followed immediately by β) is a letter string. Then, if g is the mapping that assigns to any pair of letter strings α and β their concatenation $\alpha\beta$, $\langle G,g\rangle$ is a semigroup. Likewise, if G is the set of natural numbers, $\{0, 1, 2, \ldots\}$, g stands for addition on G and e denotes the number 0, $\langle G,e,g\rangle$ is a monoid. If G is the set of the positive and negative rational numbers, g stands for multiplication, f stands for the operation taking the reciprocal value, and $e = 1$, $\langle G,e,f,g\rangle$ is the *multiplicative group of rationals.* If G is the set of all rationals, g stands for addition, f stands for multiplication by -1, and $e = 0$, $\langle G,e,f,g\rangle$ is another realization of the species of structure *group,* namely, the *additive group of rationals.* There is an unending variety of groups. I shall now propose two simple examples. Let $\langle G,e,f,g\rangle$ be any group, and let f' and g' respectively denote the restrictions of f and g to $\{e\}$. Obviously, $f(e) = e$ and $g(e,e) = e$. Therefore, $\langle\{e\},e,f',g'\rangle$ is a group. Now consider an equilateral triangle ABC with center of gravity P, and let α, β, and γ denote clockwise rotations of the triangle about P by 120, 240, and 360 degrees, respectively. Note that γ agrees then with the identity mapping, which sends each point of the triangle to itself. Put $G = \{\alpha,\beta,\gamma\}$, $f(\alpha) = \beta$, $f(\beta) = \alpha$, $f(\gamma) = \gamma$, and $\gamma = e$. Let g associate with each ordered pair $\langle\xi,\eta\rangle$ of rotations in G the composite rotation $\xi\eta$ effected by applying ξ after η. Clearly, then, $g(\alpha\beta) = g(\beta\alpha) = \gamma$, $g(\alpha\alpha) = \beta$, $g(\beta\beta) = \alpha$, and, for any rotation $\xi \in G$, $g(\gamma\xi) = g(\xi\gamma) = \xi$. Thus, $\langle G,e,f,g\rangle$ is a group.]

A *linear order* is a pair $\langle S,P\rangle$, where S is a non-empty set and P is a binary relation on S, i.e., a subset of S^2 such that for any elements a, b, and c of S, the following four conditions are satisfied:

L1. $\langle a,a\rangle \notin P$ (P is irreflexive).

L2. If $\langle a,b\rangle \in P$ and $\langle b,c\rangle \in P$, then $\langle a,c\rangle \in P$ (P is transitive).

L3. Either $\langle a,b\rangle \in P$ or $\langle b,a\rangle \in P$ or $a = b$ (P is connected).

L4. $\langle a,b\rangle \in P$ entails that $\langle b,a\rangle \notin P$ (P is asymmetric).

P is called the ordering relation and is usually symbolized by $<$ (or by $>$). One writes $a < b$ (or, alternatively, $a > b$) for $\langle a,b\rangle \in P$.

If the following conditions are substituted for L1–L4, $\langle S,P\rangle$ is said to be a *weak order.*

W1. $\langle a,a\rangle \in P$ (P is reflexive).

W2. If $\langle a,b\rangle \in P$ and $\langle b,c\rangle \in P$, then $\langle a,c\rangle \in P$ (P is transitive).

W3. Either $\langle a,b\rangle \in P$ or $\langle b,a\rangle \in P$ (P is strongly connected).

W4. Unless $a = b$, $\langle a,b\rangle \in P$ entails that $\langle b,a\rangle \notin P$ (P is antisymmetric).

The weak ordering relation P is usually symbolized by $\leq$ (or by $\geq$). We write $a \leq b$ (or, alternatively, $a \geq b$) for $\langle a,b \rangle \in P$. If P is required to be reflexive, transitive and antisymmetric, but neither connected nor strongly connected, $\langle S,P \rangle$ is a *partial order*. The three kinds of order here defined have the following, easily provable, property: If $\langle S,P \rangle$ is a linear, weak, or partial order, there is an order $\langle S,P' \rangle$, of the same kind as $\langle S,P \rangle$, such that for any elements a and b of S, $\langle a,b \rangle \in P'$ if and only if $\langle b,a \rangle \in P$. This property justifies the dual notations, ($<$, $>$) and ($\leq$, $\geq$).

[For example, if S is the set of instants discerned in a given time interval I and P stands for temporal precedence, $\langle S,P \rangle$ is a linear order. If S is the set of all events beginning during the interval I and P denotes the relation 'x began before y', $\langle S,P \rangle$ is a weak order. If S is any set, $\langle \mathcal{P}(S), \subset \rangle$ is a partial order.]

A *field* is an octuple $\langle F,F',e,f,g,e',f',g' \rangle$, subject to the following requirements:

F1. F and F' are non-empty sets.

F2. $\langle F, e, f, g \rangle$ is an abelian group (g is called 'addition'; one writes '$a + b$', rather than $g(a,b)$; the neutral element e is usually called 'zero' and denoted by 0).

F3. F' is the complement of $\{e\}$ in F.

F4. g' is a mapping of F^2 into F such that for any $a \in F$, $g'(a,e) = g'(e,a) = e$ (g' is called 'multiplication'; one writes '$a \times b$' or simply 'ab', rather than $g'(a,b)$).

F5. g' is associative (in the sense explained in G1).

F6. $\langle F', e', f', g'' \rangle$ is an abelian group, where g'' denotes the restriction of g' to F' (e' is usually called 'one' or 'unity' and denoted by 1).

F7. g and g' obey the following law of distribution: For any elements a, b, and c of F, $g'(a,g(b,c)) = g(g'(a,b),g'(a,c))$.

A field $\mathbf{F} = \langle F,F',e,f,g,e',f',g' \rangle$ is said to be *ordered* if it is divided into two mutually exclusive subsets, the *negative* and the *non-negative* elements of $\mathbf{F}$, in such a way that the neutral elements e and e' (zero and unity) are both non-negative, and that $g'(a,a)$—i.e., $a \times a$—is non-negative for any $a \in F$. It will be readily seen that if $\mathbf{F}$ is an ordered field in the sense just defined, its primary base set F is linearly ordered by the following condition:

For any $a,b \in F$, $a < b$ if and only if $a - b$ is negative (where $a - b$ stands for $a + f(b)$, i.e., $g(a,f(b))$).

Let $\mathbf{F} = \langle F, F', e, f, g, e', f', g' \rangle$ be an ordered field. If $a \in F$, I write $-a$ for $f(a)$. Put $|a| = -a$ if a is negative, and $|a| = a$ if a is non-negative. Now consider (infinite) sequences in $\mathbf{F}$, i.e., mappings of the full set of natural numbers into the primary base set F. If α_k denotes the value assigned by such a mapping to an arbitrary natural number k, we typically denote the sequence in question by $\alpha_1, \alpha_2, \ldots$ or, for greater brevity, by (α_k). The sequence (α_k) is said to be a *null* sequence if for every ε in F there is a natural number N such that for any natural number $n > N$, $|\alpha_n| < \varepsilon$. Take any given $\alpha \in F$. The sequence (α_k) is said to *converge to the limit* α, if the sequence $(\alpha_k - \alpha)$ is a null sequence. (α_k) is said to be a Cauchy sequence if for every ε in F there is a natural number M such that for any natural numbers $m,n > M$, $|\alpha_m - \alpha_n| < \varepsilon$. The ordered field $\mathbf{F}$ is said to be *complete* if every Cauchy sequence in $\mathbf{F}$ converges to a limit in $\mathbf{F}$.

If N is any positive integer, and $a \in F$, we write Na for $a + a + \ldots + a$, with a repeated N times (where '$a + a$' stands for $g(a,a)$). The ordered field $\mathbf{F}$ is said to be *Archimedean* if for any two non-negative elements $a,b \in F$, there always is a positive integer N such that $a < Nb$.

[There is an endless variety of fields, but two are of paramount importance in physics: the field $\mathbf{R}$ of real numbers and the field $\mathbf{C}$ of complex numbers.

To introduce $\mathbf{R}$, one usually takes the field $\mathbf{Q}$ of rational numbers as the starting point. Let F denote the set of rationals (all negative and non-negative, proper and improper fractions). Let e be the rational number 0, and e' the rational number 1 (or 1/1). Let g and g' stand, respectively, for addition and multiplication of rationals, f for multiplication by -1, and f' for the operation of taking the reciprocal value. The reader should verify that, in this interpretation, the octuple $\langle F, F', e, f, g, e', f', g' \rangle = \mathbf{Q}$ is an ordered (in effect, Archimedean) field. Consider Cauchy sequences in $\mathbf{Q}$. Many of them converge to a rational limit, but many more do not. For example, the sequence $((1 + 1/k)^k)$ does not converge to a rational limit. Thus $\mathbf{Q}$ is not a complete field.

The field $\mathbf{R}$ is defined from $\mathbf{Q}$ with the deliberate aim of filling the gaps that spoil the latter's completeness. Let us say that two sequences of rationals, (α_k) and (β_k), are equivalent if the sequence $(\alpha_k - \beta_k)$ is a null sequence. Let $[\alpha_k]$ stand for the equivalence class which contains the sequence (α_k). Reinterpret F to be the set of all equivalence classes of Cauchy sequences of rationals. We now define addition and multiplication on F by the rules $[\alpha_k] + [\beta_k] = [\alpha_k + \beta_k]$ and $[\alpha_k] \times [\beta_k] = [\alpha_k \times \beta_k]$. (Note that the symbols '+' and '×' inside the brackets represent addition and multiplication of rationals; the same symbols outside the brackets stand for the newly defined addition and multiplication of equivalence classes of Cauchy sequences of rationals. Of course, the definitions only make sense because it can be shown that if the sequence (α_k) is equivalent to (α'_k) and (β_k) is equivalent to (β'_k), then the sequence $(\alpha_k + \beta_k)$ is equivalent to $(\alpha'_k + \beta'_k)$, and the sequence $(\alpha_k \times \beta_k)$ is equivalent to $(\alpha_k{'} \times \beta_k{'})$.) Let e be the equivalence class of the sequence of rationals (0, 0, . . .), and let e' be the equivalence class of the sequence of rationals (1, 1, . . .). The reader should satisfy himself that the octuple $\langle F,F',e,f,g,e',f',g'\rangle$, as now interpreted, is again an Archimedean field. We call it the field of real numbers and designate it by **R**. It can be shown—though this is not the place to do it—that every Cauchy sequence of real numbers converges to a limit in **R**. As desired, **R** is a complete field. Indeed, any complete Archimedean field is isomorphic (in the exact sense explained in Section 2.8.5) with the field **R** just defined. Since isomorphic structures are mathematically indistinguishable, the field **R** of real numbers may be straightforwardly characterized as the one and only complete Archimedean field. Thus, the most pervasive structure of mathematical physics can be introduced directly, without subordinating it to the rationals in the clever, yet undeniably farfetched way we have followed. It is true of course that the equivalence classes of Cauchy sequences of rationals form a complete Archimedean field, as described above, and are therefore an instance of **R**, which is isomorphic with and thus provides an optimal representation for the complete Archimedean fields of nature.

The field **C** of complex numbers may be defined by taking as the base set F the set of all ordered pairs of reals, and defining addition and multiplication by the rules: $\langle\alpha,\beta\rangle + \langle\gamma,\delta\rangle = \langle\alpha + \gamma,\beta + \delta\rangle$, and $\langle\alpha,\beta\rangle + \langle\gamma,\delta\rangle = \langle\alpha\gamma - \beta\delta, \alpha\delta + \beta\gamma\rangle$.]

A *vector space over the field* **F**, or an **F**-*vector space*, is a 13-tuple $\langle V,F,F',\mathbf{e},\mathbf{f},\mathbf{g},e,f,g,e',f',g',h\rangle$ satisfying the following conditions:

V1. V and F are non-empty sets, whose elements are called vectors and scalars, respectively.

V2. $\langle V,\mathbf{e},\mathbf{f},\mathbf{g}\rangle$ is an abelian group (**g** is usually called ‘vector addition’; the neutral element **e** is called ‘the zero vector’ and denoted by **0**).

V3. $\langle F,F',e,f,g,e',f',g'\rangle$ is a field.

V4. h is a mapping of $F \times V$ into V (called ‘multiplication by a scalar’).

V5. For any scalar α and any vectors **v** and **w**, $h(\alpha,\mathbf{g}(\mathbf{v},\mathbf{w})) = \mathbf{g}(h(\alpha,\mathbf{v}),h(\alpha,\mathbf{w}))$.

V6. For any scalars α and β, and any vector **v**, $h(g(\alpha,\beta),\mathbf{v}) = \mathbf{g}(h(\alpha,\mathbf{v}),h(\beta,\mathbf{v}))$.

V7. For any scalars α and β, and any vector **v**, $h(\alpha,h(\beta,\mathbf{v})) = h(g''(\alpha,\beta),\mathbf{v})$.

V8. For any vector **v**, $h(e',\mathbf{v}) = \mathbf{v}$.

For the sake of perspicuity one usually writes $\mathbf{v}+\mathbf{w}$ for $\mathbf{g}(\mathbf{v},\mathbf{w})$, and $\alpha\mathbf{v}$ for $h(\alpha,\mathbf{v})$. One also writes $\mathbf{v} - \mathbf{w}$ for $\mathbf{v} + \mathbf{f}(\mathbf{w})$. Vector spaces are readily enriched with further structure. Consider the 14-tuple $\mathcal{V} = \langle V,\mathbf{R},\mathbf{R}\backslash\{0\},\mathbf{e},\mathbf{f},\mathbf{g},e,f,g,e,f',g',h,\varphi\rangle$, where $\langle V, \mathbf{R}, \mathbf{R}\backslash\{0\}, \mathbf{e}, \mathbf{f}, \mathbf{g}, e, f, g, e', f', g', h\rangle$ is a vector space over $\mathbf{R}$ or real vector space and φ is a mapping of $V \times V$ into $\mathbf{R}$, such that (in the notation just described), for any $\mathbf{u}, \mathbf{v}, \mathbf{w} \in V$, $\alpha, \beta \in \mathbf{R}$,

(i) $\varphi(\mathbf{u},\mathbf{v}) = \varphi(\mathbf{v},\mathbf{u})$,

(ii) $\varphi(\alpha\mathbf{u} + \beta\mathbf{v},\mathbf{w}) = \alpha\varphi(\mathbf{u},\mathbf{w}) + \beta\varphi(\mathbf{v},\mathbf{w})$,

(iii) $\varphi(\mathbf{u},\mathbf{u}) \geq 0$, and

(iv) $\varphi(\mathbf{u},\mathbf{u}) = 0$ if and only if $\mathbf{u} = \mathbf{0}$.

The mapping φ is called an *inner product* and $\mathcal{V}$ is said to be an *inner product space.* Instead of $\varphi(\mathbf{u},\mathbf{v})$ one writes $\langle \mathbf{u},\mathbf{v}\rangle$, or $\langle \mathbf{u}|\mathbf{v}\rangle$, or $\mathbf{u} \bullet \mathbf{v}$. The foregoing characterization of inner product spaces can be extended to complex vector spaces (i.e., vector spaces over the field $\mathbf{C}$) by substituting $\mathbf{C}$ for $\mathbf{R}$ wherever the latter symbol occurs in it, and replacing (i) with

(i*) $\varphi(\mathbf{u},\mathbf{v}) = (\varphi(\mathbf{v},\mathbf{u}))^*$ (where α^* stands for the complex conjugate of $\alpha \in \mathbf{C}$).

The following ideas will be of use in Section 5.3. Let $\mathcal{V}$ be a real or complex inner product space, as described above. The mapping of V into $\mathbf{R}$ by $\mathbf{u} \mapsto |\varphi(\mathbf{u},\mathbf{u})|^{1/2}$ is called the *norm* on $\mathcal{V}$. One writes $\|\mathbf{u}\|$ instead of $|(\varphi(\mathbf{u},\mathbf{u})|^{1/2}$. $\|\mathbf{u}\|$ is also called the *length* of the vector $\mathbf{u}$. Consider an infinite sequence of vectors, $\mathbf{u}_1, \mathbf{u}_2, \ldots \in V$. We say that $\mathbf{u}_1, \mathbf{u}_2, \ldots$ is a Cauchy sequence if for every positive real number ε there is a positive integer N such that for all integers $m,n > N$ we have that $\|\mathbf{u}_m - \mathbf{u}_n\| < \varepsilon$. The inner product space $\mathcal{V}$ is said to be a (real or complex) *Banach space* if every Cauchy sequence $\mathbf{u}_1, \mathbf{u}_2, \ldots \in V$ converges to a vector in V, i.e., if for each such sequence there is a fixed vector $\mathbf{u} \in V$ which is such that for every positive real number ε there is a positive integer N such that $\|\mathbf{u} - \mathbf{u}_n\| < \varepsilon$ for every $n > N$.

The nth Cartesian product $\mathbf{R}^n$ of the real number field with itself can be seen in a natural way as a real vector space, with the following obvious definitions of (1) vector addition and (2) multiplication by a scalar:

(1) $\langle u_1, \ldots, u_n\rangle + \langle v_1, \ldots, v_n\rangle = \langle u_1 + v_1, \ldots, u_n + v_n\rangle$;

(2) $\alpha\langle u_1, \ldots, u_n\rangle = \langle \alpha u_1, \ldots, \alpha u_n\rangle$.

The zero vector is plainly the n-tuple $\langle 0, \ldots, 0\rangle$. The standard inner product on $\mathbf{R}^n$ is given by

(3) $\langle u_1, \ldots, u_n\rangle \bullet \langle v_1, \ldots, v_n\rangle = u_1v_1 + \ldots + u_nv_n$.

This clearly yields the so-called Euclidian (or Pythagorean) norm:

(4) $\|\langle u_1, \ldots, u_n \rangle\| = u_1^2 + \ldots + u_n^2.$

The reader should verify that, with this structure, $\mathbf{R}^n$ is indeed a Banach space.

2.8.5 *Isomorphism*

Let us now define the much abused term 'isomorphism'. Consider two structures $\langle A_1, \ldots, A_r, a_1, \ldots, a_m \rangle$ and $\langle B_1, \ldots, B_s, b_1, \ldots, b_n \rangle$, with base sets $A_1, \ldots, A_r$ and $B_1, \ldots, B_s$, respectively, and a mapping **h** from $A_1 \cup A_2 \cup \ldots \cup A_r$ to $B_1 \cup B_2 \cup \ldots \cup B_s$. **h** is an *isomorphism* between the said structures if and only if (i) **h** is bijective; (ii) $r = s$ and $m = n$; (iii) for some permutation π of $\{1, \ldots, r\}$, the restriction of **h** to A_k is a bijection onto $B_{\pi k}$ $(1 \leq k \leq r)$; and (iv) the homonymous mappings induced by **h** in the echelon sets over A and by the inverse mapping $\mathbf{h}^{-1}$ in the echelon sets over B are such that for some permutation σ of $\{1, \ldots, m\}$, with inverse permutation ρ, $\mathbf{h}(a_i) = b_{\sigma i}$ and $\mathbf{h}^{-1}(b_i) = a_{\rho i}$ $(1 \leq i \leq m)$. Note that conditions (ii) and (iv) jointly imply that $\langle A_1, \ldots, A_r, a_1, \ldots, a_m \rangle$ and $\langle B_{\pi 1}, \ldots, B_{\pi s}, b_{\sigma 1}, \ldots, b_{\sigma n} \rangle$ are, in effect, two similar (r,m)–lists of structural components (see Section 2.8.4). Two structures are *isomorphic* if there exists an isomorphism between them.

[In the first example of of Section 2.8.4 I represented the boy-girl situation of Subsection (b) by the structure $\langle B, G, <_B, <_G, f \rangle$. Now, let Algernon, Angus and Marmaduke be three gluttons, such that Algernon dines on Petunia the cow and Jason the elephant, Angus devours Jonathan the rhinoceros and Edsel the whale, while Marmaduke sleeps. Let G' be the set of gluttons, B' the set of beasts, and f' the mapping that assigns to each animal the glutton who eats it. Let $<_{B'}$ be a linear order on B' in agreement with the alphabetical order of the animal species involved, and $<_{G'}$ a linear order on G' in agreement with the alphabetical order of individual names. Then, I claim, $\langle G', B', f', <_{B'}, <_{G'} \rangle$ is isomorphic with $\langle B, G, <_B, <_G, f \rangle$. To prove it, let **h** map $B' \cup G'$ onto $B \cup G$ by sending each element x of the domain to the element of the codomain which shares with x the first two letters of its name. If we choose $(1,2) \rightarrow (2,1)$ to be the permutation π, and $(1,2,3) \rightarrow (3,1,2)$ to be the permutation σ, it will be clear that **h** satisfies the conditions prescribed above for an isomorphism.]

2.8.6 *Alternative typifications*

Consider now two species of structure, $\Sigma_1 = \langle S_1, \ldots, S_m, A_1, \ldots, A_h \rangle$ and $\Sigma_2 = \langle S_1, \ldots, S_m, B_1, \ldots, B_k \rangle$, which share the same list of base sets $\mathfrak{S} = \langle S_1, \ldots, S_m \rangle$ and are governed by conditions $\mathcal{H}$ and $\mathcal{K}$, respectively. Let us denote by $\mathcal{H}'$ and $\mathcal{K}'$ the sentences formed by priming the names of the structural components $A_1, \ldots, A_h, B_1, \ldots, B_k$ in $\mathcal{H}$ and $\mathcal{K}$, respectively. Suppose that from apposite echelon sets over $\mathfrak{S}$ we can select elements $B'_1, \ldots, B'_k$ by virtue of $\mathcal{H}$, and elements $A'_1, \ldots, A'_h$ by virtue of $\mathcal{K}$, so that $\langle S_1, \ldots, S_m, B'_1, \ldots, B'_k \rangle$ satisfies conditions $\mathcal{K}'$ and $\langle S_1, \ldots, S_m, A'_1, \ldots, A'_h \rangle$ satisfies conditions $\mathcal{H}'$. Plainly, any particular instance of either type of structure also provides an instance of the other. We therefore regard Σ_1 and Σ_2 as equivalent species of structure, or, better still, as alternative typifications of what is at bottom the same kind of mathematical object, viz., the same species of structure.

I shall illustrate this idea by giving three alternative typifications of the important species of structure known as a topological space. By reflecting on them one can better understand why the identity of a species of structure cannot be made to rest on a particular formal description. The second typification is especially elegant as it characterizes a topological space by means of a simple "operation" on the power set of its base set. A brief and very lucid motivation of the other two will be found in the splendid book by Saunders Mac Lane (1986, pp. 30–33).

A *topological space* is a pair $\langle S, \mathbf{T} \rangle$, where S is a set and $\mathbf{T}$ is a subset of $\mathcal{P}(S)$—and thus a distinguished element of $\mathcal{P}^2(S)$—which satisfies the following three conditions:

T_O1. S belongs to $\mathbf{T}$.

T_O2. The intersection of any two elements of $\mathbf{T}$ belongs to $\mathbf{T}$. ($A, B \in \mathbf{T} \Rightarrow A \cap B \in \mathbf{T}$.)

T_O3. The union of any subset of $\mathbf{T}$ belongs to $\mathbf{T}$. ($\mathbf{H} \subset \mathbf{T} \Rightarrow \cup\mathbf{H} \in \mathbf{T}$.)

$\mathbf{T}$ is called a *topology* on S. An element of $\mathbf{T}$ is said to be an *open set*. If x is an element of S, any subset of S that includes an open set which contains x is said to be a *neighborhood* of x. If $A \subset S$ and A is open, the complement of A in S is said to be *closed*. The mapping $f: \mathcal{P}(S) \rightarrow \mathcal{P}(S)$ that sends each subset A of S to the intersection of all closed sets in which A is included is the *closure* of $\langle S, \mathbf{T} \rangle$. Note that T_O3 entails that $\varnothing$ is open, and so, by T_O1, S and $\varnothing$ are both open and closed. $\mathbf{T}$ is the *trivial* topology on S if $\mathbf{T} = \{S, \varnothing\}$. $\mathbf{T}$ is the *discrete* topology on S if $\mathbf{T} = \mathcal{P}(S)$.

Alternatively, a *topological space* is a pair $\langle S, f \rangle$, where S is a set and f is a mapping of $\mathcal{P}(S)$ into itself such that for any subsets A and B of S, the following four conditions are met:

T_K1. $f(A \cup B) = f(A) \cup f(B)$.

T_K2. $A \subset f(A)$.

T_K3. $f(f(A)) = f(A)$.

T_K4. $f(\emptyset) = \emptyset$.

f is said to be the *closure* of S. Any set lying on its range is said to be *closed*. If A is closed in S, the complement of A in S is said to be *open*. The collection of the open sets of S is the topology of $\langle S,f\rangle$. Neighborhoods are defined as above. It can be easily proved that if f denotes the closure of a topological space $\langle S,\mathbf{T}\rangle$, characterized by conditions T_O, f meets the conditions T_K, and that if T denotes the topology of a topological space $\langle S,f\rangle$, characterized by conditions T_K, T meets the conditions T_O.

But a topological space may be characterized in still another way. Let S be a set and ϕ a mapping of S into $\mathcal{P}^2(S)$ that assigns to each element x of S a collection $\phi(x)$ of subsets of S meeting the following conditions:

T_N1. $S \in \phi(x)$.

T_N2. $x \in U$, for every $U \in \phi(x)$.

T_N3. If $U \in \phi(x)$ and $U \subset V$, $V \in \phi(x)$.

T_N4. If $U \in \phi(x)$ and $V \in \phi(x)$, $U \cap V \in \phi(x)$.

T_N5. If $U \in \phi(x)$, there is a set V such that $x \in V \subset U$ and $V \in \phi(y)$ for every $y \in V$.

Then, $\langle S,\phi\rangle$ is a *topological space*. If $x \in S$, every element of $\phi(x)$ is a *neighborhood* of x. A subset of S is said to be *open* if it is a neighborhood of every element it contains. (Note that this condition is satisfied by $\emptyset$, for it contains no elements.) A subset of S is said to be *closed* if its complement in S is open. The collection of all open sets is the *topology* of $\langle S,\phi\rangle$. The mapping that sends each subset of S to the intersection of all closed sets that include it is the *closure* of $\langle S,\phi\rangle$. It can then be shown that if $\mathbf{T}$ is the topology of $\langle S,\phi\rangle$, $\mathbf{T}$ meets the conditions T_O; and that if f is the closure of $\langle S,\phi\rangle$, f meets the conditions T_K. On the other hand, if we choose to understand 'neighborhood' in the sense defined above after the statement of conditions T_O, and $\phi(x)$ denotes the collection of the neighborhoods of an arbitrary $x \in S$, it can be shown that $\phi(x)$ meets the conditions T_N.

[Let me illustrate the above ideas in the light of the standard topology of the Euclidian plane **E**. A set of points of **E** whose distance from a point P is less than a given real number is said to form an open disk at P. A set of points whose distance from P is not greater than a given real number is said to form a closed disk at P. For greater precision, writing $|X - Y|$ for the distance from point X to point Y, I define the open disk (P,ρ), with center P and radius ρ, as the point set $\{X \mid |X - P| < \rho\}$, and the closed disk $[P,\rho]$, with center P and radius ρ, as the point set $\{X \mid |X - P| \leq \rho\}$ $(P \in E, 0 \leq \rho \in \mathbf{R})$. A point set is open in the standard topology of the Euclidian plane if (i) it is an open disk, or (ii) it is the intersection of two open sets, or (iii) it is the union of a family of open sets. A point set is closed in that topology if (i) it is a closed disk, or (ii) it is the union of two closed sets, or (iii) it is the intersection of a family of closed sets. Let $P \in$ **E** and let ζ be a subset of **E** that includes some open disk at P; then, ζ is a neighborhood of P. A closure mapping is defined on the Euclidian plane by assigning to each point set the intersection of all closed sets that include it. It can also be characterized by defining the mapping $(P,\rho) \mapsto [P,\rho]$ on the set of open disks and extending this mapping to all subsets of **E** in such a way that conditions T_K1–T_K4 are all met. (It must of course be shown that there is a unique such extension.)]

2.8.7 Axiomatic set theory

Bourbaki's *Éléments de Mathématique* has shown that it is possible to conceive the familiar areas of mathematical inquiry as the study of different species of structure (and their subspecies). Of course, Bourbaki did not follow the naïve approach to sets adopted above. As I noted in Section 2.8.1, if we nonchalantly regard any expression of the form '$\{x \mid S(x)\}$' as the description of a set, we run into contradictions. To avoid them, Russell (1908) sought to regiment the language of science, drastically and, some would say, intemperately restricting the admissible expressions. But working mathematicians have generally preferred to deal with this problem after the manner of Zermelo (1908b), who proposed to characterize sets by an axiom system free from contradiction and not to countenance any set descriptions not warranted by this axiom system. Zermelo's axioms, subsequently perfected by Fraenkel (1922, etc.) and known thereafter as the ZF axioms, assert that certain sets exist *if* certain other sets are given (two of these conditional statements were quoted as (i) and (ii) in Section 2.8.3), and postulate that there is at least one infinite set. It is demonstrably impossible to prove the consistency of the ZF axioms without invoking premises at least as strong. But hitherto no contradictions have cropped up in axiomatic set theory. However, the ZF axioms do not by themselves suffice to characterize the mathematical concept of a set. Zermelo himself (1904, 1908a) had used one additional assumption—the so-called Axiom of Choice[77]—for proving that every set can be well-ordered.[78] Gödel (1938, 1940) proved that the ZFC axioms (Zermelo-Fraenkel axioms plus the Axiom of Choice) are consistent *if* the ZF axioms are consistent. But then Paul Cohen (1963/64) proved that the ZF¬C axioms (Zermelo-Fraenkel axioms plus the negation of the Axiom of Choice) are consistent too if the ZF axioms are consistent. Do we mean by 'sets' the objects axiomatically characterized by the ZFC system or those characterized by the ZF¬C system? Most mathematicians will not hesitate to choose the former, for far too many strong and beautiful mathematical results cannot be proved in the latter. R. M. Solovay (1970) has shown that the best known of them follow also from a weakened and in some ways more convenient version of the Axiom of Choice. However, Solovay's proposal raises the specter of some very large sets, whose existence not everyone will concede. (Cf. Section 4.2, note 7.)

A similar question arises in connection with Cantor's Continuum Hypothesis and the Generalized Continuum Hypothesis, which I shall here denote by H^c and H^g, respectively. The import of H^c and H^g can be simply explained as follows. Let us say that a set A is *strictly smaller* than a set B—and B *strictly larger* than A—if there is an injective mapping of A into B, but it is impossible to map B injectively into A. Cantor proved that every set is strictly smaller than its power set. Let us say that the power set $\mathcal{P}(A)$ of a set A is *next to* A if it does not include any subset that is both strictly smaller than $\mathcal{P}(A)$ and strictly larger than A. It is plain that if a set A has only one element, $\mathcal{P}(A)$ is next to it. However, if A is finite, but has more than one element, $\mathcal{P}(A)$ is not next to A. For example, if A has 3 elements, $\mathcal{P}(A)$ has 8 elements and includes subsets with 4, 5, 6, and 7 elements. Cantor, however, expected to prove that, if $\mathbf{N}$ denotes the set of the natural numbers, $\mathcal{P}(\mathbf{N})$ is next to $\mathbf{N}$. This is Cantor's Continuum Hypothesis H^c. The Generalized Continuum Hypothesis H^g asserts that if A is any infinite set, $\mathcal{P}(A)$ is next to A. Gödel (1938, 1940) proved that ZFH^g is consistent if ZF is consistent. But again Cohen (1963/64) proved that ZF$\neg H^c$—and hence ZF$\neg H^g$—is consistent, if ZF is consistent. When we say that a group or a topological space is a *set* endowed with such-and-such structural components, do we mean that it satisfies H^c or that it satisfies $\neg H^c$?

2.8.8 Categories

The concept of a mathematical category was first introduced by Eilenberg and Mac Lane (1945). Category theory provides a global approach to mathematics that is more flexible, closer to real mathematical work and possibly more fruitful than the exclusively set-theoretic perspective adopted by Bourbaki. While the Bourbaki group defines a species of mathematical structure by the distinctive internal properties and relations of a typical representative (e.g., a topology on an arbitrary set), the new school of thought envisions any specific field of mathematics (e.g., topology) as a system of objects (e.g., the topological spaces) characterized by the peculiar nature of the network of structurally significant mappings from one to the other (e.g., continuous functions, homeomorphisms). I will not go further into this matter.[79] Joseph Sneed and some of his followers have been using category-theoretic concepts in some of their more recent work on physical theories, but the book by Balzer, Moulines, and Sneed (1987) on which I shall base my discussion of Sneed's views in Chapter 3 describes all mathematical concepts in a strictly Bourbakian manner.

3. Theories

When speaking about physical theories in general one ought to keep in mind that they issue from a peculiar mode of thought pertaining to a particular cultural tradition. Surely, mathematical physics is the most impressive enterprise of human knowledge, and this is indeed a philosopher's chief motivation for reflecting on it. But it would be silly to expect that its methods can be fruitfully brought to bear on every conceivable subject of empirical research—for example, the causes of the U.S. trade and fiscal deficits in the 1980s or the date and authorship of the 13 letters in the Platonic corpus. Not sharing the faith in *unam sanctam catholicam scientiam* we have no business in reaching for a conception of scientific theorizing so broad that it is relevant to every field of study. To work on a single comprehensive notion of "scientific theory," capable of fitting such diverse creatures as Darwin's theory of evolution, Freud's theory of neuroses, and Schumpeter's theory of capitalist development, as well as the Salam-Weinberg theory of electroweak interactions, seems pointless to me—except, perhaps, for the sake of invidiously pronouncing unscientific any purported theory which does not comply with it.

Cultural traditions always involve some degree of self-understanding. Such self-understanding may not agree with a philosophical understanding of the same tradition—think, for example, of the Roman Catholic mass—but apart from it it is well-nigh impossible to say what the tradition is all about. When a tradition begins deliberately, as the Reformation did in the 16th century, the tradition's self-understanding is usually articulated in manifestoes. But even when such self-understanding is not made explicit, it will be present and active in the patterns of imitation and recognition without which a tradition would quickly die away.

Mindful of this, I shall look for some clues to the nature of physical theories in the writings of Galileo and Newton, who may fairly be regarded, respectively, as the modern founder of mathematical physics and the author of its first widely accepted theoretical framework. That will be the subject of Sections 3.1 and 3.2. Then I shall turn to the current handling of the question "What is a physical theory?" by Joseph Sneed and his collaborators. After sketching in Section 3.3 their structuralist explication of 'physical theory' in both the narrow and the broad sense of this expression, I examine in Section

3.4 the distinction they make between the **T**-theoretical and the non-**T**-theoretical terms of a theory **T**, and the philosophical problem—in my opinion a pseudoproblem—they raise with respect to the former. In Section 3.5 I deal with another contemporary explication of 'physical theory', put forward by the physicist Günther Ludwig, who, I think, is more sensitive than the Sneedian structuralists to questions concerning the interface between theoretical thinking and observed phenomena. Ludwig has developed an exact means of conceiving the approximative nature of empirical claims in physics which has subsequently been adopted—in a suitable adaptation—by Sneed's school. It is the subject of Section 3.6. The schemata developed by Ludwig and Sneed can be of great use for establishing and describing links between different theories. Intertheoretic relations are studied from this perspective in Section 3.7. Section 3.8 deals in particular with the relation(s) of intertheoretic reduction. Section 3.9 reviews the first three chapters and introduces the subject of the fourth.

3.1 The theory of free fall in Galileo's *Discorsi*

In the Third and Fourth Days of Galileo's *Discourse and Mathematical Demonstrations concerning Two New Sciences* (1638), Salviati introduces a new science of motion by reading from a Latin treatise Galileo had already been working on before his great astronomical discoveries of 1609.[1] In it Galileo purported to give "the elements" of "a large and excellent science [. . .] into which minds more piercing than mine shall penetrate to recesses still deeper" (EN, VIII, 190).[2] The treatise comprises three parts, respectively dealing with equable or uniform motion, motion naturally accelerated or free fall, and "violent" motion, i.e., the motion of projectiles. The second part will furnish us with a very simple example of a physical theory. Before turning to it, I must say a few words about the first part, in which Galileo laid the foundations of his algebraic handling of physical quantities.

Galileo defines uniform motion as a motion in which the spaces run through by the moveable "in any equal times whatever" are equal to one another (EN, VIII, 191). The meaning of this definition is conveniently spelled out in four "axioms." The first two concern a given uniform motion. Let s_i denote the space traversed by the moveable in time t_i ($i=1,2$). Then (i) if $t_1 > t_2$, $s_1 > s_2$, and (ii) if $s_1 > s_2$, $t_1 > t_2$. The remaining two axioms introduce the concept of speed. Let v_i be the speed with which a moveable traverses the space s_i in a fixed time. Then (iii) $v_1 > v_2$ implies that $s_1 > s_2$, and (iv) $s_1 > s_2$

implies that $v_1 > v_2$. From these axioms Galileo derives with the utmost care the familiar relations between spaces, times, and speeds characteristic of uniform motion, which culminate with the statement that "if two moveables are carried in uniform motion, the ratio of their speeds will be compounded from the ratio of the spaces run through and from the inverse ratio of the times" (EN, VIII, 196). Thus, if the quantities pertaining to one of the moveables are designated by the primed letters and those pertaining to the other by the unprimed ones,

$$v/v' = (s/s')(t'/t) \tag{1}$$

Eudoxus' theory of proportions, as explained in Book V of Euclid's *Elements*, entitled Galileo to construct the ratio between any two magnitudes of the same kind (e.g., any two lengths, such as the side and the diagonal of a square, or any two speeds). For the Greek mathematicians and their medieval and Renaissance followers it was indeed nonsense to speak of a ratio between heterogeneous quantities such as a space and a time. However, ratios as such are all of a kind, regardless of the nature of the diverse pairs of homogeneous quantities from which they are severally constructed. One could therefore form the sum, difference, product, or quotient of any two of them. Euclid equated the ratio of two areas to a ratio of volumes and also to a ratio of lengths (e.g., in Book XI, Props. 32, 34), and Archimedes equated a ratio of lengths with a ratio of times (*On Spirals*, Prop. I; see below, Section 5.3). Galileo extends this treatment to speeds and accelerations. Thus, in eqn. (1) the ratio between two speeds is equated with the product of a ratio of spaces and a ratio of times. By taking the reciprocal value of the latter, the right-hand side can also be expressed as a ratio of ratios:

$$v/v' = (s/s')/(t/t') \tag{2}$$

It is worth noting that (1) and (2) are equations between "pure" or "mathematical" quantities—"real numbers," in modern parlance—although s, t, v, and their primed counterparts denote physical quantities. Any given quantity can be represented by its ratio to a fixed quantity of the same kind, conventionally chosen as a unit. Every such ratio may be represented in turn by a straight segment which has that very ratio to the unit of length. This method of representing physical quantities is used as a matter of course throughout the rest of Galileo's treatise.[3]

Let us now look more closely into its second part. Its formal development begins with the following definition:

> We shall call that motion equally or uniformly accelerated which, abandoning rest, adds on to itself equal moments of swiftness in equal times.
>
> (Galileo, EN, VIII, 205; cf. 198)

From this definition and familiar truths of arithmetic and geometry Galileo infers exact quantitative relations between the duration of a uniformly accelerated motion and the distance traversed and the speed attained by the moving body.[4] His aim, however, is not just to make explicit the logical implications of a neatly defined concept of motion but to set up that concept as the mathematical centerpiece of a physical theory concerning a class of phenomena actually observed. Galileo asserts that the definition given above "agrees with the essence of naturally accelerated motion," i.e., with the "sort of acceleration of heavy falling bodies" actually employed by nature. This assertion is made plausible by recalling "the custom and procedure of nature herself in all her other works, in the performance of which she habitually employs the first, simplest and easiest means"; for indeed "we can discover no simpler addition and increase than that which is added on always in the same way." Ultimately, however, Galileo's claim rests on "the very powerful reason that the essentials successively demonstrated by us correspond to, and are seen to be in agreement with, that which physical experiments show forth to the senses" (EN, VIII, 197).

Replying to some unwarranted suggestions by Sagredo about "the possible cause of the acceleration of the natural motion of heavy bodies" (EN, VIII, 201), Salviati summarizes Galileo's approach as follows:

> For the present, it suffices our Author that we understand him to want us to investigate and demonstrate some attributes of a motion so accelerated (whatever be the cause of its acceleration), that the momenta of its speed go increasing, after its departure from rest, in that simple ratio with which the continuation of time increases, which is the same as to say that in equal times, equal additions of speed are made. And if it shall be found that the events that then shall have been demonstrated are verified in the motion of naturally falling and accelerated heavy bodies, we may deem that the definition assumed covers that motion of heavy things, and that it is true that their acceleration goes increasing as the time and the duration of motion increases.
>
> (Galileo, EN, VIII, 202–3)

Thus, we find in Galileo's theory of free fall a clear distinction between what, borrowing a term from Sneed (1971), we may call its *frame*, viz., the exactly defined concept of uniformly accelerated motion, and its *applications*, viz., the phenomena of falling bodies. The relation between these two sides of the theory is contained in Galileo's assertion that the naturally accelerated motions of falling heavy bodies are actual instances of his concept of uniformly accelerated motion.[5] This is the theory's *empirical claim*. Now, not only is the empirical claim of a physical theory its main, or indeed its whole, point; it is also the one feature of it which is hardest to understand. Thus, in the very simple theory we are examining, the empirical claim is not, as one might naïvely expect, that any and each of the objects described in ordinary language as "falling bodies" moves with uniformly accelerated motion. Autumn leaves and winter snow may very properly be said to fall, and yet they do not usually descend with uniform acceleration. The vulgar idea of fall reflected in common usage directs our attention to the range of phenomena which Galileo intends to bring under the sway of his theory. But it is not crudely equated by a "correspondence rule" with the exact concept of fall he put forward. Heavy bodies are supposed to move in strict agreement with the latter only when they fall unhindered, e.g., vertically in a perfect vacuum, or along a frictionless inclined plane (EN, VIII, 205). Obviously, any observed discrepancy between the laws of uniformly accelerated motion and particular instances of downward movement may be ascribed to known or unknown impediments, such as air resistance, magnetism, etc. The theory is thus well shielded against so-called falsification by experience. But it would doubtless have failed to awake anybody's interest if the phenomena roughly demarcated by our ordinary talk of falling bodies did not approach the standard of uniform acceleration as the more notorious hindrances are mitigated or removed.

I do not have to stress that our grasp of what is or is not a hindrance to free fall is open to modification by scientific thought. The important thing to note here is that, in order to judge the agreement between the proposed concept of motion and the phenomena to which one intends to apply it, one must be able to read in the latter the features of the former. These include elapsed time, space traversed, speed, and acceleration. In the exact form in which Galileo conceived of them they certainly do not belong to the prescientific—or pre-philosophical—heritage of mankind. (They have become entrenched in 20th century common sense thanks, above all, to our daily intercourse with the motor-car, a product of scientific engineering which cannot be properly used—or avoided—without some thought of acceleration and speed.) However, when Galileo wrote the *Discorsi,* the geometric notion of distance had long been a familiar one, and the development of modern time measurement was already under way. The clocks available to him—including his own heart—were probably still inadequate for timing the successive stages of a free fall. But the division of time into quite small, audibly equal parts was common practice among musicians and may have suggested to Galileo an experimental test of his understanding of free fall as uniformly accelerated motion. It can be readily proved that the spaces traversed by a uniformly accelerated body in equal times taken successively from the beginning of motion are to one another as the odd numbers from unity, i.e., as 1:3:5:7 . . . (EN, VIII, 210). Stillman Drake (1975) has published the manuscript records of Galileo's experiment and has also shown how to reproduce it. In this experiment a ball descends along a groove on a tilted plane. This is how Drake imagines Galileo's procedure:

> He tied gut frets around his grooved plane, as frets are tied on the neck of a lute, so that they are snug but can be moved as needed; to set their initial positions it sufficed to sing a march tune, release the ball on one beat, and mark its approximate positions at the following beats. With the frets roughly in place, the ball made a sound on striking the plane after passing over each one; they were then adjusted until each of those sounds was judged to be exactly on a beat. It remained only to measure their distances from the point at which the resting ball touched the plane.
>
> (Drake 1978, p. 89)

Galileo's data indicate that the inclined plane he used was about 2 meters long and had a 3% tilt. The above method cannot be applied to bodies falling on sensibly steeper planes because audible time differences have a lower bound of 1/64 s. However, Galileo's results can be extended to such cases and also to vertical fall thanks to the following assumption, introduced in the *Discorsi* right after the definition of uniformly accelerated motion:

> I assume that the degrees of speed acquired by the same moveable over different inclinations of planes are equal whenever the heights of those planes are equal.
>
> (Galileo, EN, VIII, 205)

From this assumption Galileo derives by far the greater part of his 38 propositions concerning free fall.[6] Although Sagredo describes the assumption as "a single simple principle" ("un solo semplice principi") from which "the demonstrations of so many propositions" are deduced (EN, VIII, 266), one should not equate its role in Galileo's theory with that of an axiom in an axiomatized mathematical theory. The stated assumption does not modify the concept of uniformly accelerated motion or contribute to specify it in any respect. Indeed, it is formulated in such terms that one can see no way of directly relating it to that concept. The assumption does not refer to uniformly accelerated motion as defined four lines above it but to the purported realizations of that kind of motion. It is the latter, viz., the particular instances of free fall, which can be said to proceed from a given height along diversely tilted planes. The assumption establishes an exact quantitative relation between the final speeds attained by a body falling freely from equal heights on different inclined planes. The presence in every physical theory of postulated links between its several intended applications has been emphasized by Joseph Sneed, who calls them "constraints." With a view to our subsequent examination of this idea in Section 3.3 it is convenient to take here a quick look at the use and justification of Galileo's constraint on free falls.

To my mind, its chief utility is that from it one can infer that the distance traversed by a freely falling body is proportional to the square of the time elapsed since it departed from rest, just as if it were moving with uniformly accelerated motion.[7] Thus, if Galileo's constraint holds good, freely falling bodies do instantiate the concept of uniformly accelerated motion. Galileo's spokesman Salviati proposes that the constraint be taken "as a postulate, the absolute truth of which will be later established by seeing that other conclusions, built on this hypothesis, do indeed correspond and exactly conform to experience" (EN, VIII, 208). One can easily imagine an experimental test of, say, the beautiful Proposition VI, which Galileo infers from this postulate:

> If, from the highest or lowest points of a vertical circle, any inclined planes whatever are drawn to its circumference, the times of descent through these will be equal.
>
> (Galileo, EN, VIII, 178)

To test this proposition it is enough to draw a vertical circle—a practice foreign indeed to the untutored understanding but well established in Galileo's milieu—and to let two or more metal balls roll along as many inclined planes built along any chords issuing from the circle's uppermost point. The proposition is confirmed if, whenever the balls are released together from the top of the planes, they all reach the circumference at the same time. By Salviati's hypothetico-deductive logic, this should also confirm the postulated constraint from which it follows.[8]

But Galileo apparently thought that this manner of confirmation was not enough. Perhaps his logical training made him wary of a method that seeks to verify the antecedent of a valid implication by testing the consequent. Be that as it may, the fact is that he tried twice to base the constraint in question on a broader view of the course of nature. The first attempt immediately precedes Salviati's proposal to treat the constraint as an unproven postulate. Galileo's argument can be briefly stated thus: The speed attained by a body falling freely from a given height must enable it to climb back to that same height along any frictionless plane, however tilted. Consequently, such speed must depend only on the height, not the inclination of the frictionless plane on which the body falls. This argument is voiced by Sagredo. Salviati backs it by referring to pendular motion. The bob of a pendulum returns practically to the same level from which it was released even if it is constrained to rise along a circle steeper than its downward path by making the pendulum string turn on a pin placed between the bob and the original suspension point. Salviati grants that the analogy of the pendulum might not be applicable to the case of descent on a tilted plane followed by ascent on another, because the transit from one plane to the other is unsmooth.[9]

It is at this point that Salviati proposes to treat the constraint as a postulate—*come un postulato*—to be confirmed by its testable consequences. But Galileo was obviously unsatisfied with this option, for he tried once more to establish the validity of the constraint from general principles in a passage he dictated to Viviani in the fall of 1638 and which was inserted in the 1655 edition of the *Discorsi* (EN, VIII, 214–19, footnote). The new argument turns on a consideration of "the impetus, power, energy, or let us say momentum of descent" ("l'impeto, il talento, l'energia, o vogliamo dire il momento, del descendere"). We may gather the meaning of these tentative expressions from Galileo's statement that "the impetus of descent of a heavy body is as great as the minimum resistance or force that suffices to impede and stop that descent" ("tanto esser l'impeto del descendere d'un grave, quanta è la resistenza o la forza minima che basta per proibirlo e fermarlo"). The impetus of descent of a heavy body is therefore tantamount to the effective gravitational pull (or push?) of the body. Let $F(\alpha)$ denote the impetus of descent of a body on an inclined plane which makes an angle α with the vertical. By a simple argument in statics Galileo shows that $F(\alpha) = F(0) \cos \alpha$. Galileo's constraint on free falls then follows easily, provided that one takes for granted (i) that free fall is uniformly accelerated motion and (ii) that the uniform acceleration is proportional to the "impetus of descent."[10] It is doubtful that this proof is worth the price. For if the constraint on free falls rests on (i), it cannot provide independent evidence for the theory's empirical claim. But this need not concern us here. By looking into this posthumous addition to Galileo's text I only wanted to illustrate how the neatly contrived "frame" and "constraints" that make up the explicit "core" of a physical theory swim in a less articulate stream of physical insights and ideas, which are not part of the theory proper and also lie beyond the ken of established common sense. This moving background of transtheoretical thoughts, despite its imperfect clarity, or perhaps because of it, forms a powerful link between successive physical theories.[11]

3.2 Mathematical constructs for natural philosophy

In his illuminating study of the Newtonian Revolution, I. B. Cohen (1980) discusses at length what he calls "the Newtonian style," viz., "a clearly thought out procedure for combining mathematical methods with the results of experiment and observation in a way that has been more or less followed by exact scientists ever since" (p. 16). Newton "did create what he conceived to be purely mathematical counterparts of simplified and idealized physical situations that could later be brought into relation with the conditions of reality as revealed by experiment and observation" (p. 37). It is this "possibility of working out the mathematical consequences of assumptions that are related to possible physical conditions, without having to discuss the physical reality of these conditions at the earliest stages, that marks the Newtonian style" (p. 30). These abstract descriptions evidently apply also to the procedure that Galileo followed in his theory of free fall and that led us to distinguish between the theory's conceptual frame and the host of its empirical applications. But Newton's science of motion, thanks to its broader scope and greater complexity, provides a fuller view of the relationship between physical theory and experience.

Newton's *Mathematical Principles of Natural Philosophy* comprises three books, preceded by two short introductory sections. The introductory sections—which contain the three Laws of Motion, eight definitions, six corollaries, and two cholia, including the very famous one on absolute and relative time, space, and motion—jointly establish the conceptual frame subsequently applied to the phenomena of motion in the solar system. But this application does not happen until Book III. Books I and II, though obviously composed with a view to using their results in Book III, deal not with physics but with mathematics, as Newton more than once remarks (*Principia*, pp. 266–67, 298). He explains the point of this distinction as follows:

> In mathematics we are to investigate the quantities of forces and the ratios which follow from *any conditions that may be posited.* Then, when we go down to physics, these ratios should be compared with the

> phenomena, in order to learn what conditions of forces pertain to the several kinds of attractive bodies.[12]

In agreement with this program, Book I develops the implications of the concepts put forward in the introductory sections, under several specific conditions. Such conditions involve bodies that move through free space, e.g., in the circumference of a circle, along an eccentric conic section, or on given surfaces, with an acceleration directed to an immovable center; or bodies that tend to each other with centripetal forces; or "very small bodies . . . agitated by centripetal forces tending to the several parts of a very great body," etc. Just as Euclid studied the properties of all sorts of geometrical figures regardless of their actual presence in nature, so Newton pursued his analysis of diverse dynamico-chronogeometric configurations without paying any attention to the fact that they may not exist. The choice of such configurations for study is guided by the ultimate goal of finding one or more that will provide a sufficiently accurate mathematical representation of the motions of the solar system; but it is also constrained by the necessity of understanding the mathematics of some simple unreal situations in order to be ready to solve the more complex realistic ones. The latter, of course, fit the phenomena only within some pragmatically viable margin of error. Indeed, for some—large enough—such margin, this may also be true of the former. Thus, for instance, a body moving under the action of a force directed towards an infinitely distant center is nowhere to be found in the world, yet the motion of cannonballs from one point of a battlefield to another, or that of the Earth-Moon system over a short time, can be represented tolerably well by configurations subject to that blatantly fictitious hypothesis (Newton, *Principia*, pp. 94, 117). Mathematical physics confronts the realities of nature *neglectis negligendis*.

A different but no less important aspect of the relationship between physical theory and experience is illustrated in Book II, which deals with the motion of bodies in resisting media. The purpose of this exercise is to show that if a solid body is carried around a center by a fluid vortex the radius vector joining the center to the body does not sweep equal areas in equal times. Hence, under the assumptions of Newtonian dynamics, the Cartesian vortex theory of planetary motions is incompatible with Kepler's law of areas, and therefore untenable. A series of conceivable cases are examined in Book II, not for the sake of approximating experience stage by stage, but rather in order to prove that, under the conditions that have been posited, such approximation is impossible.

I. B. Cohen (1980) uses the expression 'mathematical construct'—instead of the popular but much abused 'mathematical model'—as a generic name for the exactly thought out, more or less unreal dynamical configurations displayed in the mathematical books of the *Principia*. The abstract frame of a physical theory must be linked by such constructs to its intended applications. This can be clearly seen in Newton's transition to physics in Book III. The six "Phenomena" listed at the head of this book, for which the theory must account, are described in terms suitable to exhibit them as instances of some of the constructs presented in Book I. The terms employed are, indeed, purely kinematic, not dynamic, and are used prima facie in the same way as in pre-Newtonian astronomy.[13] However, the time measurements mentioned in the descriptions must be based, under pain of meaninglessness, on Newton's dynamical laws (Torretti 1983, pp. 12–13). A bridgehead for the theory can be established among the phenomena only by conceiving them in the theory's own terms and thus letting it, so to speak, catch hold of them.

3.3 A structuralist view of physical theories

In the humanities the term 'structuralism' may sound passé, but in science it stands to this day for the most popular and successful attempt to introduce systematic unity into the luxuriant jungle of mathematics. I refer to the Bourbaki program for reconstructing every special field of mathematics as the study of some species of structure.[14] The self-styled "structuralist view" of physical theories cultivated by Joseph Sneed and his associates ultimately rests on the realization that if theoretical physics seeks to represent and explain natural phenomena by mathematical means and there is a fairly satisfactory systematic method of conceiving all mathematical ideas, we shall attain a unified perspective on theoretical physics by bringing this same method to bear on our understanding of it. Some such realization probably motivated the logical empiricist view of physical theories as interpreted calculi or formal languages, at a time when it was still fashionable—at least among philosophers—to regard mathematical theories as uninterpreted calculi. This crass misrepresentation of mathematics must share the blame for the eventual failure of the logical empiricist program. In the late fifties, when the hopelessness of the formal language approach to the philosophy of physics was dawning on the more wakeful researchers in the field, Patrick Suppes and his pupils at Stanford had already substituted for it a Bourbakian program. Early examples of their work are the axiomatization of Classical Particle Mechanics by McKinsey, Sugar, and Suppes (1953) and E. W. Adams' doctoral dissertation on "Axiomatic Foundations of Rigid Body Mechanics" (1955; see also Adams 1959). But these writings could easily cause the reader to overlook what to us now is the key feature of the structuralist approach, viz., that a theory of mathematical physics is to be seen, not as a set of *statements* with fixed referents in nature, but as a *concept* defined by the axioms (as the concept of a topological space is defined by the axioms in Section 2.8.6), coupled with the simple but open-ended claim that such-and-such pieces or aspects of nature are "models"—i.e., instances—of that concept. This view can be extracted from Suppes' writings of the sixties (see, e.g., Suppes 1960, 1962, 1967, 1969), but I would not say it became clear—at any rate to me—before the publication of Sneed's book *The Logical Structure of Mathematical Physics* (1971).[15] Wolfgang Stegmüller promptly discovered here the conceptual means for securing the comparability of successive theories even though they are incommensurable in some Kuhnian sense. He wrote a book-length addendum to his earlier Carnapian handbook on theory and experience setting forth the terms of his "conversion" and founded the still flourishing school of Sneedian philosophy in Germany.[16] I shall base my exposition of Sneed's ideas on a mature product of his collaboration with members of that school, the lively, clear, refreshingly open-minded treatise by Balzer, Moulines, and Sneed, *An Architectonic for Science* (1987).

To begin with, let me note that Sneed and his associates do not adhere dogmatically to Bourbaki's formulation of mathematical ideas in set-theoretical terms. Indeed, in their contribution to the 1983 International Congress of Logic, Methodology, and Philosophy of Science in Salzburg, Balzer, Moulines, and Sneed (1986) describe the conceptual core of physics in terms of (mathematical) category theory, which serves well their aim of bringing out the linkage of physical theories in a global synchronic and diachronic network. In *An Architectonic for Science* they revert to their earlier Bourbakian ways mainly for didactic reasons—after all, most current textbooks introduce their respective field of mathematics as the study of a kind of structured set—and also perhaps, one may surmise, because they have not yet worked out every detail of a category-theoretic presentation.[17] These reasons are good enough for our purpose here.

When asked to name a physical theory one may mention some such imposing, protean creature as Classical Mechanics or Relativity; or one may settle for something more modest and, of course, more definite, such as, say, the (classical) theory of damped oscillations or the (general relativistic) theory of the Schwarzschild field. To resolve this ambiguity Balzer, Moulines, and Sneed represent entities of the latter sort by what they call *theory-elements*, which are woven into *theory-nets* corresponding to "physical theories" of the broader, polymorphic kind. The unit of analysis is thus the theory-element, which the authors present as "the simplest significant part of empirical science."[18] (In the rest of this section I shall make no allowance for the fact that physical theories are expected to hold only within a certain margin of error. We shall be dealing, therefore, with what Balzer, Moulines, and Sneed call *idealized* theory-elements and theory-nets. The structuralist treatment of approximation in physics is the subject of Section 3.6.)

A theory-element **T** comprises a well-defined conceptual *theory-core* **K**(**T**) and an open-ended collection of *intended applications* **I**(**T**). The intended applications of a theory are the real-life situations one seeks to grasp and understand by means of the conceptual core. Their number and variety change in the course of time, depending on the successes and failures of the theory and the creative imagination of its practitioners. It is essential to realize that, at any given moment, for any given instance of a theory-element, there are generally no necessary and sufficient conditions for membership in its family of applications **I**, so **I** is not properly a set in a mathematically acceptable sense of this word. Nevertheless, our authors define: "**T** is a *theory-element* if and only if there exist **K**(**T**) and **I**(**T**) such that **T** = ⟨**K**(**T**),**I**(**T**)⟩." It is as if they felt that vague ideas gain in precision by having their name enclosed within angular brackets. Strictly speaking, the expression on the right-hand side of the equality is nonsensical, for an indefinite object such as the one that **I**(**T**) here stands for cannot be a term in a set-theoretical ordered pair. But I do not think that much harm is made by writing things in this way.[19] On the other hand, the conceptual core **K**(**T**) is well defined, and we must now consider what Balzer, Moulines, and Sneed put into it.

As the reader will have guessed, **K**(**T**) contains the species of structure by which the intended applications are to be grasped and understood. This species of structure is equated with the class of its conceivable instances or models, and denoted by **M**(**T**). But there is more to the theory-core **K**(**T**) than just the class **M**(**T**) of its models. Our authors wish to capture in some plausible way the distinction one normally makes between the general conceptual framework within which a physical theory proposes to understand its intended domain and its full characterization of that domain by a peculiar set of laws.[20] They think they can achieve this by including in **K**(**T**), besides **M**(**T**), another species of structure $\mathbf{M}_\mathrm{p}$(**T**), instantiated by the so-called *potential models* of the theory-element. The designation is obviously intended to suggest that $\mathbf{M}_\mathrm{p}$(**T**) contains every conceivable candidate for application of the theory, while **M**(**T**) is that subset of $\mathbf{M}_\mathrm{p}$(**T**) which actually satisfies the theory's laws. However, I am not sure that the terms actually defined by Balzer, Moulines, and Sneed will live up to this suggestion.

In the case of any particular theory-core **K**(**T**), **M**(**T**) will be, of course, some species of structure defined, as usual, by listing some base sets and distinguished elements from echelon sets over them, and stating the conditions that all these structural components are required to fulfill. Our authors classify such conditions into two kinds: those that mention one and only one distinguished structural component and those that mention more than one. They take it that conditions of the former sort do no more than describe the sort of properties and relations considered in the theory and are therefore sufficient to characterize the candidates for its application. Such conditions are therefore called *characterizations.* On the other hand, the theory's laws should be expressed by conditions of the latter kind, which couple together two or more of those properties and relations. Thus, for example, in the definition of the species of structure *group* in Section 2.8.4, condition G2, coupling the distinguished components e and g, and condition G3, which links e, f, and g, are laws, while G1 is a mere characterization. Given the class of models $\mathbf{M(T)}$ of a theory-core $\mathbf{K(T)}$, the corresponding class of potential models $\mathbf{M_p(T)}$ is defined, according to Balzer, Moulines, and Sneed, by simply deleting all laws from the definition of $\mathbf{M(T)}$—i.e., by forgetting all those defining conditions that mention more than one distinguished element from an echelon set over the base sets of $\mathbf{M(T)}$.

This procedure is simple enough and, prima facie, it looks plausible. However, as one can easily show, it fails to determine uniquely the class of potential models for a given class of models. Consider again the foregoing example of the species of structure *group*—which is, of course, too simple to be the conceptual core of a viable physical theory, but is a familiar ingredient of many such cores. Let $\mathbf{M(T)}$ comprise all the objects that meet conditions G1–G3 in Section 2.8.4. Then, $\mathbf{M_p(T)}$ contains all quadruples $\langle G,e,f,g\rangle$ such that G is any non-empty set, g is any associative mapping of G^2 into G, e is an arbitrarily chosen element of G, and f is an arbitrary mapping of G into G. However, a *group* can also be defined as a pair $\langle G,g\rangle$, such that G is a non-empty set and g is a mapping of G^2 into G which meets the following three conditions:

G´1. g is associative.

G´2. For any a and b in G there is a unique x in G such that $g(a,x) = b$.

G´3. For any a and b in G there is a unique y in G such that $g(y,a) = b$.

If $\mathbf{M(T')}$ denotes the set of all objects satisfying the second definition of the species of structure *group*, it can be readily shown that for any pair $\langle G,g\rangle \in \mathbf{M(T')}$ there is a unique $e \in G$ and a unique mapping f: $G \rightarrow G$ such that $\langle G,e,f,g\rangle \in \mathbf{M(T)}$. On the other hand, given any $\langle G,e,f,g\rangle \in \mathbf{M(T)}$, g meets conditions G´1–G´3. We thus have a canonic bijection of $\mathbf{M(T)}$ onto $\mathbf{M(T')}$, which entitles us to identify these two classes of objects. But there is no such bijection of $\mathbf{M}_p(\mathbf{T})$ onto the class $\mathbf{M}_p(\mathbf{T}')$ of potential models obtained by the above procedure from $\mathbf{M}(\mathbf{T}')$, for there are no laws to delete in G´1–G´3, so that, obviously, $\mathbf{M}_p(\mathbf{T}') = \mathbf{M}(\mathbf{T}')$.

A unique $\mathbf{M}_p$, defined solely by characterizations, will, of course, be associated to any given species of structure **M** if the latter is specified by its full set of *necessary and sufficient* conditions (and not by one of its alternative sets of *sufficient* conditions, as is usual in mathematics). In each instance of **M** the necessary and sufficient conditions pick out infinitely many distinguished elements $\alpha_1, \alpha_2, \ldots$ from appropriate echelon sets $X_1, X_2, \ldots$ over the structure's base sets.[21] A species of structure $\mathbf{M}_p$ will be specified without ambiguity by (i) setting up a fixed order among the elements $\alpha_1, \alpha_2, \ldots$ (based, say, on the echelon construction of the set they belong to); (ii) taking an arbitrary family $\mathcal{F}$ of base sets, equinumerous with the family of base sets of **M**; and (iii) choosing arbitrary distinguished elements $\beta_1, \beta_2, \ldots$ from echelon sets $Y_1, Y_2, \ldots$ over $\mathcal{F}$, subject to the twofold condition that, for every positive integer *k*, the set Y_k is homologous with X_k and β_k satisfies the same characterizations as α_k. In this way, a definite set of potential models $\mathbf{M}_p$ can indeed be uniquely linked to any given set of models **M**. But the infinite list by which its typical member is "described" will hardly pass as a reasonable reconstruction of what the average physicist has in mind when he sets out to look for possible realizations of a theory.

Worse still, if $\mathbf{M}_p$ is introduced in this way for a given **M**, one will generally have that $\mathbf{M}_p = \mathbf{M}$, so that the very distinction between these two species of structure is pointless. As we saw, such is the case if **M** is the species of structure *group*, for G´1–G´3 are some of its necessary conditions. One might be tempted to argue that precisely for this reason—viz., that the species of structure *group* can be defined by means of characterizations alone, without any need for laws—group theory cannot be by itself the conceptual core of a physical theory.But this argument is no good. *Any* species of structure can be trivially defined in a similar way if only one takes care to ascend sufficiently high in the construction of echelon sets over the base sets. To see what I mean, consider once more the definition of *group* by G1–G3 in Section 2.8.4. It is tantamount to:

> A group is a pair $\langle G, \Phi \rangle$, such that G is a non-empty set and Φ maps G^2 into G^4 by $(x,y) \mapsto (g(x,y), f(x), f(y), e)$, where e, f, and g satisfy G1–G3.

This sounds rather silly, but the nice thing is that we can now replace G1–G3 by the following pure *characterizations* of Φ. (Recall that π_k designates the projection of a Cartesian product onto its *k*th component, as defined in

Section 2.8.2):
For any $a,b,$ and $c \in G$,

Γ0. $\pi_4 \circ \Phi$ is a constant function, whose value we shall denote by e.

Γ1. $\pi_1 \circ \Phi(a, \pi_1 \circ \Phi(b,c)) = \pi_1 \circ \Phi(\pi_1 \circ \Phi(a,b),c)$.

Γ2. $\pi_1 \circ \Phi(a,e) = \pi_1 \circ \Phi(e,a) = a$.

Γ3. $\pi_1 \circ \Phi(a, \pi_2 \circ \Phi(a,b)) = \pi_1 \circ \Phi(\pi_2 \circ \Phi(a,b),a)$
$= \pi_1 \circ \Phi(\pi_3 \circ \Phi(a,b),b) = \pi_1 \circ \Phi(b,\pi_3 \circ \Phi(a,b)) = e$.

Thus, it turns out, the distinction made by Balzer, Moulines, and Sneed between characterizations and laws cannot do the job for which they introduce it.[22] On the other hand, one would certainly like to establish a conceptual difference between serious candidates for the application of a physical theory and confirmed realizations of it. I shall propose a manner of doing this by means of the structuralist notion of a theory-net. But before we can go into it, I must finish the description of the theory-core **K**(**T**) of a theory-element **T**. Let me, therefore, temporarily assume that there is some sensible way of distinguishing between $\mathbf{M}(\mathbf{T})$ and $\mathbf{M}_p(\mathbf{T})$, and proceed to the definition of some further components of **K**(**T**).

One of these additional components must be included, according to the Sneedian structuralists, in every theory-core in order to solve what they call "the problem of theoretical terms." As the name suggests, this was inspired by the homonymous problem of logical empiricism (Section 2.4.3), but its import is quite different. Apparently, Sneed sought to answer Hilary Putnam's demand for an elucidation of "what is *really* distinctive" about theoretical terms in science, and was able to do so by deftly changing the accepted logical empiricist meaning of the expression 'theoretical term' (see Putnam, PP, I, p. 219; Stegmüller 1973a, pp. 30ff.; 1986, p. 32). To understand his definition of this expression, let us consider a theory-element $\mathbf{T} = \langle \mathbf{K}(\mathbf{T}), \mathbf{I}(\mathbf{T}) \rangle$ such that the species of structure instantiated by its models $\mathbf{M}(\mathbf{T})$ and its potential models $\mathbf{M}_p(\mathbf{T})$ are defined by listing real-valued functions $f_1, \dots, f_n$ defined on echelon sets over a family $\mathcal{D}$ of base sets.[23] Suppose now that we have succeeded in conceiving some intended application of **T** as a potential model in $\mathbf{M}_p(\mathbf{T})$. This entails that we have discerned in some physical process or situation S the objects we wish to represent by elements of the base sets $\mathcal{D}$ and the physical quantities that the functions $f_1, \dots, f_n$ will stand for. Now, S is a model and not just a potential model of **T** only if the functions $f_1, \dots, f_n$ satisfy the conditions that govern $\mathbf{M}(\mathbf{T})$. To ascertain whether this is indeed the case, we must measure the quantities represented by those functions. In this connection Sneed makes the following very apt and profound remark: in virtually every physical theory that can be described in this way, all established methods for measuring one or more of the functions $f_1, \ldots, f_n$ in a real physical situation or process regarded as a potential model of the theory *presuppose* that the theory has an *actual* model, i.e., that it is realized either by that very situation or process or by some other one. A function which has this property in a theory **T** is said by Sneed to be *theoretical* relative to **T**, or **T**-*theoretical.*[24] An arbitrary predicate is **T**-theoretical if its characteristic function is **T**-theoretical (see note 23). Observe that the theoreticity of terms in this sense is not only relative to a particular theory but is also dependent on the accepted methods of measurement. The invention of a new method may thus be sufficient to de-theoretize a concept (relative to a particular theory). Not all followers of Sneed are happy with such pragmatic dependence. I shall have more to say about this in Section 3.4. For the time being, let us abide by Sneed's original criterion of **T**-theoreticity, which Balzer, Moulines, and Sneed now restate "semiformally" as follows:

> A concept t is called theoretical relative to a theory **T** (or just **T**-theoretical) if and only if every determination of (a relation belonging to) t in any application of **T** presupposes the existence of at least one actual model of **T**.
>
> (Balzer, Moulines, and Sneed 1987, p. 55)

Sneed's problem of theoretical terms can now be formulated thus: Suppose that **T** is, as above, a theory-element whose models and potential models are instances of a species of structure $\langle \mathcal{D}, f_1, \ldots, f_n \rangle$, such that, for some fixed integer i $(1 \leq i \leq n)$, f_i is **T**-theoretical. Let S be a physical process or situation which is being viewed as a potential model of **T**. To establish that S is actually a model of **T** one has to measure the quantities represented by the functions $f_1, \ldots, f_n$ and verify that they satisfy the laws of **T**. But one cannot measure the quantity represented by the **T**-theoretical function f_i without assuming that there exists a physical process or situation S' which is a model of **T**. One may indeed seek out such an S', conceive it as a potential model $\langle \mathcal{D}', f'_1, \ldots, f'_n \rangle$ of **T**, and try to verify that the quantities represented by the functions $f'_1, \ldots, f'_n$ satisfy the laws of **T**. But then, in order to measure f'_i one must assume that there exists a physical process or situation S'' which is a model of **T**. For this reason, in Sneed's opinion, the statement that a certain physical situation or process is a model of a theory-element **T** sporting **T**-theoretical terms is not an empirical statement. What, then, is the empirical claim being made by someone who holds such a **T**?

The third component that Sneedian structuralists list in their theory-cores was expressly designed to answer this question. It is again a species of structure, symbolized by $\mathbf{M}_{pp}$, whose instances are said to be the *partial potential models* of the respective theory-element. We thus have that a theory-element $\mathbf{T}$ is a pair $\langle \mathbf{K}(\mathbf{T}), \mathbf{I}(\mathbf{T}) \rangle$, where the theory-core $\mathbf{K}(\mathbf{T})$ is at least a triple, comprising the species of structure $\mathbf{M}(\mathbf{T})$, $\mathbf{M}_p(\mathbf{T})$, and $\mathbf{M}_{pp}(\mathbf{T})$. $\mathbf{M}_{pp}(\mathbf{T})$ is defined by simply chopping off from $\mathbf{M}_p(\mathbf{T})$ every $\mathbf{T}$-theoretical term. Thus, if $\mathbf{M}_p(\mathbf{T})$ can be described as above in the simplified form $\langle \mathcal{D}, f_1, \ldots, f_n \rangle$, and this $(n+1)$-ple contains, say, $n-m$ $\mathbf{T}$-theoretical terms which we agree to write at the end of the list, there will be a mapping

$$\mu: \mathbf{M}_p(\mathbf{T}) \to \mathbf{M}_{pp}(\mathbf{T}) \quad \text{by}$$
$$\langle \mathcal{D}, f_1, \ldots, f_m, f_{m+1}, \ldots, f_n \rangle \mapsto \langle \mathcal{D}, f_1, \ldots, f_m \rangle \tag{1}$$

Sneedian structuralists believe that, due to the problem of theoretical terms, the empirical claim made by someone who holds such a $\mathbf{T}$ can refer only to partial potential models. The collection $\mathbf{I}(\mathbf{T})$ of intended applications must therefore be regarded as included in (the extension of) $\mathbf{M}_{pp}(\mathbf{T})$, not in $\mathbf{M}_p(\mathbf{T})$, let alone in $\mathbf{M}(\mathbf{T})$. As we noted above, $\mathbf{I}(\mathbf{T})$ is not a set but rather a "non-classical category" in Lakoff's sense,[25] clustering about one or more "prototypes"—usually the situations or processes for which the theory was first conceived—and containing a host of other cases, some of which may have been added tentatively and will eventually be dropped if it turns out that the theory cannot be successfully applied to them. But if we take the liberty to use '$\mathbf{I}(\mathbf{T})$' as if it were the name of a set, the empirical claim involved in holding $\mathbf{T}$ can be formulated in a crude first version thus:

$$\mathbf{I}(\mathbf{T}) \subset \mu(\mathbf{M}(\mathbf{T})) \tag{2}$$

Here μ is not the mapping from $\mathbf{M}_p(\mathbf{T})$ to $\mathbf{M}_{pp}(\mathbf{T})$ defined in (1), but the homonymous mapping induced by it on $\mathcal{P}(\mathbf{M}_p(\mathbf{T}))$, as was explained in Section 2.8.3.

If (2) accurately represented the empirical claim of a typical physical theory, mathematical physics would indeed be cheap. It would be fairly easy to supplement with suitable theoretical functions an isolated partial potential model of our pet theory so that it was made out to be a model of it. For example, Newton would have had little trouble in accounting for the Keplerian trajectory of the Moon around the Earth by an inverse square force law had he been free to assign any value to the mass of the Earth or to use any arbitrary gravitational constant. What made his enterprise difficult but worthwhile was the requirement that the same law—with the same proportionality constant—should also account for the Keplerian trajectory of the Earth around the Sun and for the Galilean free fall of projectiles on the Earth, and that the mass of our planet should in all three cases be the same.[26] Sneed perceived that constraints of this sort, binding together the various intended applications of a physical theory, can severely restrict the free play of invention and convention in spelling out diverse physical processes and situations as potential models of that theory, and thus make the feat of reading them as actual models of it all the more interesting. Sneed's insight enables us to understand better why a successful physical theory is held in greater esteem when it has a rich and variegated field of applications. It is not so much that it works like a master key opening many different doors, thus saving precious space in our mental key holder. It is rather that a theory which has been successfully constrained to fit, all at once, a broad variety of interconnected applications cannot be easily replaced and may long remain without peer.

Sneed has found a clever way of conceiving such constraints on physical theories by purely set-theoretical means. Note, first of all, that the constraints do not regulate the existence of an isolated model of a theory-element **T**; they preclude the *joint* existence of certain pairs or larger collections of models of **T** (e.g., two models of Classical Particle Mechanics whose domains have a particle p in common, but whose respective mass functions take different values at p). The constraints can therefore be regarded as rules of compatibility. Hence, if $\mathbf{M}_p(\mathbf{T})$ stands for the potential models of **T**, the constraints will be specified by the choice of a set **C** of subsets of $\mathbf{M}_p(\mathbf{T})$ which contains every collection of mutually compatible potential models of **T**, and nothing else. Such a set **C** necessarily meets the following two conditions: (i) every singleton in $\mathcal{P}(\mathbf{M}_p(\mathbf{T}))$ belongs to **C**, for surely each potential model of **T** is compatible with itself; and (ii) if A is a subset of $\mathbf{M}_p(\mathbf{T})$ belonging to **C** then every non-empty subset of A belongs to **C**, for surely the subsets of A will not include mutually incompatible potential models of **T** if A does not. Otherwise, **C** can be quite arbitrary. Any set $\mathbf{C} \subset \mathcal{P}(\mathbf{M}_p(\mathbf{T}))$ which satisfies (i) and (ii) can therefore be said to specify constraints for the theory **T** and may be called, following Sneed, a *constraint* for $\mathbf{M}_p(\mathbf{T})$.[27] If we now take the intersection of all the constraints postulated in this way for a theory-element **T**, we have what Balzer, Moulines, and Sneed call its *global constraint* **GC**(**T**). This is the fourth component assigned by them to theory-cores. They reckon still a fifth component, but we can safely ignore it for the time being. We may, thus, spell out the pair $\langle \mathbf{K}(\mathbf{T}), \mathbf{I}(\mathbf{T}) \rangle$—with which we initially equated the theory-element **T**— as $\langle\langle \mathbf{M}(\mathbf{T}), \mathbf{M}_p(\mathbf{T}), \mathbf{M}_{pp}(\mathbf{T}), \mathbf{GC}(\mathbf{T}) \rangle, \mathbf{I}(\mathbf{T}) \rangle$, where $\mathbf{M}(\mathbf{T})$, $\mathbf{M}_p(\mathbf{T})$, and $\mathbf{M}_{pp}(\mathbf{T})$ are species of structure such that $\mathbf{M}(\mathbf{T}) \subset \mathbf{M}_p(\mathbf{T})$ and

$\mathbf{M}_{pp}(\mathbf{T})$ is obtained from $\mathbf{M}_p(\mathbf{T})$ by deleting its **T**-theoretical distinguished structural components; and $\mathbf{GC}(\mathbf{T}) \subset \mathcal{P}(\mathbf{M}_p(\mathbf{T}))$ and satisfies the conditions on constraints.

The empirical claim (4) which we found so weak can now be duly strengthened by appealing to constraints. Not only must each intended application of a physical theory admit of being reconceived as a model of that theory, but every pair or larger collection of those intended applications should yield, when thus rethought, a collection of mutually compatible models of the theory. This amounts to saying, in the notation we have been using, that if every single intended application of the theory-element $\mathbf{T} = \langle\langle \mathbf{M}(\mathbf{T}), \mathbf{M}_p(\mathbf{T}), \mathbf{M}_{pp}(\mathbf{T}), \mathbf{GC}(\mathbf{T})\rangle, \mathbf{I}(\mathbf{T})\rangle$ is completed with **T**-theoretical terms to make a member of $\mathbf{M}(\mathbf{T})$, every pair or larger collection of them must turn out to lie in the intersection of $\mathcal{P}(\mathbf{M}(\mathbf{T}))$ with $\mathbf{GC}(\mathbf{T})$. Consequently, every subset of the set $\mathbf{I}(\mathbf{T})$ of intended applications of **T** must lie in the image of $\mathbf{GC}(\mathbf{T}) \cap \mathcal{P}(\mathbf{M}(\mathbf{T}))$ by the mapping $\mu: \mathcal{P}^2(\mathbf{M}_p(\mathbf{T})) \rightarrow \mathcal{P}^2(\mathbf{M}_{pp}(\mathbf{T}))$. (This μ is, of course, the homonymous mapping induced in $\mathcal{P}^2(\mathbf{M}_p(\mathbf{T}))$ by the mapping defined in (1)). The empirical claim of **T** should therefore be stated thus:

$$\mathcal{P}(\mathbf{I}(\mathbf{T})) \subset \mu(\mathbf{GC}(\mathbf{T}) \cap \mathcal{P}(\mathbf{M}(\mathbf{T}))) \quad (3)$$

(3) entails (2) but is not entailed by it.

To my mind, the main strength of Sneedian structuralism lies in the ease with which it can handle the plurality of theories in the intellectual arsenal of physics. Theory-elements (i.e., theories in the narrower sense noted at the beginning of this section) are bound together into theory-nets (i.e., theories in the broader sense) but also into theory-evolutions (representing the historic succession of physical theories in both senses). Indeed, in their latest work, Sneed and his associates are even speaking of a theory-*holon*, which is meant to stand for the whole growing, shifting, yet closely knit pluralistic *corpus* of science.[28] To give the holon a grip on theory-elements, Balzer, Moulines, and Sneed have added one more component to the theory-core $\mathbf{K}(\mathbf{T})$ of each theory-element **T**. They call it the *global link* of **T** and denote it by $\mathbf{GL}(\mathbf{T})$. The members of $\mathbf{GL}(\mathbf{T})$ are the potential models of **T** which satisfy the *links* of **T** with other theory-elements.[29] Our authors define such inter-theoretical links quite generally, as binary relations between the potential models of one theory-element and those of another—in other words, a link between theory-elements **T** and **T**′ is simply an element of $\mathcal{P}(\mathbf{M}_p(\mathbf{T}) \times \mathbf{M}_p(\mathbf{T}'))$—but for specific purposes they impose further requirements on them. Thus, they have been working on the concept of an *interpreting link*, by which they propose to grasp the familiar fact that in any theory-element **T** the accepted methods of measurement of some non-**T**-

theoretical function presuppose the existence of models of other theory-elements. In a fairly intuitive sense, the latter may be said to provide interpretations for the non-**T**-theoretical terms of **T**.[30] I shall not go here into the labored—though quite elementary—definition of an interpreting link by Balzer, Moulines, and Sneed (1986, pp. 301–2). Let me note, only, that such a link establishes an antisymmetric relation between an interpreted and an interpreting theory-element, which naturally sets up a directed path from one to the other. Whether such paths can or cannot form loops in the theory-holon is a moot question (Moulines 1984). I shall touch again on interpreting links in the next section. But first I must round off this sketch of the Sneedian view of theories with a brief description of the idea of theory-nets, which, as I announced above, can provide us with a workable distinction between models and potential models.

As I indicated earlier, the notion of a theory-net is intended to capture the complexity of a physical theory in the standard broader sense in which one says that Classical Mechanics is a theory and Classical Electrodynamics is another one. When a theory in this sense is brought to bear on a particular problem, one takes into account some general principles of the theory—e.g., the Newtonian Laws of Motion or the Maxwell equations—plus one or more special laws relevant to that problem, while forgetting, and at times even contradicting, other laws that one would certainly consider in solving other problems. For example, in figuring out the trajectory of an artillery shell across a battlefield one will normally ignore the varying distance from the shell to the center of gravity of the Earth-shell system—although the shell's acceleration depends on it according to Newton's Law of Gravity—and will be content to apply Galileo's Law of Free Fall (in which the said acceleration is a constant). This sort of situation can be readily understood in structuralist terms if we regard a grand theory as consisting of many theory-elements which share the same basic concepts, given by some class of potential models, but differ in their models and intended applications. Sneedian structuralism also assumes that all the theory-elements of a grand theory will share the same class of partial potential models, and thus the partition of their terms into theoretical and nontheoretical. On the other hand, they will normally differ in their global link and their global constraint. In view of such considerations, Balzer, Moulines, and Sneed define the following antisymmetric relation of *specialization* between theory-elements:

> A theory-element $\mathbf{T}_1$ is said to be a *specialization* of a theory-element $\mathbf{T}_2$—abbreviated $\mathbf{T}_1 \sigma \mathbf{T}_2$—if and only if (i) $\mathbf{M}_p(\mathbf{T}_1) = \mathbf{M}_p(\mathbf{T}_2)$ and $\mathbf{M}_{pp}(\mathbf{T}_1) = \mathbf{M}_{pp}(\mathbf{T}_2)$, and (ii) $\mathbf{M}(\mathbf{T}_1) \subset \mathbf{M}(\mathbf{T}_2)$, $\mathbf{GC}(\mathbf{T}_1) \subset \mathbf{GC}(\mathbf{T}_2)$, $\mathbf{GL}(\mathbf{T}_1) \subset \mathbf{GL}(\mathbf{T}_2)$, and $\mathbf{I}(\mathbf{T}_1) \subset \mathbf{I}(\mathbf{T}_2)$.

A *theory-net* is a collection of theory-elements partially ordered by specialization. Typically, a theory-net will be a tree, issuing from a single theory-element of great generality and progressively specializing along several branches. This tree structure of theory-nets—i.e., of physical theories in the broad ordinary sense—is often reflected in the diagrams drawn by textbook authors to explain the way their successive chapters depend on one another: Usually, the first four or five chapters should all be read in numerical order to understand everything that follows, but from then on the book branches off into mutually independent chapters or short sequences of chapters.

I shall now assume that such a single-trunk tree structure is essential to the rational reconstruction of grand theories as theory-nets. Any physical theory in the broad sense has a framework of basic concepts and principles shared by all its subtheories. A classic example of this is the material presented in the introductory section of Newton's *Principia* under the title "Axioms or Laws of Motion." It includes the concepts of space and time, force, mass and momentum, and *all three* Laws of Motion. With this alone you cannot solve any physical problems, but any problem amenable to a Newtonian solution must be brought under them. I put the framework of a grand theory into the first element of the pertinent theory-net. The following simple postulate ensures that it is inherited by every other element in the net:

> A theory-net contains a theory-element $\mathbf{T}_F$, the net's *framework* element, such that (i) $\mathbf{M}(\mathbf{T}_F) = \mathbf{M}_p(\mathbf{T}_F)$ and (ii) for any theory-element $\mathbf{T}$ in the net, $\mathbf{T}_\sigma\mathbf{T}_F$.

The difficulty we had in distinguishing the models from the potential models of a theory-element is thereby solved. I make no such distinction for the framework element $\mathbf{T}_F$, since, as we saw above, none can be properly made for a theory-element considered in isolation. But a difference between models and potential models automatically arises in any nontrivial specialization $\mathbf{T}$ of $\mathbf{T}_F$, for then surely $\mathbf{M}(\mathbf{T})$ is a proper subset of $\mathbf{M}(\mathbf{T}_F) = \mathbf{M}_p(\mathbf{T}_F) = \mathbf{M}_p(\mathbf{T})$. Thus, the distinction between models and potential models is available where it can be of real use, i.e., when the grand theory is brought to bear on actual physical problems. Such problems must be analyzed in terms of the framework, but can be solved—if at all—only in terms of one of its nontrivial specializations.[31]

3.4 T-theoretical terms

In Section 3.3 I presented Sneed's characterization of **T**-theoretical terms (relative to a theory-element **T**) and explained the problem he raised in connection with them. Balzer, Moulines, and Sneed (1987) retain that earlier characterization as a launchpad for more formal attempts. However, they allow the so-called problem of theoretical terms almost to vanish out of sight—indeed, it is not even listed in their subject index! On the other hand, the "problem" is still prominent in the latest addition to Stegmüller's *Theorie und Erfahrung* (1986). Here the author raises the specter of an "epistemological circle" and solemnly warns us that "*ein epistemologisches Zirkel ist nichts Triviales. Er ist etwas Furchtbares. Genau so furchtbar wie eine Antinomie*" (1986, p. 42; his italics.) As we saw in Section 3.3, the Sneedian solution to this fearsome problem consists in assuming that every physical theory **T** contains non-**T**-theoretical terms supplied by other theories or by common sense, and that the empirical claim of **T** involves such non-**T**-theoretical terms only. In the present section I shall first explain why I believe that in any acceptable formulation of a basic (or "framework") theory-element $\mathbf{CM_F}$ for classical mechanics *every* primitive term will be $\mathbf{CM_F}$-theoretical. Obviously, this conclusion need not hold if a different definition of '**T**-theoretical' is substituted for Sneed's. Ulrich Gähde (1983) made such a substitution and succeeded in proving that *position* and *time* are non-$\mathbf{CM_F}$-theoretical in *his* (Gähde's) sense. I shall next examine Gähde's proposal. Finally, I shall take a look at the new approach to **T**-theoreticity sketched in the paper presented to the 1983 Salzburg congress by Balzer, Moulines, and Sneed (1986). Though seemingly accessory to the mainstream of this book, my critical remarks on these developments of the Sneedian school will corroborate the need for intellectual creativity in physics and lend additional support to the stance I have taken against foundationism.

In my view, a basic theory-element for classical mechanics (in a Newtonian formulation) must contain primitive terms for *mass* and *force*, *position* and *time*, bound together by Newton's three Laws of Motion. Sneedian structuralists, however, do not share this view. They choose for this purpose the axiomatization of Classical Particle Mechanics by McKinsey, Sugar, and Suppes 1953, which they designate by **CPM**.[32] Now, **CPM** includes Newton's First and Second Laws (although the former has been collapsed into the latter as the limiting case in which the force $\mathbf{F} = \mathbf{0}$) but excludes the Third, which is relegated to a specialization of **CPM** that Sneedians call Newtonian Classical Particle Mechanics (**NCPM**).[33] The primitive physical concepts of **CPM** are the force function **F**, the mass m, and the position function **r**. Time is not regarded as a physical quantity of **CPM**, but only as the parameter on which **F** and **r** depend, and which is required to range over an arbitrary open interval *T* of the auxiliary mathematical structure **R**.[34] Although it is clear to me that without the Third Law one cannot make sense of the Newtonian description of motion (see note 39), I shall grant here for the sake of the argument that the basic theory-element of Classical Mechanics need only contain the First and the Second Law.

Sneed and his followers have little difficulty in showing that the force **F** and the mass *m* are **CPM**-theoretical concepts. Gähde (1983), in particular, has provided detailed structuralist descriptions of some simple standard methods for measuring mass and force, in the light of which it is evident that these methods can do their job only if the processes of measurement are understood as models of **CPM** (indeed, of appropriate specializations of **CPM**). I wish to claim, further, that the position function of **CPM** is also **CPM**-theoretical. I state Newton's Second Law of Motion in the following familiar form (where a dot over the name of a function signifies differentiation of that function with respect to time):

$$\mathbf{F} = m\ddot{\mathbf{r}} \tag{1}$$

The position function **r** maps particles-at-an-instant into a three-dimensional real vector space; though not, indeed, into *any* such space, but into one attached to an inertial frame of reference. (One may, no doubt, define a position function suitable for **CPM** which is referred to a non-inertial frame, but the Second Law cannot then be stated as in eqn. (1) but must explicitly mention the angular velocity and the translational and angular acceleration of that frame with respect to the inertial frames—see, for instance, Landau and Lifshitz 1960, p. 128, eqn. 39.7.) Thus, it is impossible to measure the position function **r** in an intended application of **CPM** without assuming the existence of an inertial frame. Now, an inertial frame is, of course, a physical system of which Newton's First Law is true. Consequently, by Sneed's criterion of theoreticity, the position function **r** whose second time derivative occurs in eqn. (1) is **CPM**-theoretical. Sneed and his associates have failed to see this because, distracted by McKinsey et al., they neglect the inescapable connection between every intended application of Classical Mechanics and the inertial frames.[35] I cannot refrain from emphasizing that the quantity with respect to which the position function **r** has been differentiated twice in eqn. (1) is not just any real variable, freely ranging over any real interval, but is *time*, which cannot be arbitrarily assigned to physical events, but must be read on true clocks. Only on this condition is eqn. (1) a law of Classical Mechanics. This obvious, yet essential point, which earlier expositions of

Sneedian structuralism forgot to mention, is openly acknowledged by Balzer, Moulines, and Sneed (1987, p. 51). They insist, however, that not all true clocks are *mechanical.* "Astronomical, hydrological, physiological, biological, and other kinds of 'clocks' have been either found or devised since antiquity which do not presuppose the laws of mechanics." Hence, they conclude, the quantity with respect to which **r** is differentiated in eqn. (1), is non-**CPM**-theoretical. Apparently, they overlook the fact that the readings of a non-mechanical clock can yield admissible values of time, in the sense of **CPM**, only if that clock is known—or assumed—to agree with real or ideal clocks governed by the laws of mechanics. The prototype standard clock proposed for Classical Mechanics by W. Thomson and P. G. Tait (1867) in Britain and by C. G. Neumann (1870) in Germany—viz., a free particle moving along a graduated ruler—is rather hard to come by. So in the heyday of classical physics the standard timekeeper was the freely rotating rigid top, whose angular velocity is constant according to classical rigid body mechanics.[36] The handiest such top, our own planet Earth, is neither perfectly free nor perfectly rigid, but its motion can be duly corrected for expected deviations from the ideal standard. (The Earth is gradually losing angular momentum due to the tidal friction caused by the gravitational pull of the Moon and the Sun. But even in the short run, while angular momentum remains practically constant—and due to its very constancy—the angular velocity of the Earth varies slightly yet noticeably when sizable masses of water rise as vapor into the tropical atmosphere or settle down as ice around the poles. Note that such effects are quite small and could only be appreciated *in the light of Newtonian mechanics.*) Hence, until fairly recently the unit of time was defined as 1/86,400 of the mean solar day of the year 1900. Its redefinition in 1967 as 9,192,631,770 periods of the radiation corresponding to the transition between the two hyperfine levels of the ground state of the cesium-133 atom is one more indication that Classical Mechanics no longer has the standing it once possessed.

We can therefore conclude that all the primitive terms of **CPM** are **CPM**-theoretical in Sneed's sense. Hence, in the basic theory-element of Classical Mechanics, the forgetful mapping μ defined in (1) of Section 3.3 will cut off *all* the structural components of the potential models, except the bare base sets. Therefore, a partial potential model of **CPM** consists solely of unstructured sets of suitable cardinality.[37] An empirical claim for **CPM** of the form of (3) in Section 3.3 would then be utterly trivial. Does this imply, as Stegmüller suggests, that Classical Mechanics cannot make any empirical claim without falling into a vicious circle or an infinite regress? Ludwig Lange (1885), who inquired into the kinematic foundations of Newtonian mechanics in the wake of Neumann, was well aware of the dynamical assumptions involved in measurements of time and distance in that theory. He showed, however, that we can make empirical claims on behalf of the theory provided that one true clock and one inertial frame stand at our disposal. To secure them, all that we need are a rigid body and three free particles moving relatively to it, not on the same plane, in accordance with Newton's First Law.[38] For those three particles, of course, the First Law holds only by convention. But the statement that every other free particle moves according to the First Law when referred to the agreed frame and clock is surely a testable empirical claim. We have here no trace of circularity or regress but only a useful reminder that the mathematical description of physical phenomena is not available as a matter of course—so to speak, as a gift from nature—but can proceed only on conditions that we ourselves lay down.[39]

A proof that time and position are non-**CPM**-theoretical can be readily given if the criterion of **T**-theoreticity is suitably modified. This was accomplished by Ulrich Gähde in the doctoral dissertation, *T-Theorizität und Holismus* (1983), he wrote under Stegmüller.[40] His aim, indeed, was not to show this, but rather to establish a more stable and workable partition of scientific terms into theoretical and nontheoretical (relative to a given theory) than the one put forward by Sneed (1971). However, in his list of desiderata for an improved criterion of **T**-theoreticity, Gähde mentions that it ought to classify the position function of **CPM** among the non-**CPM**-theoretical terms, in agreement with current "intuitions" ("vorliegende intuitive Vorstellungen"—Gähde 1983, p. 103). Gähde is dissatisfied with Sneed's original criterion because, in his view, it burdens us with a Sisyphean task. A function f is **T**-theoretical by Sneed's criterion if *every* accepted procedure for measuring values of f presupposes the existence of a model of **T**. (As I have already noted, any concept k can be classified as **T**-theoretical if its characteristic function χ_k is **T**-theoretical.) Hence, Gähde says, in order to investigate "in a formally precise setting" the **T**-theoreticity—as determined by this criterion—of the concepts of a particular theory-element **T**, one would require a formal reconstruction not only of **T** itself but also of *all* the procedures accepted for measuring each function of **T**. Since there are many such procedures, and new ones are being introduced all the time, their formal reconstruction can never be completed.

According to Gähde, a satisfactory criterion of **T**-theoreticity should, in the first place, establish a sharp, unambiguous distinction between the **T**-theoretical and the non-**T**-theoretical functions of a theory-element **T**. Its formulation should involve a most precise and detailed grasp of the theory-dependence of the measurement procedures for functions of **T**. It should classify force and mass as **CPM**-theoretical and position as non-**CPM**-theoretical (Gähde does not mention time). Finally, it ought to be decidable de facto and not just "in principle." Gähde first develops a criterion of **CPM**-theoreticity meeting the above requirements, which he then generalizes to fit any theory-element. He considers some well-known (ideal) methods of measuring **CPM** mass and shows that they do not merely presuppose Newton's Second Law, but also some additional law which the Sneedian school assigns to **CPM**-specializations (a weak form of the Third Law is assumed in measuring mass ratios by inelastic collisions; a special case of Hooke's Law, in measuring them by undamped oscillations).[41] Thereupon he makes the essential remark leading to his definition of **T**-theoreticity: Every model of **CPM** is "the mathematical description of a concrete mechanical system referred to a given inertial frame (or, more precisely: to a coordinate system Σ 'affixed' to such a frame). [. . .] Descriptions of the same mechanical system referred to different inertial frames exhibit one remarkable feature, which is very essential for what follows: such a change of reference frame does not affect the validity of Newton's Second Law nor of the relevant special law" (1983, p. 92). In other words, the special laws presupposed by the methods of mass measurement examined by him are invariant under Galilean transformations, no less than the Second Law.[42] Gähde stresses that analogous "invariance requirements play a decisive role in every physical theory" (1983, p. 125). Armed with this insight, Gähde produces a definition of '**T**-theoretical' in structuralist idiom, first for **T** = **CPM**, then for arbitrary **T**, which has been designed to deliver the following result: A function of **CPM** is **CPM**-theoretical if and only if it cannot be uniquely determined by methods of measurement which presuppose the axiom of **CPM**—i.e., Newton's Second Law of Motion—alone, but is uniquely determined by methods of measurement which presuppose that axiom plus one or more special laws invariant under Galilean transformations. Force and mass are shown to be **CPM**-theoretical by such a criterion, but position plainly is not. This is so not because Newton's Second Law is not *required* in order to determine the values of the position function, but rather because it is *insufficient* for that purpose, since "the position function is always underdetermined by Newton's Second Law plus any Galilei invariant special laws" (Gähde 1983, p. 119).

Gähde's definition of '**T**-theoretical' is so cumbersome that he finds it necessary to present it twice, first for the special case of **CPM**, then again for the general case. However, like other such ponderous brainchilds of the Sneedian school, it is quite elementary. I shall sketch it here, not so much for its own sake, as for the light it throws on other notions we have been dealing with. Although I shall not copy Gähde's symbols and shall not provide literal translations for his German terms, I shall seek to portray his concepts faithfully. (Readers who are not particularly interested in the subject may skip the next four paragraphs without a significant loss of continuity.)

Let me begin by defining some simple set-theoretical notions. A family of objects $\{\alpha_k\}_{k\in\lambda}$ indexed by a set λ is a non-empty set whose members have each been labelled by an element of the non-empty set λ.[43] Thus, for example, an n-tuple $\langle\alpha_1, \ldots, \alpha_n\rangle$ is a family indexed by $\{1, \ldots, n\}$, and if, for each integer k in this set of indices, $\alpha_k \in A_k$, the Cartesian product $A_1 \times \ldots \times A_n$ can be regarded as a set of such families. We are already acquainted with the kth projection π_k which maps the Cartesian product $A_1 \times \ldots \times A_n$ onto its kth component A_k by $\langle\alpha_1, \ldots, \alpha_n\rangle \mapsto \alpha_k$. If $\mathcal{F}$ denotes any set of families indexed by the same set λ, I can define, for each subset $\sigma \subset \lambda$, the σ-projection π_σ on $\mathcal{F}$ by $\{\alpha_k\}_{k\in\lambda} \mapsto \{\alpha_k\}_{k\in\sigma}$. π_σ sends each indexed family z in $\mathcal{F}$ to a family z' indexed by σ, which is obtained by "chopping off" every member of z labelled by an index not in σ. If I wish to cut off from each family in $\mathcal{F}$ precisely those members indexed by labels in σ, I simply resort to the $(\sigma\backslash\lambda)$-projection on $\mathcal{F}$ (where '$\sigma\backslash\lambda$' stands for the complement of λ relative to σ). Rather than use for this mapping the unwieldy symbol $\pi_{\sigma\backslash\lambda}$, I shall denote it by ρ_σ and call it the σ-complementary projection on $\mathcal{F}$. Note that there are exactly $2^{\text{Card}\,\lambda}$ distinct such complementary projections on $\mathcal{F}$, and that $\rho_\varnothing$, in particular, is simply the identity on $\mathcal{F}$.[44] Unless $\sigma = \varnothing$ or $\sigma = \lambda$, ρ_σ is not an injective mapping, and therefore an inverse ρ_σ^{-1} does not exist. However, for each value x in its range, ρ_σ—like any other mapping—unambiguously fixes the set of arguments it sends to that value, that is, the fiber of ρ_σ over x. Thus, ρ_σ determines a "fiber map" from its range $\rho_\sigma(\mathcal{F})$ into the power set $\mathcal{P}(\mathcal{F})$ by $x \mapsto \{y \mid \rho_\sigma(y) = x\}$. I shall denote this map by φ_σ. A typical element of the range of φ_σ is a non-empty subset of $\mathcal{F}$, consisting of many families $\{\alpha_k\}_{k\in\lambda}$, all of which share the family members labelled by indices in $\sigma\backslash\lambda$ but differ from one another as to some family member labelled by indices in σ.[45] The foregoing piece of freshman set theory is all we shall need to penetrate Gähde's definition of **T**-theoreticity.

Let us now consider an arbitrary theory-element **T**. Since we are looking for a criterion of **T**-theoreticity, we may not proceed as if we could identify the partial potential models of **T**. The theory-core **K(T)** must therefore be completely specified by **T**'s models, potential models, and global constraint. Let us put, therefore, $\mathbf{K(T)} = \langle\mathbf{M(T)}, \mathbf{M_p(T)}, \mathbf{GC(T)}\rangle$.[46] If **T** stands for a real physical theory, there are generally two kinds of operations on structural components of $\mathbf{M_p(T)}$ leading from one instance of $\mathbf{M_p(T)}$ to another: (i) scale transformations, whereby new units of measurement are substituted for those employed in the construction of a given $\mathbf{m} \in \mathbf{M_p(T)}$, and (ii) transformations belonging to **T**'s characteristic symmetry group, such as the Galilean group of Newtonian mechanics alluded to above, or the Lorentz group of Special Relativity. (From a structuralist standpoint, such symmetries can be regarded primarily as subgroups of the permutation group of one—

or several—of the base sets of each $\mathbf{m} \in \mathbf{M}_p(\mathbf{T})$, which in turn induce homonymous mappings on the echelon sets over the base sets. With mild abuse of language, one may therefore describe them as mappings from $\mathbf{M}_p(\mathbf{T})$ to $\mathbf{M}_p(\mathbf{T})$.) I shall say that two potential models of **T**, $\mathbf{m}_1$ and $\mathbf{m}_2$, are equivalent modulo a choice of units, or U-equivalent, if $\mathbf{m}_1$ is obtained by operating a scale transformation on $\mathbf{m}_2$; and that they are equivalent modulo a symmetry of **T**, or $\Sigma(\mathbf{T})$-equivalent, if $\mathbf{m}_1$ is obtained by operating on $\mathbf{m}_2$ a transformation of **T**'s characteristic symmetry group. Plainly, both U-equivalence and $\Sigma(\mathbf{T})$-equivalence are genuine equivalences—i.e., reflexive, symmetric, and transitive binary relations—which partition $\mathbf{M}_p(\mathbf{T})$ into equivalence classes. I denote the U-equivalence class of $\mathbf{m} \in \mathbf{M}_p(\mathbf{T})$ by $[\mathbf{m}]_U$ and its $\Sigma(\mathbf{T})$-equivalence class by $[\mathbf{m}]_{\Sigma T}$. It should be clear that if **m** is a potential model of **T**, either $[\mathbf{m}]_U \subset \mathbf{M}(\mathbf{T})$ or $[\mathbf{m}]_U \cap \mathbf{M}(\mathbf{T}) = \varnothing$. Now let **T**´ denote a specialization of **T**. I say that **T**´ is *true* to **T**'s symmetry group or $\Sigma(\mathbf{T})$-true if and only if, for every $\mathbf{m} \in \mathbf{M}_p(\mathbf{T}')$, either $[\mathbf{m}]_{\Sigma T} \subset \mathbf{M}(\mathbf{T}')$ or $[\mathbf{m}]_{\Sigma T} \cap \mathbf{M}(\mathbf{T}') = \varnothing$.

If **S** is a species of structure given by a collection $\mathcal{D}$ of base sets and a list $\alpha_1, \ldots, \alpha_n$ of distinguished structural components, any instance of **S** can plainly be regarded as a family $\{\mathcal{D}, \alpha_1, \ldots, \alpha_n\}$ indexed by $\{0,1, \ldots, n\}$. So the set-theoretical notions introduced in the last paragraph but one are immediately applicable to **S**. Put $\mathbf{S} = \mathbf{M}_p(\mathbf{T})$ and take any $\mathbf{m} \in \mathbf{M}_p(\mathbf{T})$. A distinguished structural component α_r of **m** ($r \in \{1, \ldots, n\}$) is **T**-theoretical by Gähde's criterion if and only if for some $\sigma \subset \{1, \ldots, n\}$, the following conditions are fulfilled:

(i) $r \in \sigma$.

(ii) For every $x \in \rho_\sigma(\mathbf{M}_p(\mathbf{T}))$, $\varphi_\sigma(x) \cap \mathbf{M}(\mathbf{T})$, if not empty, contains members of more than one U-equivalence class.

(iii) There is a $y \in \rho_\sigma(\mathbf{M}_p(\mathbf{T}))$ such that for some $\Sigma(\mathbf{T})$-true specialization **T**´ of **T**, $\varphi_\sigma(y) \cap \mathbf{M}(\mathbf{T}')$ contains members of one and only one U-equivalence class.[47]

Note that conditions (ii) and (iii) hold for the entire subfamily $\{\alpha_k\}_{k \in \sigma}$, which, by condition (i), contains α_r. Condition (ii) says that in any model of **T** the terms labelled by indices in σ are not determined up to a scale transformation by the laws of **T** and the remaining terms of that model. Condition (iii) says, on the other hand, that the terms indexed by σ are unambiguously determined (modulo a scale transformation) by the laws of some $\Sigma(\mathbf{T})$-true specialization **T**´ of **T**, in some model $\mathbf{m}' \in \mathbf{M}(\mathbf{T}') \subset \mathbf{M}(\mathbf{T})$, if the terms of **m**´ not indexed by σ are given. Obviously, α_r may not satisfy the said conditions when one considers it all by itself (viz., by putting $\sigma = \{r\}$) and yet qualify as **T**-theoretical jointly with other terms of **m**.[48] Of course, α_r should test positive for **T**-theoreticity in Gähde's sense if conditions (ii) and (iii) are met by *any* set of integers $\sigma \subset \{1, \ldots, n\}$ that contains r. Hence, α_r may qualify as **T**-theoretical on several counts, i.e., as a member of several distinct groups of terms. Each such group marks out a set of terms of **m** as possibly non-**T**-theoretical—viz., the group's complement relative to $\{\alpha_1, \ldots, \alpha_n\}$. Of course, a term α_r is actually non-**T**-theoretical according to Gähde's criterion only if it is not **T**-theoretical on any count (i.e., only if conditions (ii) and (iii) are not satisfied by any $\sigma \subset \{1, \ldots, n\}$ that contains r).

Gähde's concept of **T**-theoreticity will hardly pass for an explicans of the notion adumbrated by Sneed. At any rate, non-**T**-theoretical terms in Sneed's and in Gähde's sense do not—pace Balzer (1986, p. 87)—always function in the same manner "at the pragmatic level." Take, for example, Newtonian force. The vector-valued function **F** of 20th-century standard textbooks is **CPM**-theoretical by Gähde's criterion. But 19th-century textbooks conceived Newtonian force in the guise of the real-valued functions F_ξ, F_η, and F_ζ, the components of force in three mutually perpendicular directions. These three functions are non-**CPM**-theoretical by Gähde's criterion, for any model **m**´ of any specialization **T**´ of **CPM** can surely be transformed by a simple rotation R of the reference frame into another model R(**m**´) of **T**´ such that $F_\xi \neq R(F_\xi)$. On the other hand, by Sneed's criterion, F_ξ, F_η, and F_ζ are no less **CPM**-theoretical than **F**. This example also suggests that Gähde's criterion does not mark an important epistemological distinction between the concepts of a physical theory—as Sneed's surely does in every case in which it partitions such terms into two non-empty classes. Nevertheless, I think that we ought to be grateful to Gähde for showing us how the properties discernible in a concept of a physical theory may depend on the way it is grouped with other concepts of the same theory, for working out the set-theoretical means for describing such groupings and their consequences, and for suggesting that a definite assignment of partial potential models is not required for the structuralist specification of a theory-element (see note 46).

In their Salzburg paper, Balzer, Moulines, and Sneed (1986) propose still another definition of '**T**-theoretical'. Here the property of being '**T**-non-theoretical' is assigned a positive characteristic, and '**T**-theoretical' is used as an abbreviation for 'non-**T**-non-theoretical', i.e., the condition of items lacking the former property. The new definition turns on the concept of an *interpreting link* between two theory-elements $\mathbf{T}_1$ and $\mathbf{T}_2$, or, more precisely, between their respective theory-cores $\mathbf{K}_1$ and $\mathbf{K}_2$ (for the sake of brevity, I shall write $\mathbf{K}_i$ for $\mathbf{K}(\mathbf{T}_i)$). The theoretical items in the potential models of such a theory-core **K** are the distinguished structural components that are *not* affected by any interpreting link of **K**; the nontheoretical items are those whose values are correlated in some way, by interpreting links of **K**, with the values of distinguished structural components of potential models of other theory-cores (Balzer, Moulines, and Sneed 1986, pp. 301–2). We need not go here into the labored—though again quite elementary—definition of an interpreting link. The following informal description will suffice for our purpose:

> A $\mathbf{K}_1/\mathbf{K}_2$ link is an interpreting link for $\mathbf{K}_2$ when models of $\mathbf{K}_1$ serve as acceptable means of measuring or determining the values of components in potential models of $\mathbf{K}_2$. More precisely, an interpreting $\langle\alpha_1, \ldots, \alpha_r\rangle$- $\mathbf{K}_1/\langle\beta_1, \ldots, \beta_s\rangle$-$\mathbf{K}_2$ link is a link that allows us to infer something interesting about values of the components $\langle\beta_1, \ldots, \beta_s\rangle$ of at least some potential models of $\mathbf{K}_2$ from knowledge of the values of the components $\langle\alpha_1, \ldots, \alpha_r\rangle$ in models of $\mathbf{K}_1$.
>
> (Balzer, Moulines, and Sneed 1986, p. 295)[49]

This characterization at once suggests that the theoretical terms of Sneed 1971 could now be reclassified as nontheoretical items of self-interpreting theories. (Put $\mathbf{K}_1 = \mathbf{K}_2$ in the above quotation. The α's may—but need not—differ from the β's. A **K**/**K** link of this sort would be involved, for example, in the familiar use of the electromagnetic properties of some electromagnetic systems for measuring electromagnetic properties of other such systems.) However, the authors dismiss this suggestion offhand. *"There are neither uninterpreted nor self-intepreting formal structures in empirical science"* (Balzer, Moulines, and Sneed 1986, p. 297; my italics).[50] This raises again the specter of circularity or fathomless regress, for it follows at once that the net of interlinked physical theories "is unbounded with respect to interpreting links or that there are closed 'chains' of interpreting links." But the authors "do not find either 'horn' of this dilemma obviously unacceptable."[51]

The italicized statement is, of course, a truism if by 'formal structure' we mean the lifeless skeleton of a theory divested of thought. But that is not the guise in which theories of mathematical physics make their appearance in intellectual history. Nor will the skeletonized theories recover their senses merely by being coupled to other formal structures in ordered pairs. It will be said that the proposed couples are not symmetrical, that in a $\mathbf{K}_1/\mathbf{K}_2$ interpreting link properties of known models of the interpreting theory $\mathbf{K}_1$ are employed for ascertaining properties of *potential* models of the interpreted theory $\mathbf{K}_2$. But, if there are no self-interpreting theories, $\mathbf{K}_1$ is in and of itself no more than just another uninterpreted structure, which can reach its physical models only through an interpreting link with still another theory. Yet heaping structures on structures by an endless succession of interpreting links will not give a single one among them a hold on reality.

This is not to deny that such interpreting links between physical theories can and do exist. But the light they provide often flows in a direction opposite to that indicated by Sneed and his associates. Consider the following example—the first that comes to my mind. In relativistic cosmology, the "expansion of the universe" that has taken place since the emission of an extragalactic light-signal is measured by the redshift, i.e., by the displacement towards the red end of the spectrum of the position of some known pattern of spectral lines. The redshift is established by comparing the position of those lines in the spectrum of the light-signal in question with their standard position as predicted by atomic theory and recorded in the spectra of light emitted from terrestrial sources. Insofar as the spectrum of the light-signal can be regarded as a model of atomic theory, we can infer from it a particular value of a function pertaining to relativistic cosmology. But no sensible person would say that relativistic cosmology is herewith interpreted by atomic theory. On the contrary: by furnishing a geometrodynamic account of the anomalous position of the spectral lines in the extragalactic light-signal, relativistic cosmology can properly be said to interpret a phenomenon that, without its assistance, we could not view as a model of our atomic theory.[52]

I surmise that, despite Sneed's fearless and uncompromising stand for the primacy of concepts in physics, there lurks in his writings an antiintellectualist distrust of the enlightening power of thought. This would account for the curt rejection of self-interpreting theories in the Salzburg paper. Perhaps for the same reason the Sneedian school has hitherto evaded the central philosophical question concerning the application of a physical theory, namely, how does a dull, confusing stream of events come to be understood as an illustration of such a theory and thus as a model of a mathematical structure? In the next section we shall see how Günther Ludwig has grappled with this question.

3.5 To spell the phenomena

> Erscheinungen zu buchstabieren, um sie als Erfahrung lesen zu können.
>
> KANT, *Prolegomena*, §30

Günther Ludwig (1978) describes a physical theory *PT* as a combination of three factors: a *mathematical theory MT*, a *domain of reality W*, and a set of *mapping principles* (—) that link the latter to the former. Schematically: *PT* = *MT*(—) *W*. Ludwig is no doubt aware that in history these three factors actually grow together. Mathematical theories do not lie ready in a Platonic smorgasbord for the physicist to choose the one that suits him best. As is well known, some of the greatest mathematical ideas were born in the struggle for a better grasp of physical phenomena. The *MT* of a *PT* is laboriously developed in the flickering light of the *W* it is meant to apprehend. However, in order to understand the workings of a *PT* we must, for definiteness, fix one of the three factors and show how the other two are shaped with respect to it. The *MT* is the only likely candidate for this role. In fact, the mathematical part of an established or a dead physical theory can be given a neat and stable formulation, which in turn provides the means for outlining the theory's intended domain of application, whereas the latter cannot be properly referred to without the conceptual resources of the mathematical part.

Ludwig's understanding of mathematics is Bourbakian, like Sneed's. The *MT* of a *PT*, as conceived by him, is therefore typically a theory—stronger than Zermelo-Fraenkel set theory—that characterizes a species of structure in Bourbaki's sense. Ludwig, however, does not take the informal set-theoretic approach to mathematical structures that Sneed learned from Patrick Suppes but remains beholden to the old faith regarding the possibility and desirability of expressing all mathematical discourse in a strictly regimented formal language. "A mathematical theory is defined as a collection of symbols according to certain rules of the game," he writes.[53] We need not go to this extreme; but in order to follow his exposition we must equate the *MT* of a *PT* not with a concept whose extension comprises the models of a certain Bourbakian species of structure but rather with a set of statements, namely, the statements that are true of just those models. Ludwig expects this set to be axiomatized, i.e., to include a distinguished finite or at least computable subset of which all statements in the set are logical consequences.

According to Ludwig, the domain of reality *W* of a physical theory *PT* is partly determined by the *MT* and the mapping principles. However, it must contain a part, a *fundamental domain G*, that does not depend on *MT*. *G* stands for the processes and situations that are the factual basis of *PT*. "The fundamental domain *G* of *W* does not consist of any *statements* about physical processes but of the given processes themselves. Just as the text of a book lies there beforehand [*vorliegt*], so *G* presents a domain of real states of affairs. Because of the similarity with the text of a book, any *given pieces* [*vorliegende Teilstücke*] of *G* will be called *real-texts* [*Realtexte*] of the theory *PT*" (Ludwig 1978, p. 13). Despite its illustrious pedigree, the book metaphor may seem farfetched, for there is a deep difference between the two sorts of thing that it brings together. Anybody who has mastered the English words 'book' and 'page' will be able to tell at a glance where a page begins and where it ends, and what features of it represent the text. On the other hand, mastering Ludwig's expression 'real-text' gives no such grip on the physical states of affairs which it is meant to designate. There are no universal conventions for singling out the situations and processes to which physical theories are applicable and for distinguishing in them the relevant text from the indifferent background. However, if we take a closer look into Ludwig's writings, we see that this gap between real texts and what he calls real-texts does not matter much in the end. For the fragments of physical becoming on which, according to him, a *PT* can rest are not just any crudely delineated real-texts that happen to be there but only what he calls "normalized" (*genormte*) real-texts. A *normalized real-text* of a *PT* is a physical state of affairs that is analyzed into factors, a finite number of which have been labelled with names—usually letters—of the language in which the respective *MT* is expressed. The analysis and the labelling clearly suffice to delimit the situation or process which is to be the object of inquiry. And Ludwig's description of it as a kind of text is somehow vindicated by the standard use of letters as labels. But a normalized real-text cannot be established simply by pointing, as some of Ludwig's statements might suggest. The analysis of observed things and events requires a conceptual grasp of them, without which one cannot even properly be said to *observe* them. Ludwig does not deny this (although, in my judgment, he does not sufficiently stress it). He explicitly notes that the real-texts of a given *PT* may have to be normalized with the aid of other, earlier physical theories. But he emphatically maintains that the *PT* to whose fundamental domain *G* a certain real-text belongs can have no say in its normalization. "How one reads [. . .] the real text *cannot* [*darf nicht*] depend on the *PT* under consideration" (Ludwig 1978, p. 40). "We shall use the expression *fundamental domain G* to denote those facts that can be specified in advance of a particular *PT* and which are mapped into the mathematical framework *MT* of *PT*. [. . .] The fundamental domain consists of all that can be determined before the application of the theory" (Ludwig 1983, p. 3). I shall argue that to exclude every physical theory from the required intellectual elaboration of its own factual basis is neither justified nor altogether consistent with some of Ludwig's more valuable insights. But I must first say something about the mapping principles that connect the *MT* of a *PT* with its fundamental domain *G*.

It is not unreasonable to equate the mathematical part of a *PT*, for the purposes of a philosophical discussion, with the axiomatized theory MT_Σ of a Bourbakian species of structure Σ (Ludwig 1983, p. 3), but it would be utterly unrealistic to attribute a comparable degree of definiteness to the theory's fundamental domain *G*. Ludwig emphasizes that *G* does not denote a set, "for nothing has been defined that might correspond to the 'elements' of a set." Since new real-texts are continually being added by new experiments, "we can collect 'all' real-texts only notionally [*nur begrifflich*], precisely in the notion of the fundamental domain *G*" (Ludwig 1978, p. 15). I take this to mean that *G* refers to the real-texts of *PT* distributively, not collectively, for they do not form a definite collection. The mapping principles will therefore link *MT* severally with each given particular normalized real-text.

How do the mapping principles perform their job?

In the first place, they pick out certain monadic and polyadic predicates of *MT*, which Ludwig calls iconic terms (*Bildterme*) and iconic relations (*Bildrelationen*), respectively.[54] Ludwig assumes that if *R* stands for an iconic relation, any sentence formed by substituting names (constants) for the variables x_k in an open formula of the form $Rx_1 \ldots x_n$ must be required by syntax to carry the name of a real number in the place of x_n. This requirement may seem needlessly restrictive, insofar as not every physical observation must be a measurement in the ordinary sense. However, as I indicated in note 23, any monadic or polyadic predicate *P* is uniquely associated with a real-valued function on the set of objects to which *P* can be meaningfully ascribed. This function, called the *characteristic function* of *P*, takes the value 1 at every object of which *P* is true and the value 0 at every object of which *P* is not true. Hence, when I verify the presence of H_2S by its fetid smell, I may be said to measure the characteristic function of the respective qualitative predicate and to obtain the value 1.

In the second place, the mapping principles provide rules by which one or the other iconic term is predicated of the several names assigned to factors in the real-text.

In the third place, they provide rules by which the iconic relations are predicated of those names and of certain particular real numbers, "the so-called quantitative measurement results," which, I presume, must somehow turn up in the real-text.

Ludwig (1978, p. 34) stresses that all these rules must be applied "on the basis of the normalized and read present real-text" ("aufgrund des vorliegenden, genormten und gelesenen Realtextes"). I take this to mean that the rules must be adjusted to each particular real-text or family of related real-texts; that they are, so to speak, tailored to fit them.

Let *A* be the conjunction of all statements made about a given normalized real-text pursuant to the said rules. Since the names used for labelling the real-text belong to the language MT_Σ, *A* is a statement of that language, which can be added to MT_Σ to form a stronger theory $MT_\Sigma A$. If $MT_\Sigma A$ is consistent, we say that by means of the mapping principles employed, MT_Σ provides a usable (*brauchbare*) description of the real-text in question. If no inconsistency has been found in any of the theories $MT_\Sigma A$ formed as I have said by applying to all hitherto investigated normalized real-texts of *PT* the rules prescribed for them by the mapping principles, we say that MT_Σ together with the utilized mapping principles provides a usable description of the *entire* fundamental domain *G*, or, more briefly, that "*PT* is a *definitively usable theory*" ("eine *endgültig brauchbare Theorie*"—Ludwig 1978, p. 42).

The construction of the full domain of reality *W* of a physical theory *PT*—of which, as the reader will recall, the fundamental domain *G* is only a part—proceeds by including "hypotheses" in the theories $MT_\Sigma A$ by which the real-texts of *PT* are consistently described. In Ludwig's parlance, a hypothesis is a statement in which an *n*-adic predicate of MT_Σ ($n \geq 1$) is predicated of *n* names of MT_Σ, one or more of which are not labels of factors in a given real-text but stand for conceived objects (*gedachte Gegenstände*). Here are some examples of such conceived objects: (*a*) an unrecorded instant between two consecutively recorded ones; (*b*) the invisible trajectory of an uncharged elementary particle joining the photographed tracks of two charged particles;[55] (*c*) the orbit of the hypothetical planet Vulcan, as computed from Newton's laws and the observed orbits of Mercury and the other known planets. Ludwig's classification of hypotheses and his carefully grounded assignment of certain classes of them to the job of determining the domain of reality of a *PT* are interesting and difficult matters on which, unfortunately, I cannot dwell here (Ludwig 1978, §10, pp. 116–94).

Let us now return to the question I left pending when I went on to speak about the mapping principles: Is Ludwig right in denying that a *PT* can play a role in the spelling and reading of its real-texts? To my mind, this denial stems from the same sort of foundationist scruples that mired Sneed and his followers in the bogus problem of *T*-theoretical terms. Balzer, Moulines, and Sneed (1986, p. 297) went so far as to assert that there are no self-interpreting formal structures in empirical science—although any such structure can interpret another one![56] Ludwig has furnished us with an obvious reply to that assertion: strictly speaking, a formal structure cannot interpret itself (or anything else, for that matter) but a physical theory consists of a mathematical structure together with mapping principles and is therefore inherently self-interpreting. Seasoned physicist that he is, Ludwig wastes no time on the theoretical/nontheoretical conundrum. Yet he seems to fear that if he grants the physical theory a say in the articulation of its own real-texts, its factual basis will be undermined. Such would no doubt be the case if the reading of the real-texts stood to the theory's axioms as the premises to the conclusion of some perverse form of inference. The inductivist school of philosophy has long sought for the rules of a "logic" that would warrant just this kind of "reasoning." But Ludwig will have none of it:

> The much debated problem of "incomplete induction" is avoided in actual physics, because one simply *does* theoretical physics *without worrying* about the question of "incomplete induction."[57] Theoretical physics, as it in fact exists, *gives no prescriptions* for the establishment of axioms in *MT* but leaves it to an "intuition" of some sort. The General Theory of Relativity is one of the most striking illustrations of the fact that the axioms of a theory are in no way derived from experience pursuant to the principle of incomplete induction. On the contrary, the starting point for the establishment of the axioms of the General Theory of Relativity was Einstein's intuitive grasp of the connection between spacetime structure and gravitation.
>
> (Ludwig 1978, p. 45)

Hence, from Ludwig's own standpoint there is no compelling reason why a physical theory should not contribute to the delimitation and articulation of its real-texts. And indeed Ludwig appears to grant as much when he equates "the *precise* structure of a real-text" with what is said about it pursuant to the mapping principles (in the statements conjoined into the statement *A*, in the notation employed above).[58]

Ludwig is perfectly right to stress that the real-texts from which a physical theory draws its sustenance have usually been worked over by other theories and by untutored common sense. A physical theory does not come to life in full armor in the midst of an intellectual vacuum. As Ludwig aptly puts it: "The possibility of knowing certain facts and processes 'immediately,' i.e., without any *PT*, is an everywhere noticeable basis of all experiments; e.g., the state of a counter is accepted as a fact that is not to be analyzed further. No scientific, let alone physical, criteria are employed to establish such facts" (1978, p. 14). The presence of such common, readily identifiable items in real-texts that different physical theories understand differently constitutes a powerful link between those theories, which it would be foolish to ignore. But this alone does not warrant the foundationist claim, made by Ludwig on the same page, to the effect that "all *PT*'s can be built together upon the simplest possible 'immediately' readable pieces of real-texts, [. . .] such as we acknowledge as facts in our everyday behavior, without any reflection or philosophical inquiry—e.g., that there is a chair in the room, a cup of coffee on the table, a stone that shattered a window-pane." If prephilosophical common sense had furnished a coherent and cogent reading of the real-texts of everyday life, I doubt very much that anybody would have ever felt a need for mathematical physics. And I have yet to see how such a reading can be obtained merely by putting together a mosaic of small unreconstructed pieces of pre-scientific experience. Physicists attain to their innovative understanding of phenomena through a deliberate effort to read them and even to spell them in new ways.

Ludwig has proposed a familiar example to illustrate his requirement that "what is being mapped [by the mapping principles of a *PT*] must be expressed in terms of experiment and experience without the need for the application of a new theory."

> For example, it is possible to specify the position of the planets without requiring the use of Newton's mechanics and Newton's law of gravitation. The position of the planets at different times provides the experimental material which can be compared to Newton's theory, that is, which is to be mapped into the mathematical framework *MT* of Newton's theory.
>
> (Ludwig 1983, p. 3)

As I see it, this is an excellent illustration of the opposite view, for Ludwig's requirement is here violated in several respects. *First,* the timing of planetary positions must be done with good clocks, and for the purposes of Newtonian theory, only the direct or indirect agreement of a clock with Newton's laws can warrant its goodness. (This point was discussed in Section 3.4.) *Second,* to test the Newtonian laws in their standard form, the planetary positions must be referred to an inertial frame, the criteria of inertiality being provided by Newton's theory. Thus, for example, it is only on the strength of this theory that Foucault's pendulum demonstrates that the fixed stars define an inertial frame relative to which the Earth is rotating. *Third,* before the discovery of the aberration of light by Bradley (in 1728) and the measurement of the parallax of a nearby star by Bessel (in 1838), the only cogent reason for regarding the center of gravity of the Solar System—and not, say, the center of the Earth—as a point at rest in an inertial frame was the successful dynamical explanation of planetary motions by Newton's theory. In other words, if *MT* stands for the mathematical framework of Newtonian Gravitation Theory (*NGT*) and *A* and *B* denote, respectively, the standard quasi-heliocentric and a geocentric description of planetary motions to within suitable margins of error, then, in Newton's time, the only serious ground for accepting *A* and not *B* as a description of the real-text consonant with the mapping principles of *NGT* was that the theory *MTA* is consistent, and the theory *MTB* is not (see Newton, *Principia*, Bk. III, Phenomenon V).

These three points sufficiently support my allegation that, at least in Ludwig's own example, the theory decisively intervenes in the reading of the real-text. The example raises still another question: Why is "the experimental material which can be compared to Newton's theory" confined to the Sun and planets, while the fixed stars are mercilessly chopped off from the real-text? In the light of current astronomical data, a Newtonian could answer more or less as follows: The entire Solar System undergoes an acceleration directed towards the center of gravity of the Galaxy; as the distance to that center is enormous, the acceleration is very nearly constant; hence, by Corollary VI to the Laws of Motion, the Solar System can be treated as an isolated mechanical system. Obviously, in this reply it is the theory that draws the boundary around its real-text. On the other hand, I find no evidence that Newton ever raised the question himself. Since both the force of gravity and the intensity of light obey an inverse square law, he might have reasoned that in the Sun's presence the gravitational influence of the fixed stars is about as insignificant as their light, for surely the latter should be roughly proportional to their masses.[59] In the Corollary to Proposition XII of Book I of the *Principia*, Newton explicitly equates the center of gravity of the Solar System with the motionless center of the world (*centrum mundi*). The fixed stars are not mentioned in this context; but if Newton believed, with Kepler, that they are more or less evenly distributed in a spherical shell about the center, he could have invoked Proposition LXX of Book I to prove that the gravitational force exerted by the fixed stars at any point in the interior of that shell is null. Be that as it may, with these speculations I only wished to suggest that the real-text in Ludwig's example cannot be circumscribed without some theoretical assumptions. The same is true, a fortiori, of less simple cases.

3.6 Approximation and idealization

For simplicity's sake I have hitherto ignored what is perhaps the most remarkable feature of the relationship between the mathematical theories of physics and the actual situations to which they are purportedly applicable. Roughly stated, it is this: on the one hand, physical states of affairs agree only *approximately* with the mathematical theories that physics brings to bear on them, so that such theories can furnish, at best, only an *idealized* representation of physical reality; on the other hand, mature physical theories include a *precise quantitative estimate* of their respective degree of approximation (or imprecision). This feature distinguishes the conception of knowledge embodied in modern physics from everything that went before it and probably accounts to a large extent for the estrangement of modern scientific thinking from the majority of mankind. Before I try to elucidate it, I shall illustrate it with a very simple example, which may also serve to clarify some of Ludwig's ideas about mapping principles and real-texts presented in Section 3.5.

Let *PT* denote Galileo's theory of free fall (Section 3.1). Let MT_Γ stand for elapsed from the release of the freely falling ball *x* at *y* until its arrival at *z*. The mapping principles must also specify rules (methods) for measuring the distance *d* from *p* to *q*, the distance *d´* from *p* to *q´*, and the times *t* and *t´* that the freely falling ball *b* takes to go from *p* to *q* and *q´*, respectively. The mapping principles will therefore prescribe the following reading *A* of the above real-text:

$$Fa \mathbin{\&} Gb \mathbin{\&} Hp \mathbin{\&} Hq \mathbin{\&} Hq' \mathbin{\&} Jap \mathbin{\&} Jaq' \mathbin{\&} Sapqd \mathbin{\&} Sapq'd' \mathbin{\&} Tbpat \mathbin{\&} Tbpq't' \qquad (1)$$

In MT_Γ, (1) must entail that $d/d' = (t/t')^2$. However, it is extremely unlikely that the numbers *d, d´, t,* and *t´*, measured on the real-text in accordance with the mapping principles, will satisfy this equation. The theory $MT_\Gamma A$ is therefore inconsistent and *PT* is unusable for the real-text proposed. How come, then, that we still regard Galileo's theory of free fall as eminently applicable to experimental situations of the kind described? The answer to this question is obvious: We do not expect the measured values of physical quantities to satisfy the equations of mathematical physics exactly, but only to within a suitable margin of imprecision.

Similar inconsistencies arise in any theory *MTB* if the iconic relations that go into a reading *B* of an appropriate real-text are drawn, without qualifications, from a mathematical theory *MT*. But that is not the way iconic relations are used in physics. Their actual occurrence in the reading of real-texts is fairly well described if, following Ludwig, we say that the mapping principles of a physical theory associate a *blurred* (*verschmierte*) relation with each iconic relation drawn from the pertinent mathematical theory, and prescribe the attribution of such blurred relations to items discerned in the real-text. The construction of the blurred relation $\tilde{R}xy$ associated with a given relation *Rxy* between a list *x* of objects falling under iconic terms and a real number *y* is quite simple. The mapping principles should associate with R a real-valued function ε on the real numbers, depending on R but also on the nature of the real-text in question and the method by which the relevant measurements are to be performed on it. Then, $\tilde{R}xy$ holds if and only if there exists a real number z such that Rxz and $|y - z| < \varepsilon(z)$.

The procedure can be readily applied to the above example of Galileo's theory of free fall. For simplicity's sake, let us associate constant functions $\varepsilon_s = \sigma$ and $\varepsilon_t = \tau$ with the iconic relations S and T, respectively. We define the blurred relations $\tilde{S}$ and $\tilde{T}$ by

$$\begin{aligned} \tilde{S}xyzu &\Leftrightarrow \exists w(Sxyzw \,\&\, |u - w| < \sigma) \\ \tilde{T}xyzu &\Leftrightarrow \exists w(Txyzw \,\&\, |u - w| < \tau) \end{aligned} \qquad (2)$$

We now claim that the measured values of distances and times sustain only these blurred relations with the appropriate points, balls, and inclined planes; so we substitute in (1) $\tilde{S}$ for S and $\tilde{T}$ for T. Let $\tilde{A}$ denote the statement obtained by this substitution. Conjoined with MT_Γ, $\tilde{A}$ does not entail—like A—that $d/d' = (t/t')^2$ but only that there are real numbers u, u', v, and v', such that $u/u' = (v/v')^2$, which satisfy the following four inequalities: $|d - u| < \sigma$, $|d' - u'| < \sigma$, $|t - v| < \tau$, and $|t' - v'| < \tau$. For a sensible choice of σ and τ this requirement is normally fulfilled, and $MT_\Gamma\tilde{A}$ is consistent.

The blurring of real-valued functions in physics can be demonstrated very clearly, as we have just seen; but the physicist blurs all sorts of relations. We can conceive such blurring with great generality by means of some simple set-theoretical notions. Let S be any set and consider the binary relations on S, i.e., the subsets of the Cartesian product $S \times S$. If F is a binary relation on S, the *inverse* relation $-F$ is the set $\{\langle x,y\rangle \mid \langle y,x\rangle \in F\}$. If F_1 and F_2 are binary relations on S, the *composition* $F_1 + F_2$ is the binary relation $\{\langle x,y\rangle \mid \exists z(z \in S \,\&\, \langle x,z\rangle \in F_1 \,\&\, \langle z,y\rangle \in F_2)\}$. I shall write $2F$ for $F + F$, and generally $(n + 1)F$ for $nF + F$. The *diagonal* $\Delta(S)$ is the set $\{\langle x,x\rangle \mid x \in S\}$. If F is a binary relation on S which includes $\Delta(S)$ and is identical with its inverse $-F$, F is said to be an *entourage* of $\Delta(S)$. Note that, according to this definition, $\Delta(S)$ is an entourage of itself and is also the intersection of the set of all its entourages. If F is an entourage of $\Delta(S)$ and $\langle x,y\rangle \in F$, I write $|x - y| < F$. The reader should verify that if x, y, and z are any elements of S, and F and F' are any entourages of $\Delta(S)$, then (i) $|x - x| < F$; (ii) $|x - y| < F$ if and only if $|y - x| < F$; and (iii) $|x - y| < F$ and $|y - z| < F'$ only if $|x - z| < F + F'$. Consider now an n-ary relation R picked out by the mapping principles of a physical theory from the pertinent mathematical theory. Let

D_R denote the domain of definition of R, i.e., the set of all the n-tuples to which R can be meaningfully attributed. Let F be an entourage of $\Delta(D_R)$. F determines a blurred relation $\tilde{R}_F$ associated with R by the condition:

$$\tilde{R}_F x \Leftrightarrow \exists w (Rx \ \& \ |x - y| < F) \qquad (3)$$

(where the variables x and y range over the set of n-tuples D_R). Following Ludwig, I call F an *imprecision set* (*Unschärfemenge*) for R. $\tilde{R}_F$ is the relation R *as blurred by* F. Note that the relation R as blurred by the diagonal $\Delta(D_R)$ is equivalent to R itself.

The familiar saying that a physical theory represents this or that class of phenomena only *approximately*, or to such-and-such *approximation*, can thus be given a precise sense by associating definite imprecision sets with all iconic relations involved in the reading of the relevant real-texts according to the mapping principles of the theory. We also have to make sense of the notion that some approximations are better or worse than others, and of the belief, common among physicists, that every approximation ought to be liable to gradual improvement. Ludwig achieves this by requiring that any imprecision set A_R associated in some application with an iconic relation R should belong to a uniformity U_R on R's domain of definition D_R.[60] By a *uniformity* on a set S, I mean a non-empty set U such that

U0. If $A \in U$, A is an entourage of $\Delta(S)$.

U1. If $A \in U$ and B is an entourage of $\Delta(S)$ such that $A \subset B$, $B \in U$.

U2. If $A \in U$ and $B \in U$, their intersection $A \cap B \in U$.

U3. If $A \in U$, there is in U a set B such that $2B \subset A$.

A uniformity on a set is weakly ordered by inclusion.[61]

Now, if A and B are imprecision sets for a given iconic relation R, and A is a proper subset of B (i.e., if $A \subset B$ but $A \neq B$), it is reasonable to say that R is approximated *more closely*—and hence *better*—by the blurred relation $\tilde{R}_A$ determined by A than by the blurred relation $\tilde{R}_B$ determined by B, because the condition $A \subset B \ \& \ A \neq B$ implies that $\tilde{R}_B x$ holds good whenever $\tilde{R}_A x$ does but not vice versa. In other words, the said condition implies that $\tilde{R}_B x$ is weaker than $\tilde{R}_A x$. But why should the imprecision sets for an n-ary iconic relation R constitute a uniformity on D_R? One readily sees that any such imprecision set A_R must be an entourage of the diagonal $\Delta(D_R)$. For the blurred relation $\tilde{R}_A$ determined by A should certainly be true of any n-tuple x of which R itself is true; and if $\tilde{R}_A x$ is true whenever Ry is true, then surely $\tilde{R}_A y$ is true whenever Rx is true. So the imprecision sets for R satisfy condition U0. But what about the remaining three conditions on uniformities, U1–U3? U1 is satisfied by all conceivable imprecision sets for R if and only if given a blurred relation $\tilde{R}_A$ which approximately renders R, any blurred relation $\tilde{R}_B$ which approximates R less closely, and is therefore strictly worse than $\tilde{R}_A$, *also* renders R approximately. This certainly sounds plausible. U2 is satisfied if and only if whenever two blurred relations $\tilde{R}_A$ and $\tilde{R}_B$ occur among the approximate renderings of R, the conjunctive relation $\tilde{R}_A$ & $\tilde{R}_B$ also occurs among them. Again this sounds plausible. But to vindicate U3 is not so easy. What U3 says, in effect, is that for every conceivable inexact approximation of the iconic relation R there is another conceivable approximation which is "twice as good" and yet still falls short of exactness. I do not see how one could be sure that this is true in every case. Ludwig (1978, p. 55) grants that U3 is an "idealization." Yet it does express well the physicist's belief in successive, gradually improving approximations. One might argue on its behalf that, in a situation in which a blurred relation $\tilde{R}$ is known to be the best conceivable approximation of an iconic relation R drawn from some *MT*, but $\tilde{R}$ is not equivalent to R, *MT* should be replaced by another theory *MT*′, perhaps involving $\tilde{R}$ instead of R. When a better approximation of R is precluded by the failure of U3, an unbridgeable gap is revealed between the theory and its real-texts. R cannot then be held as a goal to reach for, and therefore ought to be discarded.[62]

The picture of a physical theory that I have sketched, after Ludwig, in Section 3.5 should therefore be completed to include, besides the mathematical theory *MT*, the domain of reality *W*, and the mapping principle (—), a family of uniformities $\{U_k\}$, one for each iconic relation R_k employed in the reading of real-texts. The mapping principles that govern the reading of a given real-text in a given set of circumstances must then pick out, from every relevant uniformity, an imprecision set by which to blur the respective iconic relation. The choice of the appropriate imprecision set depends on several factors. Foremost among them is the precision—or, rather, the imprecision—of the instruments used for observation. All instruments have a limited power of resolution. This is clearly noticeable when an instrument produces numerical readings. There is always then a lower bound below which the readings either are no longer consistent (among themselves and with those of like instruments) or fail to make any further distinctions. In other words, below that bound the instrument does not discriminate reliably, or does not discriminate at all. The instrument's inaccuracy thus contributes to fix the imprecision set by which any relation observed with it will be blurred. For example, let $Dxyz$ state the relation that holds between the points x and y and the real number z, when z is the distance in meters between x and y. Suppose that the procedure used for measuring distances in the real-text under consideration accurately discerns up to 1 part in 10,000. We should then substitute for D the blurred relation $\widetilde{D}$ defined by:

$$\widetilde{D}xyz \Leftrightarrow \exists w(Dxyw \ \& \ |z - w| < 0.0001w) \tag{4}$$

(In this example I have made no special allowance for the inevitable imprecision in the identification of points, for it is normally absorbed into the "error" in the measured value of the distance function.)

A second factor governing the selection of imprecision sets for the reading of a real-text arises from the fact that experimental situations do not exactly satisfy the conditions in which a given physical theory is strictly applicable. Predictions that are meant to hold for isolated physical systems will be tested on systems that are not fully isolated. Moreover, the need for mathematical tractability imposes streamlined conditions which the available real-texts only approximately fulfill. Thus, Galileo's theory of free fall applies to a ball falling freely in a total vacuum on a perfectly flat, frictionless inclined surface. Since no such things are available, the blurring of the relations S and T, as in (1), must be made to depend not only on the accuracy with which points and events can be identified and distances and times can be measured between them, but also on the discrepancy between the ideal conditions assumed by the theory and those actually prevailing in the real-texts.

A third motive for blurring the iconic relations of a physical theory stems from the theory's mathematics. This usually postulates ordinary or partial differential equations governing the distribution in space and time of the physical quantities of interest but is unable to provide exact solutions of those equations except under quite special, generally unrealistic initial and boundary conditions. Let me illustrate this point by quickly recalling the classical reading of the motion of Mercury as a test of General Relativity. In its now standard form, it employs the Schwarzschild solution of the Einstein field equations. This is the exact solution for a static, spherically symmetric, asymptotically flat gravitational field in vacuo.[63] Its application to this partic-ular case therefore assumes that the Sun is perfectly symmetric and is the only extant source of gravity and that Mercury is a freely falling test particle of negligible mass. Under these assumptions, General Relativity almost exactly predicts the part of Mercury's perihelion advance that Newtonian kinematics and dynamics do not account for, and which amounts to less than 0.8% of the total perihelion advance recorded in the laboratory frame. The remaining 99.2% is due to the motion of our terrestrial frame in the Schwarzschild field of the Sun and to the presence in that field of significant gravitational sources, viz., Venus, Jupiter, and the other planets. Since an exact solution of the many-body problem is not available in Einsteinian (or indeed in Newtonian) dynamics, the effect of these additional sources must be handled by approximation methods as a perturbation of the Schwarzschild field.[64] One would generally expect that any imprecision set selected on grounds of mathematical impotence will be included in the imprecision set that must be assigned to the same predicate because of the inaccuracy of the instruments of observation and the imperfect adequacy of the experimental conditions. But one can occasionally choose to work with a cruder mathematics than the given real-text would warrant, because it is easier to do so and a higher precision is not needed. Philosophers tend to think that the pragmatic acceptance of inexactness is the mark of engineering, not science. But engineering, in the guise of experimental design, pervades contemporary physics.

The approximative nature of physical theories accounts for much of the traditional importance of continuity in mathematical physics. Let me explain this by means of an abstract example. Suppose that the *MT* of a *PT* entails that a certain relation R holds for an n-tuple in its domain of definition D_R if certain conditions obtain. Such conditions can always be conceived as a particular instance of a relation C on a suitable set of m-tuples D_C (for sufficiently large m). D_R and D_C are topologized in the manner explained in note 61 by the uniformities associated with R and C, respectively. If the link between C and R is determined in *MT* by a continuous mapping $f: D_C \to D_R$, then, for every imprecision set ε in the uniformity of R, there is an imprecision set δ in the uniformity of C such that $\widetilde{C}_\delta$ entails $\widetilde{R}_\varepsilon \circ f$ (In the symbolism introduced earlier, $\widetilde{C}_\delta$ stands, of course, for C as blurred by δ, etc.) Since only blurred relations can be read in the real-texts of *PT*, the entailment of unblurred R by unblurred C would be of purely academic interest if f were not continuous (in the topologies induced in D_R and D_C by the respective uniformities).

One often hears that physical theories represent their referents approximately, through idealized "models."[65] Such a manner of speaking easily suggests that the physicist on the one hand faces a definite object (a system, a process, etc.) which he handles and observes, while on the other hand he contemplates, so to speak, with the mind's eye a simplified image of that object, which imperfectly mirrors it. This suggestion may cause one to overlook the fact that the referents of a physical theory owe their segregation and articulation to the intellectual efforts of the theorizer. We tend to forget this because scientific thought never arises in a conceptual void but follows upon and is prompted by an earlier grasp of things. But if physical theorizing were to abide by that former grasp, while seeking merely to approximate it with its "models," it would indeed be futile and redundant. *What* is observed and handled is not to be decided outside the bounds of human discourse; and within those bounds there are no privileged oracles to which a physicist must unconditionally submit. The above remarks concerning approximation and continuity indicate that idealization in physics runs farther and deeper than the talk of "models" would imply. Approximation by a conceptual structure only makes sense if the objects approximated by it are incorporated with it in a broader structure, which is also conceptual. If every attempt to understand nature by means of so-called mathematical—i.e., exactly defined—concepts amounts to an idealization, then, in physics, idealization is pervasive; for the blurred relations in terms of which the phenomena are actually read by a physical theory are no less mathematical than the unblurred relations that go into their definitions.

3.7 On relations between theories

By judiciously combining Ludwig's approach with some ideas of Sneed we can draw a sketch of the internal organization of physical theories that will be useful for studying their mutual relations. Both Sneed and Ludwig build their prototype of a physical theory around a mathematical core. I am reluctant to identify this core with a set of sentences of a formal or an informal language, let alone with a set of uninterpreted formulae of a so-called calculus. I propose therefore that we regard, with Sneed, the mathematical "heart" (should not one rather say the "brain"?) of a physical theory as a *concept* by which the physical phenomena to which the theory purportedly applies are to be grasped. Sentences ("theorems") are still required for spelling out the implications of bringing something under that concept, but many alternative sets of sentences can perform this task equally well. The Sneedian "applications" of the theory crystallize naturally enough around its Ludwigian "real-texts." Sneed's notion of "constraints" nicely captures how a theory's "texts" are bound together into a coherent "domain of reality." I have argued that Sneed's emphasis on a physical theory's "nontheoretical" terms, and on the "partial models" which can be fully described by means of such terms, is only a barren remnant of the foundationist quest for certainty, and should be ignored. On the other hand, I do not dispute the usefulness of Ludwig's distinction between a theory's "fundamental domain" of observable "real-texts" and its completion by the theory's "hypotheses" into a full "domain of reality." This distinction somehow parallels Sneed's contrast between partial and full models; and indeed Ludwig also shows a deep and, to my mind, unwarranted distrust of creative understanding when he requires a theory to describe its real-texts in borrowed terms—whereas it is often the failure of earlier science and plain common sense to furnish a cogent reading of such texts that motivates the development of new theories. I propose that we simply dismiss this requirement of Ludwig's. Not only is it up to a physical theory to say what is or is not to be seen in its real-texts; but, moreover, the frontier between the latter and the conjectured "domain of reality" beyond them will shift as the theory becomes gradually entrenched and some of its initially adventurous hypotheses come to be accepted as observation data (typically because the hypothesized entities are being detected as a matter of course with instruments built to the theory's specifications). In Ludwig's picture the crowning feature of a physical theory is the "mapping principles." A counterpart to them is entirely missing in Sneed's scheme of things. Such "mapping principles" are hard to explicate and probably impossible to codify, but they come closest to expressing the intellectual feat by which the theoretical physicist seizes on a fragment of reality and articulates it as a concrete illustration of an abstract mathematical structure. The philosopher who ignores them—or who, worse still, reduces them to arbitrary lexical conventions—is reaching only for the mindless cadaver of a physical theory.[66]

From our present standpoint it seems natural that physics should develop many theories, both successively and simultaneously, proffering different structures to deal with the various real-texts we can discern all about us. According to the foregoing sketch such theories can be quite disparate, for although each one of them must cut out its real-texts from the same all-engulfing stream of life, each theory does it after its own manner and completes them by means of its peculiar hypotheses, thereby building its own "domain of reality." In actual fact, however, all the known theories of physics are mutually related, either directly or through the mediation of others. In this section and the next I shall describe and illustrate several types of relations between physical theories. But before proceeding to do so I must briefly comment on the common grounds on which such relations can rest.

Two theories, T_1 and T_2, can seek to account for the same phenomena; or, in the jargon we have chosen, they may share (at least some of) their real-texts. This seems too obvious to require any discussion; but it is important to see its import. Believers in senseless reference will sense no difficulty here (cf. Section 2.6). For them any object or collection of objects—things, processes, events, qualities, quantities, etc.—can be unambiguously referred to by a person, independently of how she conceives it. In their view, you can know precisely what your finger is pointing at, without having to grasp it, say, as a lake, or a surface, or a glitter. But those of us who have outgrown such blissful innocence must carefully consider how the real-texts of two conceptually different theories can yet be said to be the same. If T_1 and T_2 were two comprehensive, mutually exclusive conceptual schemes such that the "conversion" from one to the other would send the convert into "a different world," there could be no question of identifying a real-text of T_2 with one of T_1.[67] But physical theories have no such global scope, for they are not intended as *Weltanschauungen* but are more like little hard clear-cut gems strewn in the magma of human understanding. Metaphors aside, it is clear that every physical theory is developed by people who already have some grasp, however unsatisfactory, of their environment, and particularly of the matters on which the new theory will be brought to bear. Today such "pretheoretical" understanding is nourished to a considerable extent by earlier physical theories, some of which may appear to be in dire need of replacement and repair (otherwise, what inducement would there be for creating a new one?) but which still guide the selection and the rough preliminary scanning of the real-texts whose detailed articulation is the appointed task of the new theory. The continuity of thought on the way towards an improved reading of those texts certifies and in a sense constitutes their sameness. Such continuity depends, in turn, on the persistent use, throughout the entire development, of the same auxiliary theories for designing the experiments, calibrating the instruments, interpreting and controlling the results, etc., and also, indeed, on the pervasive intellectual support provided by common sense. Because conceptual change is never total but must occur—in order to *be* a change—against a backdrop of conceptual permanence, it will not normally bring about a complete breakdown of reference.

Andy Pickering (1984a) analyzes a somewhat unusual example which nicely illustrates how a new physical theory is apt to retain and yet remake an inherited real-text.[68] Until the late sixties, the available theories of weak interactions (Fermi 1934; Feynman and Gell-Mann 1958; Sudarshan and Marshak 1958) sought to account for so-called charged-current events, in which two weakly interacting particles respectively acquire and lose one electric charge (supposedly through the mediation of a charged particle emitted by one of the interacting particles and absorbed by the other), but precluded the existence of neutral-current events, in which no such exchange of electric charge occurs between the interacting particles. Talk of 'currents'—charged and neutral—followed the analogy with the quantum field theory of electromagnetic interactions between charged particles, viz., quantum electrodynamics (QED). "In QED, the interaction between electric currents is mediated by an electrically neutral particle, the photon. Thus the 'interaction current' carried by the photon is an electrically neutral current.

In 1933, Fermi suggested that the weak interactions have a field-theoretic form analogous to that of QED, except that the interaction current [. . .] would in this case carry non-zero electric charge" (Pickering 1984a, p. 89n.). The exclusion of weak neutral currents agreed well with experience, for in point of fact only weak charged currents had been observed. However, the unified theory of electromagnetic and weak interactions independently proposed by Steven Weinberg (1967) and Abdus Salam (1968) predicted the existence of weak neutral currents. A few years later the prediction was tested and confirmed by an experiment carried out in the new giant bubble chamber Gargamelle at CERN (Hasert et al. 1973a, 1973b; 1974). Gargamelle holds 18 tons of liquid Freon (CF_3Br) surrounded by some 1,000 tons of metal walls and ancillary equipment. The liquid is held under pressure, on the verge of boiling. When pressure is relaxed, small bubbles form along the path of every charged particle that happens to be crossing the Freon tank. By synchronizing the firing of a particle beam with pressure relaxation in the tank and shutter release in suitably placed cameras, one obtains a photographic record of the trajectories of charged particles. High-energy neutrinos fired at Gargamelle's belly should weakly interact either with leptons (such as electrons) or with hadrons (specifically, neutrons). Hasert et al. claimed to have detected a single instance of weak neutral-current interaction between neutrinos and atomic electrons in a total of over 700,000 photographs. They also reported having found, in a total of 290,000 photographs, about 400 charged-current and 100 neutral-current neutron-neutrino interactions. Pickering's discussion is concerned only with the latter. Charged-current neutron-neutrino interactions are readily recognized because the incoming neutrino turns into a charged lepton, whose track is recorded in the photograph. But the photographic evidence of a neutron-neutrino *neutral*-current interaction can consist only in the tracks of the spatter of charged hadrons generated by the neutron's decay (upon interacting with the energetic neutrino). However, a spatter of the same type may arise if the neutron collides with another neutron issuing from a *charged*-current event outside the Freon tank. Stray neutrinos are bound to produce such charged-current events in the bubble chamber's walls and equipment. By a sophisticated chain of statistical inferences, Hasert et al. filtered the hundred or so "genuine" hadronic neutral-current events claimed by them from the background noise attributable to external charged-current events. Such procedures are of course normal in experimental science and illustrate the difficulties in establishing the real-texts on which one's theory is brought to bear. The question that, following Pickering, we must now ask is why, if the Gargamelle collaboration succeeded in detecting one hadronic weak neutral-current event for every four charged-current events, no such events had been observed before. The answer is quite simple: according to pre-1967 theories, neutral currents were not supposed to exist; hence, no elaborate statistical computer program was developed for plucking neutral-current events from under the rubble of hadronic showers caused by the fallout from external charged-current events, and all recorded showers of this sort were discounted as noise and were therefore excluded from the real-texts under consideration. Indeed, even when someone expressed an interest in such events he would assume as a matter of course that they were due to strong neutron-neutron, not weak neutron-neutrino, interactions (see Myatt 1969, p. 146; quoted by Pickering 1984a, p. 99). Confronted with the same experimental records—readily identifiable by common sense, assisted perhaps by library science—a post-1975 physicist would doubtless draw a different line between the relevant and the irrelevant. Curiously enough, in a detailed study of CERN neutrino runs from 1963 to 1965, E. C. M. Young (1967) had given a careful estimate of the neutron-induced background, implying that the latter could account for most, but not all, the observed events of the neutral-current type. "Left unaccounted for were around 150 neutral-current type events, to be compared with around 570 positively identified charged-current events: a ratio of roughly one neutral-current event to every four charged-current events—the same as that which was later reported from Gargamelle" (Pickering 1984a, p. 99; Pickering notes that "Young did not state this conclusion explicitly, but it follows directly from his background estimate and the counts of various types of event on film"). Of course, you may say that the weak neutral current was always there, but scientists were prevented by their prejudices from seeing it. But you ought then to keep in mind that what now enables them to see what they formerly overlooked should in your idiom also be called a "prejudice"—although at this point a "usable" one.

The foregoing example illustrates particularly well how reference is refocussed through conceptual change. This phenomenon, however, should be discernible everywhere, though camouflaged, so to speak, by the ordinary imprecision of our thinking. (If one knows only roughly what one is talking about, one may reconceive it and still claim that one is speaking about "the same thing.") Anyway, it should be clear by now that physical theories can share facts only insofar as they also share some thoughts, and that concepts provide a less shifty ground than raw data for building bridges between those theories. When dealing with the conceptual side of physics one may distinguish between (i) the well-defined mathematical structures that are the backbone of particular physical theories and can be shared by several such theories, and (ii) certain lasting ideas that guide the formation or selection of such structures and inspire the mapping principles that bind them to experience. Towards the end of Section 3.1 I alluded very briefly to the importance of such general ideas as an overarching link between theories. Two of them, probability and necessity, will be subject of the following chapters. I shall therefore confine myself here to a few remarks on shared structures.

Two theories, T_1 and T_2, can have the same mathematical structure ($MT_1 = MT_2$, in Ludwig's symbolism) and yet be different because they refer to different domains of reality ($W_1 \neq W_2$). Such structural identity can be very beneficial, as advances in one area can then be used without delay in the other. Indeed, I surmise that analogies in physics—whose supposedly explanatory value was much talked about in the 1960s—owe their significance chiefly to this possibility of extending to a new field, through a shared structure, the mathematical methods for solving problems which are already available elsewhere. In principle it could also happen that T_1 and T_2 share both their structure and their domain and still $T_1 \neq T_2$, because they differ in their mapping principles. But this situation, of which one can find mutatis mutandis plenty of instances in mathematics,[69] would be pretty farfetched in physics.

A philosophically more interesting case occurs when our two theories seek to account for roughly the same phenomena, and their mathematical structures, though different, are more or less closely related to one another. This case is quite common and many examples of it come promptly to one's mind. The real number field **R** is an ingredient of the mathematical core of every physical theory. The mathematical structures of both Newton's and Einstein's theory of gravity can be built around the concept of a linear connection on a four-dimensional differentiable manifold. The currently fashionable gauge theories of elementary particle physics share the notion of a fiber bundle acted on by a Lie group. Each theory has its own distinctive group, and hence a different structure, but their comparability is ensured by the conceptual affinity of all such structures.

If the theories T_1 and T_2 deal with the same real-texts—in some appropriate, duly qualified sense of 'same'—one of them, say T_2, can be a *specialization* or an *extension* or a *restatement* of the other; but T_2 can also be proposed as a *substitute* for T_1. A few examples will clarify these four types of relations between theories, while illustrating the role of shared structures.

A physical theory can be made more specific by imposing additional conditions on its mathematical structure. Thus, the diverse classical force laws, such as Newton's Law of Gravity or Hooke's Law, specify the rather general theory that Sneed calls Classical Particle Mechanics (CPM), which is characterized by Newton's Second Law of Motion. Without such specifications the general theory is indeed no more than a vacuous scheme, and in ordinary scientific talk one would not say that the several force laws generate as many specific theories, distinct from CPM, but rather that they develop the latter and make it applicable. Nor would anyone say that the diverse solutions of the Einstein field equations obtained by imposing conditions on the stress-energy tensor generate new theories of gravity, distinct from General Relativity, although they evidently specify this theory in the same sense in which the force laws can be said to specify CPM. In fact, if $\mathbf{T} = \{T_1, \ldots, T_n\}$ is a set of physical theories, in Ludwig's sense, partially ordered by specification, Sneed and his collaborators would represent each T_i $(1 \leq i \leq n)$ by a theory-element, in their sense, so that the partially ordered set **T** would amount to what they call a theory-net. As I noted in Section 3.3 their theory-nets correspond better than their theory-elements to what are ordinarily recognized as distinct physical theories.

Let us consider the more interesting case in which a physical theory can be said to be an extension of another one. I wish this to be understood strictly. The extended theory T_2 should agree *exactly* with the restricted theory T_1 in dealing with the kind of situations for which T_1 was intended (or is, at any rate, still usable), but it must also be applicable in other cases. This is achieved if the mathematical structures of T_1 and T_2 are such that the former can be conceived as a special case of the latter. 19th century theories of electrodynamics aimed at being extensions, in this strict sense, of the earlier theory of the electrostatic potential. Note that in any nontrivial case such as this one the extended theory T_2 is not obtained by simply relaxing the conditions on the mathematical structure of T_1 so as to enlarge its set of models. Of course, according to my characterization of theory extension, the models of T_1 must be contained in a proper subset of the set of models of T_2. But the latter will meet specifications of its own which cannot be conjectured by examining the structure of T_1. By merely relaxing the requirement that all charges be at rest you will not go very far towards guessing the Maxwell equations and the Lorentz Force Law from Poisson's Equation and Coulomb's Law. There are instances, however, in which the mathematics of a physical theory, viewed under the proper light, may suggest how to proceed to a required expansion of that theory.

Consider the transition from Special to General Relativity. The Principle of Equivalence first stated by Einstein in 1907 implies that an inertial frame is physically equivalent to a nonrotating frame falling freely in a uniform gravitational field. This in turn implies that inertial frames cannot coexist with nonuniform gravitational fields and must therefore be infinitely remote from all sources of gravity. Now, the flat Minkowski spacetime geometry of Special Relativity guarantees that an inertial frame endowed with Einstein time can be constructed everywhere. Einstein's Principle of Equivalence entails therefore that the spacetime geometry cannot be flat unless the gravitational field is uniform. This may suggest a link between gravity and the Riemannian curvature of spacetime. With the benefit of hindsight, we find the suggestion fairly obvious; but it also seemed natural to Einstein's mathematician friend Marcel Grossmann, who in 1912 directed Einstein's attention to Riemannian geometry. The suggestion, however, did not furnish Einstein with the field equations establishing the required link between the spacetime curvature and the distribution of gravitational sources: he did not light on them until November 1915. The flat Minkowski geometry is an exact solution of the field equations in the special case in which spacetime is wholly empty—though not the only such solution. It also agrees with the structure of the tangent space at each point of the spacetime manifold. As a consequence of this, any non-rotating freely falling frame can be regarded as inertial to within the approximation in which the local gravitational field is uniform. In this qualified sense, Special Relativity may be said to be locally valid in the curved spacetime of General Relativity.

It sometimes happens that two physical theories, while ostensibly differing from one another in their conceptual makeup, nevertheless yield exactly the same predictions for every conceivable input of data (conceivable, that is, in terms of either theory). In such cases it seems reasonable to regard the younger theory, call it T_2, as a mere restatement of the older theory T_1. It may be the case that T_1 and T_2 have been developed independently, and that their equivalence comes to their authors as a pleasant or unpleasant surprise. A rather striking example of such unexpected concurrence of theories was provided by Schrödinger's discovery that his Wave Mechanics is mathematically equivalent to Heisenberg's Matrix Mechanics, although their guiding motives were, in a way, diametrically opposite (Schrödinger 1926). Or T_2 may have been deliberately worked out in order (i) to assist in the solution of problems which had proved intractable for T_1, (ii) to reach a better understanding of the purport of T_1, or (iii) to make it easier to compare T_1 with still a third theory that is meant to replace it. Hamilton's version of Classical Mechanics—as an alternative to Lagrange's—is an example of (i); (ii) was the primary aim of Minkowski's restatement of Einstein's Special Relativity as a spacetime theory; and the four-dimensional formulations of Newton's theory of gravity (Cartan 1923; Friedrichs 1927; Havas 1964), which place it squarely at the side of Einstein's General Relativity, exemplify (iii). I surmise that in all cases of this kind T_1 and T_2 will be found to share one and the same mathematical structure, which each theory typically characterizes by a different set of sufficient conditions, while its necessary conditions are never fully spelled out. (Hence the apparent structural differences between T_1 and T_2.)

My surmise is easily confirmed in the case of the formulations of Special Relativity by Einstein (1905b) and Minkowski (1908, 1909). Einstein referred the laws of nature and the description of phenomena to inertial frames furnished with coordinate systems linked to one another by Lorentz transformations. Minkowski's spacetime can be constructed as follows from any such inertial frame endowed with Einstein time coordinate t and Cartesian coordinates x, y, and z. Let T denote the domain of t and let S denote the domain of x, y, and z. Form the Cartesian product $T \times S = M$. If π_τ and π_σ denote the projections of M onto T and S, respectively, the composite mappings $t \circ \pi_\tau$, $x \circ \pi_\sigma$, $y \circ \pi_\sigma$, and $z \circ \pi_\sigma$ are the coordinates of a global chart on M, whose very existence suffices to make M into a real four-dimensional differentiable manifold. Endow M with the Riemannian metric η, whose components with respect to the chart $(t \circ \pi_\tau, x \circ \pi_\sigma, y \circ \pi_\sigma, z \circ \pi_\sigma)$ are $\eta_{tt} = -c^2$, $\eta_{xx} = \eta_{yy} = \eta_{zz} = 1$, $\eta_{ik} = 0$ if $i = k$ (where c stands for the vacuum speed of light measured in the units employed in defining t, x, y, and z). $\langle M, \eta \rangle$ is Minkowski spacetime. On the other hand, if the Minkowski spacetime $\langle M, \eta \rangle$ is given, Einstein's inertial frames can be readily carved out of it. Take a congruence K of parallel timelike straights in M. There is a unique foliation H of M into spacelike hyperplanes orthogonal to the straights in K.[70] Say that two points in different hyperplanes of H are *isotopic* if they lie on the same straight of K. Say that two points in different straights of K are *simultaneous* if they lie on the same hyperplane of H. Plainly, both isotopy and simultaneity are equivalences. Equate the *distance* between two equivalence classes P and Q of isotopic points with the spacelike separation between any point p of P and the one and only point of Q which is simultaneous with p. The set S of all equivalence classes of isotopic points is made thereby into a Euclidian space. Equate the *time interval* between two equivalence classes P and Q of simultaneous points with the proper time elapsed along the segment joining any point p of P and the one and only point of Q which is isotopic with p. The set T of all equivalence classes of simultaneous points thereby embodies Einstein time. S and T are to be regarded as the domains of space and time coordinates adapted to a particular inertial frame. In this way, each global inertial frame in Einstein's sense is determined by a unique congruence of timelike straights in the Minkowski spacetime $\langle M, \eta \rangle$.

The common structure underlying both Lagrangian and Hamiltonian dynamics has been laid bare by the modern differential-geometric approach to classical dynamics. Roughly, the idea is as follows: The motion of a system with n degrees of freedom can be represented by a curve γ, parametrized by time, in a real n-dimensional differentiable manifold, the system's *configuration space* M. Lagrangian dynamics calculates γ from the behavior of the Lagrangian of the system, which is a real-valued function on the tangent bundle *TM*. Hamiltonian dynamics calculates γ from the behavior of the Hamiltonian of the system, which is a real-valued function on the cotangent bundle *T*M*. A given Lagrangian *L* determines a Hamiltonian *H*, and vice versa, if and only if certain conditions are met; viz., given *L*, if and only if a certain mapping of *TM* into *T*M*, sometimes called the fiber derivative of *L*,[71] is a diffeomorphism (i.e., an isomorphism of the differentiable structures); or, given *H*, if and only if the homologous mapping of *T*M* into *TM*—the fiber derivative of *H*—is a diffeomorphism. Note that if either fiber derivative is a diffeomorphism, so is the other and each is the inverse of the other. (See, for instance, Abraham and Marsden 1978, Chapter 3, "Hamiltonian and Lagrangian Systems".)

Strictly equivalent physical theories, in the sense we have been considering, agree exactly on all their predictions from any conceivable data input. One may also wish to consider theories that agree on all *verifiable* predictions from any *available* set of data. Let us say that two such theories are observationally equivalent. Observationally equivalent theories may be quite different in their respective mathematical setup, but they must share some structural features if they are to make predictions from the same data. In view of the inherent inaccuracy of empirical data, the agreement required for observational equivalence need only be approximate. Moreover, it can vanish as the methods of observation evolve. However, Clark Glymour (1977) found application, in the context of General Relativity, for a stronger, irrevocable relation of observational equivalence between mathematically inequivalent structures. The equivalence in question rests on the fact that in a relativistic spacetime all data employed in a physical calculation must issue from the calculator's causal past.[72] Consequently, if two non-isomorphic models of General Relativity are such that the causal past of any arbitrarily chosen timelike curve in either model is isomorphic with the causal past of a timelike curve in the other, then, according to the theory, it is impossible to decide on empirical grounds with which model the world agrees. For reasons of mathematical expediency, Glymour's discussion turns on the isomorphism of *chronological*—not *causal*—pasts in nonisomorphic spacetimes.[73] Two spacetimes M and M' are said by Glymour to be *observationally indistinguishable* (o.i.) if for any future-directed, future inextendible, timelike curves γ in M and σ' in M', there are curves of the same type γ' in M' and σ in M, such that the chronological past $I^{-}[\gamma]$ of γ is isometric with $I^{-}[\gamma']$, and $I^{-}[\sigma]$ is isometric with $I^{-}[\sigma']$.[74] Glymour constructs several pairs of o.i. spacetimes such that the underlying manifold of one member of a pair is a covering space of that of the other, or that both have a common covering space.[75] But Malament (1977b) gives other examples not subject to this constraint. Malament also introduces a weaker, asymmetric relation of observational indistinguishability between spacetimes, which is probably more significant than Glymour's from an epistemological point of view. A spacetime M is *weakly observationally indistinguishable* (w.o.i.) from a spacetime M' if for every point x in M there is a point x' in M' such that the chronological past $I^-(x)$ of x is isometric with the chronological past $I^-(x')$ of x'. If M is w.o.i. from M', an observer in M is unable to tell at any time of his life whether he lives in M or in M', even if he happens to be immortal. Malament lists several important global properties of M which might not be shared by M' even if the former is w.o.i from the latter.[76] He remarks that the notion of weak observational indistinguishability

> seems a straightforward rendering of conditions under which observers could not determine the spatio temporal structure of the universe. Yet, and this is the most interesting, the condition of weak observational indistinguishability is so widespread in the class of space-times as to be of epidemic proportions.
>
> (Malament 1977b, p. 69)

For a confessed metaphysical realist like Glymour the discovery of observationally indistinguishable spacetimes in General Relativity must have come as a shock. But it also deserves the attention of those of us who scorn that philosophical creed, for it helps to undermine the popular opinion that science aims for the description of reality in the metaphysical realist's sense.[77]

Finally, let us consider the relation between a theory T_1 and a structurally different theory T_2 that is intended to displace it and replace it. Obviously, we shall speak of displacement or replacement only if T_1 has in science a place of its own in which it has become entrenched by its success in accounting for its domain of reality. An inducement to replace such a "well-placed" theory T_1 can then arise from "anomalies"—i.e., from real-texts purportedly within the scope of T_1 that it has persistently failed to read satisfactorily—or from the dismissal of some other theory associated with it that has gone bankrupt even while T_1 retains an air of solvency. Examples of anomalies abound in the literature of science. Let me mention, without further comment, the measurements of specific heats at low temperatures, which remained intractable until the innovative work of Einstein (1907; but see Kuhn 1978, p. 212). The most remarkable instance of guilt by association in the history of physics is presumably that of Newton's theory of gravity, which at the turn of the century could boast of a predictive precision unmatched by any other physical theory and hardly showed any anomaly worth mentioning, and yet had to be replaced because the Newtonian kinematics on which it rested crumbled under Einstein's criticism (prompted by the quest for a viable electrodynamics of moving bodies).[78]

To provide an alternative to the established theory T_1, T_2 should of course give a satisfactory account of the real-texts on which T_1 based its success. Although T_2 will propose a different conception of those real-texts, it will usually proceed from the same analysis on which their normalization in terms of T_1 depends (see Section 3.5). In other words, although T_2 will give a new reading of them, it will—at least initially—retain T_1's spelling. This should provide ample grounds for comparison between them, unless, indeed, T_2 issues from a successful criticism of the very elements of T_1's approach to phenomena; in which case such a comparison will not be called for (see Section 2.5).

3.8 Intertheoretic reduction

Philosophers of science have often used the expression 'intertheoretic reduction' to refer to the relationship between (*a*) an entrenched physical theory T_1 and (*b*) a different theory T_2 that replaces T_1 on the grounds that T_2 accounts for every observation that supports T_1 and has still other advantages over the latter. Among such advantages one may count, for example, that T_2 accounts better than T_1 for new, improved observations within T_1's domain or that it also accounts for observations outside T_1's domain or that it is "simpler" or more "elegant" or "deeper" than T_1.

However, not all philosophers understand intertheoretic reduction in the same way. In the heyday of logical syntax and the formal mode of speech, the relation was defined for pairs of theories expressed, with partly disjoint vocabularies, in the same formal language. It was then said that T_2 *reduced* T_1 if every theorem of T_1 could be derived in the language common to both theories from theorems of T_2 conjoined with so-called bridge laws, which stated, by means of the predicates of one theory, necessary, or sufficient, or necessary and sufficient conditions for predicates of the other. Ernest Nagel (1949) noted that lest the reduction be trivial, the bridge laws cannot just follow from definitions and the laws of logic but must be scientific hypotheses liable to confirmation and disconfirmation. This implies in effect that unless the predicates of T_1 also belong to T_2 (the case of "homogeneous reduction" that Nagel leaves aside as unproblematic), T_1 is being reduced, not to T_2 alone, but to a stronger theory, viz., the union of T_2 and the bridge laws. It implies moreover (in the case of "inhomogeneous reduction" considered by Nagel) that the predicates of T_1 and T_2 which occur in the bridge laws make sense even when they are torn out of their native theoretical context and employed in statements that belong to neither theory. According to the philosophical tradition favored by Nagel at the time, such would indeed be the case of "observational"—as opposed to "theoretical"—predicates.

But Nagel's own examples—viz., 'kinetic energy' and thermodynamic 'temperature'—can hardly qualify as "observational." The bridge laws can be dispensed with, however, if one gives up thinking that the "reduced" or "secondary" theory T_1 must somehow be derivable from the "reducing" or "primary" theory T_2. Kemeny and Oppenheim (1956) took this stance. They relinquished all idea of linking the structures of the reduced and the reducing theory and defined intertheoretic reduction relative to a set of observational data. As they put it, T_1 is reduced by means of T_2 relative to observational data O if (i) the vocabulary of T_1 contains terms not in the vocabulary of T_2, (ii) any part of O explainable by means of T_1 is explainable by T_2, and (iii) T_2 is at least as well systematized as T_1.[79]

Kemeny and Oppenheim's definition of data-relative intertheoretic reduction plainly presupposes that the observation data O can be identified and described apart from the theories T_1 and T_2 which are supposed to share and explain a non-empty subset of those data. The inadequacy of this crude understanding of observation was one of the main grounds of Feyerabend's critique of the very idea of intertheoretic reduction in one of his seminal contributions to the doctrine of intertheoretic incommensurability (Feyerabend 1962). Within the following decade, some epistemologists of the new generation, better schooled in physics and its history than their philosophical mentors, tried to salvage some of the earlier work on intertheoretic reduction (while at the same time acknowledging the complexities of the relation) and sought to classify the wide variety of cases that had customarily been swept under that omnibus term.

Schaffner (1967, p. 142) pointed out that even in the exemplary case of Maxwell's reduction of optics to electrodynamics, the secondary theory, viz., the wave theory of light developed by Fresnel and his followers, did not agree exactly with all the optical consequences that can be drawn from the identification of visible light with a certain range of solutions of the electromagnetic wave equation. Mid-19th-century optics had to be somewhat modified in order to annex it to classical electrodynamics. Mindful of such examples, Schaffner offered a characterization of intertheoretic reduction according to which what can, under certain conditions, be derived from the primary theory T_2 (together with suitable "reduction functions") is usually not the secondary theory T_1 but a *corrected version* of it T_1^*. In other words, T_1 is reduced to T_2 if T_2 yields a theory T_1^* that can pass for a modified version of T_1. Thus, Schaffner's proposal substitutes for the problems hitherto encountered in explicating reduction the problem of ascertaining when and how a physical theory can be regarded as a "correction" of another one. According to Schaffner, T_1* should provide "more accurate experimentally verifiable predictions than T_1 in almost all cases [. . .], and should also indicate why T_1 was incorrect (e.g. crucial variable ignored), and why it worked as well as it did"; T_1* should produce "numerical predictions which are 'very close' to T_1's" and "the relations between T_1* and T_1 should be one [*sic*] of strong analogy."[80]

However, as Sklar (1967, p. 111) aptly noted, "to say that A is at best an approximation to B is to say that A and B differ, and that they are, in fact, incompatible." One must therefore distinguish between "reductions in which the reduced theory is *retained* as correct subsequent to the reduction" and "those reductions in which the reduced theory is instead replaced by the theory which it reduces." Sklar observes that, to avoid misunderstandings, "it might be better to restrict the extension of the term 'reduction' so as to exclude replacements." However, he does not embrace this proposal, because in ordinary scientific parlance the replacement of one theory by another is often called 'reduction', at least when "the former theory is retained 'as a useful instrument of prediction', despite its known falsity as a scientific theory" (Sklar 1967, p. 117).

It surely is a good thing to remain faithful to ordinary usage in philosophical discourse, provided that one is not thereby mired in confusion. Now, in the present case, as Nickles (1973a) remarked, intertheoretic 'reductions' spoken about in physics do not always proceed in the same direction as in epistemological talk. Physicists might indeed say that Maxwell reduced optics to electrodynamics or that Boltzmann and Maxwell reduced thermodynamics to (statistical) mechanics. But they will also say that Einstein's kinematics reduces to Newton's in the limit $(v/c)^2 \to 0$ and that Quantum Mechanics reduces to Classical Mechanics in the region of large quantum numbers. Obviously, in the last two cases, it is the "primary," replacing theory that is said to be reduced to the "secondary," replaced theory. Nickles believes that this difference in usage arises from a genuine duality in meaning, and that we are dealing here with two distinct relations between physical theories, which he denotes by 'reduction$_1$' and 'reduction$_2$'. 'Reduction$_1$', which Nickles classifies into "exact" and "approximative," can still be explicated along Nagel's lines as involving some manner of derivation of the reduced from the reducing theory. But 'reduction$_2$' is nonderivational. Unfortunately, Nickles does not define it; but he explains that "for example, a successor theory reduces$_2$ to its predecessor (not *vice versa*) if applying an appropriate operation (for example, a mathematical limit) to some equations of the successor yields the formalism of the predecessor" (Nickles 1973b, p. 588 n.

19). Now, it is clear that, if T_2 is to replace the established theory T_1 as the standard account of a certain range of phenomena *P*, and T_1 accounts satisfactorily for some subset *P′* of *P*, T_2 must, for the values of its parameters which suit the cases in *P′*, yield predictions that do not differ from those of T_1 by more than the admissible experimental error. To meet this requirement, T_2 need not have any structural connection with T_1. The approximate agreement between them in those experimental situations in which T_1 is still successful does not *presuppose* a mathematical transformation that reduces some equations of T_2 to equations of T_1. However, the existence of such a transformation may be necessary for *proving* that the required agreement will be reached in *all* relevant cases. More importantly perhaps, such formal links can be very useful when *searching* for the theory T_2, and can serve as a guide for its construction. Indeed, as Nickles (1973b, p. 585) ably remarks, once we recognize that a theory's domain of phenomena is organized and structured by it, we are bound to see "that even the bland requirement that future theories must give an equally good account of the present theory's domain will quite probably impose constraints on the internal structure of the new theory" so that "in this manner, structure may be subtly transmitted from theory to theory via the domain."[81]

Throughout the years Sneed and his associates have paid considerable attention to intertheoretic reduction (Sneed 1971, pp. 216–48; Stegmüller 1973a, pp. 144–52, 249ff.; 1986, 128–36; Balzer and Sneed 1977/78; Moulines 1984; Sneed 1984; Balzer 1984; Balzer, Moulines, and Sneed 1987, pp. 252–84, 306–20). Balzer and Sneed (1977/78) conjectured that every interesting relation between theory-elements would be a composite of specialization, theoretization, and reduction. (We have already met specialization in Section 3.3; a theory-element $\mathbf{T}_1$ is a *theoretization* of another such element $\mathbf{T}_2$ if all non-$\mathbf{T}_1$-theoretical concepts of $\mathbf{T}_1$ belong to $\mathbf{T}_2$, either as $\mathbf{T}_2$-theoretical or as non-$\mathbf{T}_2$-theoretical concepts.) Balzer, Moulines, and Sneed (1987) have added intertheoretic approximation and equivalence to that list.[82] They observe, quite rightly, that "in real-life science reduction plays a much more modest role than general discussions tend to assume," for "many would-be cases of reduction on a closer scrutiny reveal themselves as cases of approximation" (1987, p. 252). They note that intertheoretic reduction has been understood in at least two ways. First, there is what they propose to call *historical* reduction: an earlier theory **T** reduces to a new theory **T*** "in a way that transmits the main achievements of the preceding theory so that they can also be regarded as achievements of **T***." Then, there is *practical* reduction, useful in dealing with difficult problems. In such cases, it is quite common to proceed as follows:

> One tries to simplify ("to reduce") the theory used by consciously omitting some "parts" of it which for the problem at hand do not distort the "correct" solution too much so that the simpler theory will produce a solution approximately equal to the "correct" one. By using the simpler theory one then solves the problem in a "coarse" version and thus obtains an approximate solution of the original problem. In some cases, the solution thus obtained may be equally satisfactory from the empirical point of view.
>
> (Balzer, Moulines, and Sneed 1987, p. 253)

Our authors believe, however, that both sorts of reduction can be brought under the same roof, so that one is formally conceived as the converse of the other. They study two examples in some detail and then propose a fairly weak general concept of reduction. This comes in two shapes: direct reduction of one theory-element to another and reduction of a theory-element $\mathbf{T}_1$ to another theory-element $\mathbf{T}_2$ through a specialization of the latter to which $\mathbf{T}_1$ reduces directly. To give a taste of what our authors are up to, I shall paraphrase their definition of direct reduction.

Let $\mathbf{T}_1$ and $\mathbf{T}_2$ be two theory-elements and let ρ denote a binary relation between potential models of $\mathbf{T}_1$ and $\mathbf{T}_2$. Set-theoretically speaking, $\rho \subset \mathbf{M}_p(\mathbf{T}_1) \times \mathbf{M}_p(\mathbf{T}_2)$. We require, further, that every potential model of $\mathbf{T}_2$ should stand in this relation with some potential model of $\mathbf{T}_1$. We say that ρ *directly reduces* $\mathbf{T}_2$ to $\mathbf{T}_1$ —abbreviated $\mathbf{T}_2\ \rho\ \mathbf{T}_1$— if and only if

(i) for every pair $\langle x_1, x_2 \rangle \in \rho$, if x_1 is a model of $\mathbf{T}_1$, then x_2 is a model of $\mathbf{T}_2$;

(ii) for every set X_1 in the global constraint of $\mathbf{T}_1$, if every element of X_1 has the relation ρ to some potential model of $\mathbf{T}_2$, then the set $\rho(X_1)$ formed by the potential models of $\mathbf{T}_2$ that are ρ-related to elements of X_1 belongs to the global constraint of $\mathbf{T}_2$;

(iii) for every $\langle x_1, x_2 \rangle \in \rho$, if x_1 belongs to the global link of $\mathbf{T}_1$, then x_2 belongs to the global link of $\mathbf{T}_2$;

(iv) for every intended application $y_2 \in \mathbf{I}(\mathbf{T}_2)$, there is an intended application $y_1 \in \mathbf{I}(\mathbf{T}_1)$ and a pair $\langle x_1, x_2 \rangle \in \rho$, such that $\mu_1(x_1) = y_1$ and $\mu_2(x_2) = y_2$ (here μ_i denotes the appropriate "cutoff" function for $\mathbf{T}_i$ as defined in (1) of Section 3.3).

The relation between theory-elements thus defined is rather loose. We shall have that $\mathbf{T}_2\ \rho\ \mathbf{T}_1$ whenever (α) every potential model of $\mathbf{T}_2$ is matched by at least one potential model of $\mathbf{T}_1$, (β) every potential model of $\mathbf{T}_2$ that is matched by a model of $\mathbf{T}_1$ is in effect a model of $\mathbf{T}_2$, and (γ) for every intended application of $\mathbf{T}_2$ there is an intended application of $\mathbf{T}_1$ such that a potential model of $\mathbf{T}_1$ embracing the latter matches a potential model of $\mathbf{T}_2$ embracing the former. The matching itself can otherwise be quite arbitrary. It is understood, of course, that conditions (ii) and (iii) must also be fulfilled. But these conditions, which are meant to ensure compatibility between the global constraint and the global link of $\mathbf{T}_1$ and $\mathbf{T}_2$ insofar as these theory-elements are related by ρ, do not strike me as particularly severe.

3.9 Recapitulation and preview

Having come all this way, it is time that I recapitulate the main points I believe I have made in the three foregoing chapters.

In Chapter 1, I emphasized the need for general concepts in the observation of particulars. Upon performing an observation of any object (in the widest sense, which includes things, processes, events, states, etc.), the observer will always grasp it as an object of such-and-such a narrower or broader sort. Moreover, the observer's judgment of the epistemic import of an observation often depends on his understanding of the physical interaction from which, in his view, the observation issues. This rule, which is increasingly but not always satisfied by ordinary personal observations of our surroundings, admits no exception in the case of the impersonal observations with artificial receivers which are the main source of scientific data. One cannot ascribe a definite objective meaning to a click or a pointer reading except on the strength of some conception of the physical process leading to it. Thus, although our understanding of the facts of observation is measured by its greater or lesser success in fitting them together, it cannot be simply gathered from them; it must already be at work in the factories where such facts are framed.

Now, our human understanding—at any rate, according to our current grasp of it—has not stood there ready-made from the beginning of time, but has been taking shape in many, not necessarily consistent or coherent, ways in the course of history. As I mentioned in Chapter 2, the dramatic changes in our grasp of natural phenomena that took place especially in the 17th and in the early 20th century have led some writers to deny the continuity of science and to question the comparability of its several stages. In the same chapter I discussed and dismissed some of the replies that can be made to those writers, based on the distinction between observational and theoretical terms and on the doctrine of senseless—i.e., thoughtless—reference. I then went on to sketch a different answer to their problem. If there is genuine novelty in intellectual history, there must indeed be rupture. However, the break with the past will not be catastrophic if the work of innovation engages, not the entire broad, loose, quivering web of ideas and intelligent practices that makes up "the human understanding" at any given time and place, but only this or that limited segment of it. Rupture occurs more readily where the web is tighter, where the ideas and their links are neater and trimmer and steadier, and hence less flexible. I can think of no better example of this than the successive theories of mathematical physics. But these distinguished subsystems of the modern mind have never been anywhere near replacing the fuzzier, lazier, shiftier, but enormously resilient patterns of thinking by which we live most of our lives. The radiant heights of reason rise from and communicate across the duller lowlands. It is, indeed, perverse to expect that the humbler everyday modes of thought can fix the meaning of physical theories and set standards of acceptance for them. In fact, as Plato's Socrates was fond of showing, common discourse is not even clear as to its own meaning. But it does furnish a common ground that different theories may approximately conceive, so that, for all their conceptual diversity, they do in effect somehow concern "the same things" (at least, according to commonsense standards of sameness).

The present chapter has dealt with the structure of physical theories and has probed into their mutual relations. Rather than tax the reader with a new scheme of analysis (and myself with the toil of devising it), I have resorted to the proposals made by Joseph Sneed and Günther Ludwig, which, put together, sufficed for my needs. What I judge most valuable in Sneedian structuralism—and in the Suppesian tradition to which it succeeded—is the notion that the intellectual core of a physical theory is a *concept* and not a set of *statements*. To hold a theory amounts then to claiming that certain natural realities can suitably be thought of as instances of that concept (within a certain agreed margin of imprecision), and can be handled as such. On this view, the theory's integrity is not compromised by changes in the list of realities for which such a claim is made (or by changes in the imprecision allowed). Moreover, there is no contradiction in making the same claim for conceptually discrepant theories with regard to commonsensically identical realities (possibly to within different margins of error). To my mind, Ludwig's most important and original contribution lies in his study of the relationship between the intellectual core of a physical theory and its chosen domain of reality, whereby the latter is articulated and, in part, constructed in the light of the former. While endorsing Sneed's and Ludwig's central ideas, I have criticized and rejected the former's assertion that the core structure **M**(**T**) of any given physical theory **T** must include so-called non-**T**-theoretical terms (instances of which can be produced without presupposing that there is an actual instance of **M**(**T**) itself); as well as the latter's suggestion that a physical theory may not redress the received analysis of the "real texts" that it sets out to understand. In my opinion, both contentions stem from an implicit denial of the enlightening powers of thought and of its ability to put things differently from how they have been seen before. Such a denial, if consistently upheld, leads of course to the positivist's faith in self-classifying sense data, or else to an infinite regress.

I now propose to examine the concept of probability as an illustration of what I have said about the concepts of physics and their crystallization in theories. By thus looking at a full-fledged example from real life, I expect to find support for some of the ideas that I have put forward, but also to qualify and refine them. On the other hand, if those ideas are found applicable to the concept of probability, they might in turn throw some light on the much disputed question of its nature and scope.

Obviously, probability can only serve my present purpose if it behaves in physics like other physical concepts; not quite, perhaps, like the more specific ones, such as *viscosity* or *impedance* but rather in the manner of such ubiquitous concepts as *energy* or *entropy*. This would entail that no matter what we think about its peculiarities, the concept of probability actually enters into our intellectual grasp of some physical systems, processes, or states of affairs and contributes to fix whatever we take them to *be*. That is, of course, the view held, in some way or another, by all "objectivist" philosophers of probability, from Cournot, Ellis, Fries, Boole, Venn, and Peirce, through von Mises, Reichenbach, Braithwaite, and Popper, right down to our contemporaries, Hacking, Mellor, Gillies, Giere, van Fraassen. But since the late twenties they have been opposed by the growing and increasingly vocal "subjectivist" school, led, until his death in 1985, by the mathematician Bruno de Finetti. To him probability was not a feature of anything found in the physical world but only a means of quantifying our personal prevision of uncertain future or unknown events. De Finetti's arguments look strong and persuasive from an empiricist standpoint due to the failure of earlier attempts to "coordinate" the "theoretical" concept of probability with the "observational" concept of frequency, but also due to de Finetti's own success in proving how, under certain conditions, subjective probability estimates of unknown events must approach, lest they become incoherent, the recorded frequency of known events of the same kind ("the same," that is, in the subject's estimation). However, the case against objective probability loses much of its point if we do not equate the *objective* with the *given* but think of it as a work of creative understanding.

4. Probability

The general purpose of this chapter was explained at the end of Section 3.9. Now I shall briefly indicate the content of its several sections. In Section 4.1 the meaning of the Greek word πιθανός—which Cicero rendered as *probabilis*—is explained and contrasted with the quantitative concept of facility of occurrence we meet in a short paper by Galileo. Section 4.2 presents the species of structure *probability space* as a suitable way of articulating Galileo's concept. Section 4.3 sketches the early development of the modern idea of probability in the 17th and 18th centuries. Section 4.4 deals with the conception of probability as a limiting frequency, in the version of Richard von Mises. Section 4.5 comments on Bruno de Finetti's denial of objective chance and his interpretation of the mathematical concept of probability as a form of subjective "prevision." Section 4.6 studies several versions of the current understanding of chance as an objective property of physical systems—dubbed 'propensity' by Popper. Section 4.7 examines a different understanding of chance, irreducible to the former, which I believe can be documented by actual scientific practice.

4.1 Probability and the probable

Our word 'probability' is derived by nominalization from the predicate 'probable', which is the modern form—written, though not spoken, in the same way also in Spanish and in French—of the Latin 'probabilis, e'. The latter is derived from the verb 'probo' ('to pronounce good', 'to approve', 'to prove') and applies primarily to something that admits or deserves approval. However, Cicero used 'probabilis' as the Latin equivalent of πιθανός, a term employed by the Greek philosopher Carneades to characterize what we would call *likely* or *verisimilar* sense appearances; and there can be little doubt that the modern acceptation of 'probable' comes from this usage.[1] Now, πιθανός, a cognate of πείθω ('to persuade'), is predicated in classical Greek of *persuasive* persons (Thucydides, 3.36, 6.35) and their *captivating* manners (Xenophon, *Memorabilia*, 3.10.3); but also of *plausible* arguments, of *credible* reports (Herodotus, 1.214; *specious*, 2.123), and even of statues that by resembling the shapes of the living look as if alive (Xenophon, *Memorabilia*, 3.10.7). As I said, Carneades (ca. 150 B.C.) applied the term πιθανός to sense appearances. Sextus (ca. 200 A.D.) explains his usage as follows: By virtue of their relation to the thing that appears (τὸ φανταστόν), sense appearances (φαντασίαι) can be true—if they agree with the thing—or false—if they disagree (ἀληθὴς μέν ὅταν σύμφωνος ᾖ τῷ φανταστῷ, ψευδὴς ὅταν διάφωνος). Depending on its relation to the person who experiences it (τὸν φαντασιούμενον), a φαντασία can be one that appears to be true or one that does not appear to be true (ἡ μέν ἐστι φαινομένη ἀληθὴς ἡ δὲ οὐ φαινομένη ἀληθής). The former is called πιθανὴ φαντασία, the latter ἀπίθανος φαντασία (Sextus, *Adv. Math.*, 1.168–69). Having no means of ascertaining which sense appearances are true and which are not, we must, according to Carneades, accept those that appear to be true and base our decisions on them. In this way, he and his followers "use probability as a guide of life" (τῷ πιθανῷ προσχρῶνται κατὰ τὸν βίον—Sextus, *Hyp. Pyrrh.*, 1.231). It is not easy to see a connection between the property of apparent truth, predicable of present appearances, and our concept of probability, predicable of future or hidden events. Yet Cicero, reporting on Carneades, unhesitatingly and—I am tempted to say—unwittingly provided for this extended application of *probabilis*, his rendering of πιθανός. He notes that the sage must often follow the probable, which is neither grasped nor perceived nor assented to be true but is verisimilar; for, if he did not accept it, life would become impossible ("multa sequitur probabilia, non comprehensa, neque percepta, neque assensa, sed similia veri, quae nisi probet, omnis vita tollatur"). He then gives the following example: "When the sage boards a ship, has he grasped and perceived in his mind that the trip will proceed as planned? Impossible. But if he now departs from this place on a seaworthy vessel [*probo navigio*], with a good helmsman, by this lovely weather, for Puteoli, which is only three miles away, it will seem probable to him [*probabile videatur*] that he will arrive there safe and sound" (*Lucullus*, 31). Cicero does not explain how or why, under the stated circumstances, the pending arrival of the sage at Puteoli takes on the semblance of truth—which is what 'probabile' was supposed to mean. But it is clear that his use of the term in this passage substantially agrees with, and may well have led to, our ordinary, prescientific understanding of 'probable'.

This qualitative Græco-Roman notion of the probable is a far cry from numerical probabilities as they are understood and used in modern science. The probability of an event is now represented by a real number not smaller than 0 or greater than 1; or, if that should seem impossibly precise, by a subinterval within that interval. Given some such numbers, we add them up or multiply them by one another in order to calculate further probabilities. Not only are low but positive probabilities assigned to improbable events, but events that no ancient writer would have termed probable or improbable may now sport a middling probability (e.g., the event that a card drawn at random from a standard pack will be red). Of course, the Greeks graded the probable. The adjective πιθανός actually occurs in the comparative—πιθανότερος—or the superlative—πιθανότατος—in several of the classical references I gave above. Indeed Sextus tells us about Carneades' interesting views on 'the concurrence of appearances' (ἡ συνδρομὴ τῶν φαντασιῶν), which increases their credibility. Such concurrence results from the fact that "no sense appearance is ever a simple nature, subsisting in isolation, but they hang from each other as links in a chain."[2] If in a set of connected appearances none "distracts us by appearing false, but all consistently appear true, we believe more [μᾶλλον πιστεύομεν]" (Sextus, *Adv. Math.*, 7.177). This is sufficient evidence that Greek philosophers had some conception of degrees of belief. But I find no trace of their ever having sought to quantify, let alone to compute, probabilities.

The earliest crystal-clear example of probability calculations known to me is contained in a short paper of uncertain date, written by Galileo for a patron with an interest in dicing, which was first published in 1718.[3] Galileo discusses "the fact that in a dice game certain numbers [*punti*] are more advantageous than others." There is, he says, one very obvious reason for this, namely, "that some are more easily and more frequently [*più facilmente e più frequentemente*] made than the others." Galileo is talking about games played by throwing two or more dice on a table. The numbers at issue are made by adding the points displayed on the uppermost faces of the dice when they come to rest on the table. As every crapshooter knows, in a moderately long series of throws with two ordinary, well-balanced dice, a 7 is made in this way much more often than, say, a 2. Consequently, if the same odds were given on either number, there would be a considerable advantage in betting repeatedly on the former. Galileo refers in one breath to the greater frequency with which some numbers turn up—which is, of course, both necessary and sufficient to make them more advantageous—and the greater "ease" or "facility" with which such numbers are made. Unless Galileo's phrase is redundant and the ease in question is just a metaphor for the frequency, he must be speaking here of facilities—or difficulties—inherent in each act of throwing the dice, whose cumulative effect gives rise to the actual frequencies. In other words, he must be suggesting that 7's come up more frequently than 2's in a hundred throws of a pair of dice because in each throw it is easier to make a 7. The drift of Galileo's subsequent argument favors this interpretation. The greater frequency and ease with which some numbers are made depends, he says, "on their being able to be made up with more variety of numbers [*più sorti di numeri*]," i.e., of summands contributed by each die. Now, this claim might seem to clash with a fact mentioned by Galileo and presumably known to his patron, viz., that in the game with three dice "long observation [*la lunga osservazione*] has made players consider 10 and 11 to be more advantageous than 9 or 12." In effect, each of these four numbers can be formed in exactly six different ways from three positive integers smaller than 7. (Thus, for instance, 9 = 1 + 2 + 6 = 1 + 3 + 5 = 1 + 4 + 4 = 2 + 2 + 5 = 2 + 3 + 4 = 3 + 3 + 3.) Galileo overcomes this objection by figuring out how many alternative kinds of throw can be made *with the same ease* with three dice and then ascertaining what proportion of them would yield each of the numbers in question. He notes that a single die, when thrown, can equally well come to rest—"può indifferentemente fermarsi"—on any one of its six faces. Therefore, he concludes, two dice can fall "indifferently" in any one of 36 alternative ways, and three dice in any one of 216. Of these 216 kinds of throw, one makes a 3, three make a 4 (namely, the throws that display 2 on the ith die and 1 on each of the other two, where i ranges over {1,2,3}), and so on. In particular, it is found that 10 and 11 are each made by 27 distinct kinds of throw, whereas 9 and 12 are each made by 25. Hence, by Galileo's reasoning, a 9 should be 25 times easier to make than a 3, but only 25/27 times as easy as a 10.

But we are less concerned with the details of Galileo's arithmetic than with the very concept of facility on which his reasoning turns. We must ask, in the first place, about its relationship with the probable, in the ordinary qualitative sense we considered earlier. For, as the reader will have noted, the facility of events measured in accordance with Galileo's criteria will agree in every relevant case with the quantity we call probability. On the other hand, it is clear that Galileo's results are based, in his view, on the actual physical properties of dice and dice throwing, and not on the effects that their appearances might have on a gambler's mind. If the dice are loaded or handled by a trickster, Galileo's conclusions will not be applicable to the game. In point of fact, the word 'probability' does not occur in Galileo's paper, nor, for that matter, does the earliest printed textbook of the probability calculus, Christiaan Huygens' *De ratiociniis in aleae ludo* (1657), employ that word to designate its subject. However, sixty years later, the new connotation of the term must have been familiar at least to buyers of mathematical books, for Abraham de Moivre used it in the title of his classic, *The Doctrine of Chances; or, A Method of Calculating the Probability of Events in Play* (1718).[4] This semantic evolution is not at all surprising, for what is known to happen easily is also judged to be likely. As Leibniz wrote with characteristic pointedness, at some time between the two years just mentioned: "Quod facile est in re, id probabile est in mente."[5] Anyone who has been persuaded by Galileo's reasoning will be more confident that a 10 will come forth rather than a 9 in the next good throw of three fair dice. The objective facility of events, if measurable, naturally induces a measure on their subjective plausibility. This, however, does not entail that the latter is a quantity in its own right, whose numerical value can be meaningfully estimated even where it is not backed up by a measurable facility, e.g., in horse racing or in weather forecasting.

We must also consider the connection between Galileo's metricized facility of events and their comparative frequency. It should be clear by now that they are not equivalent, even if they roughly go together. The computed value of the facility of events of a given class usually does not and often cannot agree with their observed frequency. There is no way that 1/216 can be the relative frequency of 3's in a million throws with three dice, for 1,000,000 is not divisible by 216. Nor does Galileo invoke the observed frequency of throws with a single die to establish his assertion that the die will stop *indifferently* in any one of six possible positions. Apparently he saw it as an obvious consequence of the die's symmetry, of the absence of any dynamically significant *difference* between its faces. But he did not seek to explain how the greater ease of a particular kind of outcome in a single throw translates into a greater relative frequency in a long run of throws. We shall return to this question more than once. At this point, I only wish to suggest that the men who first substituted symmetric dice for the irregular *tali,* or knuckle-bones, commonly used in antiquity for gambling and soothsaying must have had a dark deep-seated conviction that frequency flows from facility. They must have noticed that any given *talus* would show in the long run a bias for this or the other face, and blamed its shape for it. Surely, when they set out to make cubic, well-balanced dice, they somehow expected that the alternative outcomes, which, by symmetry, are physically indifferent, would take place more or less equally often.

4.2 Probability spaces

Galileo's analysis of the three-dice game can be read as a simple application of the theory of probability spaces, as I shall now show. Readers familiar with the theory may proceed at once to Section 4.3, after taking a glance at the definitions of italicized terms.

A *finite probability space* is a pair $\langle S, \mathbf{p} \rangle$, where S is a finite set, and $\mathbf{p}$ is a mapping of the power set $\mathcal{P}(S)$ into the real interval [0,1] that meets the following two requirements:

FPS 1. $\mathbf{p}(S) = 1$.

FPS 2. If A and B are two disjoint subsets of S, then $\mathbf{p}(A \cup B) = \mathbf{p}(A) + \mathbf{p}(B)$.

It follows at once that $\mathbf{p}(\varnothing) = 0$ and that if A' is the complement of A in S, $\mathbf{p}(A') = 1 - \mathbf{p}(A)$. S is called the *sample space*; the elements of $\mathcal{P}(S)$ are called *events*; $\mathbf{p}$ is known as the *probability function* and its value at a given event A is the *probability* of A.

If A is an event such that $\mathbf{p}(A) \neq 0$, the *conditional* probability $\mathbf{p}(B|A)$ of event B given A (also referred to as the probability of B conditional on A) is customarily defined by the equation

$$\mathbf{p}(B|A) = \mathbf{p}(A \cap B)/\mathbf{p}(A) \tag{1}$$

The following easy corollary of (1) is the simplest form of *Bayes' Theorem*:

$$\mathbf{p}(A|B) = \mathbf{p}(B|A)\mathbf{p}(A)/\mathbf{p}(B) \tag{2}$$

Events A and B are said to be mutually independent if and only if

$$\mathbf{p}(A \cap B) = \mathbf{p}(A)\mathbf{p}(B) \tag{3}$$

Clearly, if A and B are mutually independent and $\mathbf{p}(A|B)$ is defined (that is to say, if $\mathbf{p}(B)$ is not equal to 0), $\mathbf{p}(A|B) = \mathbf{p}(A)$. Events $A_1, \ldots, A_n$ are mutually independent *by pairs* if eqn. (3) holds for every pair of distinct events selected among them. They are *all* mutually independent if the probability of the intersection of all n events equals the product of their probabilities. The *product space* of two finite probability spaces, $\langle S_1, \mathbf{p}_1\rangle$ and $\langle S_2, \mathbf{p}_2\rangle$, is the finite probability space $\langle S_1 \times S_2, \mathbf{p}\rangle$, where the sample space is the Cartesian product of the sample spaces S_1 and S_2, and the probability function $\mathbf{p}$ is defined by the following simple stipulation: If A is a subset of S_1 and B is a subset of S_2,

$$\mathbf{p}(A \times B) = \mathbf{p}_1(A)\mathbf{p}_2(B) \tag{4}$$

Let D be the set of all the different throws that can be made with one perfectly symmetric die. Let a_k denote the throw that makes the number k. A probability function $\mathbf{p}$ is readily defined on $\mathcal{P}(D)$ by assigning to each singleton $\{a_k\}$ $(1 \leq k \leq 6)$ a positive real number less than 1, proportional to the facility of the throw a_k. Because of the die's symmetry, $\mathbf{p}(\{a_h\}) = \mathbf{p}(\{a_k\})$ for all six values of the indices h and k. Since $\Sigma_i \mathbf{p}(a_i) = \mathbf{p}(D) = 1$ (by FPS 1), $\mathbf{p}(\{a_k\}) = 1/6$ for all k.

Now form the Cartesian product $D^3 = D \times D \times D$. D^3 contains the 216 triples $\langle a_h, a_i, a_k \rangle$ $(1 \leq h,i,k \leq 6)$. In the following pages I shall denote by $\mathbf{P}$ the probability function defined by eqn. (4) on $\mathcal{P}(D^3)$ when the said function $\mathbf{p}$ on $\mathcal{P}(D)$ is given. Set

$$A_r = \{\langle a_h, a_i, a_k \rangle \mid h + i + k = r\} \qquad (3 \leq h \leq 18) \qquad (5)$$

It will be readily seen that $\mathbf{P}(A_r)$ is exactly proportional to Galileo's estimate of the facility with which the number h can be made with three dice. Therefore, if the probability function is interpreted as a representation of physical facility, the three-dice game as analyzed by Galileo is a model of the finite probability space $\langle D^3, \mathbf{P} \rangle$.

We saw above that Galileo implicitly based his expectations concerning the long-run relative frequency of events of a certain kind on the given—perceived, calculated, postulated—facility of occurrence of a single event of that kind. The mathematical theory of probability spaces vindicates his tacit inference. To be precise, the theory warrants an inference from the given single-case probability to the *probability* of a certain long-run frequency. By lengthening the run the latter can be made arbitrarily close to 1; whence, if probability is interpreted as facility, one can be *practically* certain that the frequency in question will eventually obtain.

The following exercise will show what I mean. Let $\langle D^3, \mathbf{P} \rangle$ be as above. To simplify the notation I set $T = D^3$. A sequence σ of n throws with three dice may be identified with an element of the Cartesian product T^n, that is, with an ordered n-tuple of ordered triples $\langle a_h, a_i, a_k \rangle$. We are interested in the proportion of the throws in which a given number r turns up. They are represented by the triples $\langle a_h, a_i, a_k \rangle$ in σ such that $h + i + k = r$. In eqn. (5) I introduced the designation A_r for the set of all such triples. Forget the order in which the triples occur in the finite sequence σ and count the number of those belonging to A_r. Dividing that number by n, we obtain the relative frequency $\mathbf{f}_n(\sigma, r)$ with which the number r is made in a given sequence σ of n throws. As σ ranges over T^n, $\mathbf{f}_n(\sigma, r)$ ranges in turn over all integral multiples of $1/n$, from 0 to 1. Little can therefore be said about a relationship between $\mathbf{f}_n(\sigma, r)$—for fixed r but arbitrary σ—and the facility $\mathbf{P}(A_r)$ with which the number r is made with three dice. Now, T^n is the sample space of a finite probability space constructed from T by repeated use of stipulation (4). Let $\mathbf{P}_n$ denote the probability function of that space. It makes sense to ask for the probability of those sequences $\sigma \in T^n$ for which $\mathbf{f}_n(\sigma,r)$ comes close to $\mathbf{P}(A_r)$. According to our interpretation of probability as facility, that probability will measure the ease with which a series of throws with the stated property is apt to happen. Specifically, let us consider the sequences $\sigma \in T^n$ for which $\mathbf{f}_n(\sigma,r)$ comes within some fixed distance ε of $\mathbf{P}(A_r)$—or, in other words, for which $\mathbf{f}_n(\sigma,r)$ falls within an interval of length 2ε centered at $\mathbf{P}(A_r)$. Let us denote the set of all such sequences by $B_{r,n,\varepsilon}$, where the indices *r, n,* and ε name the three parameters required for identifying the set. Clearly,

$$B_{r,n,\varepsilon} = \{\sigma \mid \sigma \in T^n \;\&\; |\mathbf{f}_n(\sigma,r) - \mathbf{P}(A_r)| \leq \varepsilon\} \tag{6}$$

It should not be too hard to see that for large n and, say, $\varepsilon = 0.01$, $\mathbf{P}_n(B_{r,n,\varepsilon})$ is fairly close to 1. Moreover, it can be proved that for any ε, no matter how small, $\mathbf{P}_n(B_{r,n,\varepsilon})$ converges to the limit 1 as n goes to infinity. This merely states, in terms adapted to the present example, a theorem proved by Jacques Bernoulli (1713, pp. 228ff.), generally known as the (weak) law of large numbers.[6] To recover the theorem's full generality, let $\langle T,\mathbf{P}\rangle$ be an arbitrary finite probability space and let A_r stand for any subset of $\mathbf{P}$, labelled by the—now meaningless—index *r*. Then $\mathbf{f}_n(\sigma,r)$ is simply the proportion of instances belonging to A_r in the ordered n-tuple σ of elements of T.

Bernoulli's proof is sketched by Stigler (1986, pp. 67ff.). A modern proof that purports to follow the drift of Bernoulli's argument is given by Renyi (1970, pp. 195–96). The following remarks may clarify the import of such proofs to readers who are not willing to work their way through them. Let us go back to our example. For simplicity's sake, let A_r be the set of all throws with three dice that make an even number. $\mathbf{P}(A_r)$ is then the probability of making an even number with three dice, which—on Galileo's assumptions—happens to be exactly 1/2. The set T^n of possible n-tuples of throws with three dice contains 216^n distinct elements. If $\mathbf{P}_n$ is defined in accordance with eqn. (4), every such n-tuple is just as probable as any other. Thus, $\mathbf{P}_n(B_{r,n,0.01})$ is simply the proportion, among all such n-tuples, of those containing between $0.49n$ and $0.51n$ even-valued throws. For large n, that proportion comes close to 100%. Modern proofs of the weak law of large numbers show further that for every choice of positive real numbers δ and ε, no matter how small, there exists a large integer N such that, for every $n > N$, δ is greater than the proportion of alternative n-tuples of throws with three dice containing less than $(0.5 - \varepsilon)n$ or more than $(0.5 + \varepsilon)n$ even-valued throws. In other words, they show that for every δ and ε there is an N such that, for every $n > N$, there are more than $216^n(1 - \delta)$ such n-tuples of throws in which the proportion of even-valued throws comes within ε of 0.5. Since for fixed n all such n-tuples are equally probable (by eqn. (4)), the choice, for any given ε, of a convincingly small real number δ and a suitably large integer n will make the *probability* overwhelming that any arbitrarily chosen n-tuple contains a *proportion* of even-valued throws within ε of the *probability* of making an even number in a single throw.[7]

Let us reflect for a moment on the meaning of Bernoulli's Theorem. We have assumed that the probability function $\mathbf{P}$ on subsets of T measures the comparative facility with which different sorts of throw can be made with three dice. Bernoulli's Theorem says that, if eqn. (6) defines the typical term of an infinite sequence of subsets of the Cartesian products T^n, the corresponding sequence of values of the appropriate probability functions $\mathbf{P}_n$ converges to the limit 1.[8] The theorem is a logical consequence of the definition of probability spaces. Therefore, it is a logical truth that *if* $\mathbf{P}_n$ measures the comparative facility of events drawn from the sample space T^n, *then* the kind of outcome described by (6) is extremely easy, and hence almost inevitable, in a very long sequence of throws with three dice. However, it is *not* a logical truth that $\mathbf{P}_n$ measures facility in the set of sequences represented by T^n *whenever* $\mathbf{P}$ measures facility in the set of throws represented by T. Specifically, our initial assumption that $\mathbf{P}(A_r)$ is the comparative facility of making the number r with three dice does not entail that the comparative facility of making r in n such throws a number of times within εn of $\mathbf{P}(A_r)\,n$ is given by $\mathbf{P}_n(B_{r,n,\varepsilon})$. Nevertheless, a condition can be stated under which this interpretation of $\mathbf{P}_n$ will be necessary, given our interpretation of $\mathbf{P}$. *If* that condition holds in a particular case for every number n of trials, no matter how large, *then* Bernoulli's Theorem expresses a logical truth concerning the relationship existing in that case between the known—perceived, calculated, or postulated—facility for obtaining a certain kind of outcome in a single trial and the facility with which any given proportion of outcomes of that kind would be achieved in a long run of trials.

Recall Galileo's analysis of the three-dice game. He took for granted, on obvious grounds of symmetry, that the facility of the throw a_k with one die is the same for all six possible values of the index k. Therefore, he concluded, the facility of the throw $\langle a_h, a_i, a_k \rangle$ with three dice is the same for all 216 possible combinations of the indices h, i, and k. This is tantamount to our assumption that $\mathbf{P}$ measures the comparative facility of throws with three dice. But Galileo's conclusion would not be valid if the comparative facility of throws with any one of the three dice could in any way be influenced by the throws made with the other two. Now, Galileo believed that no such influence would be exerted, in spite of the fact that, as the game is usually played, the three dice noticeably interact inside the cup from which they are thrown. He must have thought, again on grounds of symmetry, that their clash and clatter cannot modify the comparative facility with which the several numbers are made with each die. A similar consideration can obviously justify the interpretation of $\mathbf{P}_n$ as a measure of facility. Given our interpretation of $\mathbf{P}$, $\mathbf{P}_n$ measures the comparative facility with which different sorts of sequences of n throws can be obtained with three dice *if*, but *only if*, the facility of any throw in such a sequence is not affected by the nature of the other throws.[9]

This condition is not easily met in a prolonged game of dice. The dice may become biased as they wear out by repeated throwing; the thrower may develop a habit of shaking the cup in a peculiar fashion and have some preferred ways of placing the dice in the cup, etc. Moreover, no real die is perfectly symmetric, and a slight bias, too small to be of any consequence in a single throw, becomes significant in the long run. As is well known, casinos fight such contingencies by rotating croupiers and often changing dice, presumably in the expectation that their different biases will compensate one another. By such measures, the management seeks to ensure that games on the premises provide tolerably good models of the theory of probability, and thereby to prevent enterprising clients from developing a successful gambling system that could ruin the house.

The situation here is reminiscent of other applications of mathematical structures to the description and explanation of physical realities. If the system under consideration actually satisfies the necessary and sufficient conditions of a particular species of structure, it is an instance of that species, and the "bonds of necessity"[10] will constrain it to follow the structure's laws. The instantiation of probability spaces by games of chance is somewhat peculiar and perplexing because there are no physical agencies at hand to enforce such constraints. Indeed, as I have suggested, the disclosure of causal links between some elements of the physical model may even preclude the application of the theory. But surely, this only shows that physical necessity is not always fettered to causality.

For future reference, I shall give here Kolmogorov's definition of probability spaces. A Kolmogorov probability space, or, as I shall say hereafter, a *probability space,* is a triple $\langle S, \mathcal{F}(S), \mathbf{p} \rangle$; where S is an arbitrary set, known as the *sample space;* $\mathcal{F}(S)$, the *field of events,* is a Borel field of subsets of S; and $\mathbf{p}$, the *probability function,* is a mapping of $\mathcal{F}(S)$ into the real interval $[0,1]$. As a Borel field of subsets of S, $\mathcal{F}(S)$ must meet the following requirements: (α) $S \in \mathcal{F}(S)$; (β) if A belongs to $\mathcal{F}(S)$, then $S \setminus A$, the complement of A in S, also belongs to $\mathcal{F}(S)$; (γ) $\mathcal{F}(S)$ contains the union of any finite or countably infinite collection of elements of $\mathcal{F}(S)$. The probability function $\mathbf{p}$ is subject to the following conditions:

KPS 1. $\mathbf{p}(S) = 1$.

KPS 2. If A is the union of a finite or countably infinite collection $[A_\lambda]_{\lambda\in\Lambda}$ of pairwise disjoint elements of $\mathcal{F}(S)$, then $\mathbf{p}(A) = \Sigma_{\lambda\in\Lambda}\,\mathbf{p}(A_\lambda)$.

When forming the product space of two probability spaces we must bear in mind that the Cartesian product of two Borel fields is generally not a Borel field. We shall therefore define the *product space* of the probability spaces $\langle S_1,\mathcal{F}(S_1),\mathbf{p}_1\rangle$ and $\langle S_2,\mathcal{F}(S_2),\mathbf{p}_2\rangle$ as the probability space $\langle S,\mathcal{F}(S),\mathbf{p}\rangle$, where

(i) $S = S_1 \times S_2$;

(ii) $\mathcal{F}(S)$ is the smallest Borel set that contains $A \times B$, for every $A \in S_1$ and every $B \in S_2$; and

(iii) for every such A and B, $\mathbf{p}(A \times B) = \mathbf{p}_1(A)\mathbf{p}_2(B)$.

That the function $\mathbf{p}$ defined by (iii) on $\mathcal{F}(S_1) \times \mathcal{F}(S_2)$ can be extended in one and only one way to a probability function on $\mathcal{F}(S)$ is a nontrivial theorem.[11]

Let us briefly discuss the difference between the KPS axioms and the FPS axioms given at the beginning of this section. A probability function $\mathbf{p}$ satisfying KPS 2 is said to be *completely additive* (or σ-additive). On the other hand, if $\mathbf{p}$ satisfies FPS 2 but not KPS 2 it is said to be *finitely additive.* Complete additivity is obviously a stronger requirement than finite additivity. KPS 2 is irrelevant if the sample space S is finite, for then there are no countably infinite collections of subsets of S on which to test its strength. Thus, any finite probability space $\langle S,\mathbf{p}\rangle$, governed by the axioms FPS, can be readily identified with the probability space $\langle S,\mathcal{P}(S),\mathbf{p}\rangle$, governed by the axioms KPS. Note, however, that the Kolmogorov axioms KPS do not presuppose that the probability function $\mathbf{p}$ is defined on the entire power set $\mathcal{P}(S)$ of the sample space S, but only on some Borel field $\mathcal{F}(S) \subset \mathcal{P}(S)$. That the domain of $\mathbf{p}$ is thus limited somehow compensates for the strength of KPS 2. The limitation is necessary if the Axiom of Choice holds for infinite sets.[12] For then it is generally not the case that a completely additive probability function on a Borel field $\mathcal{F}(S)$ of a sample space S can be extended to a completely additive function on $\mathcal{P}(S)$.[13]

If the probability function is defined only on a proper subset $\mathcal{F}(S)$ of the power set of the sample space S, the "events" of the probability space are only the members of that subset. De Finetti and others have objected that this is an utterly artificial stipulation, adopted for mathematical convenience, without the faintest philosophical motive. To ensure that a probability can be assigned to every subset of a sample space, de Finetti discarded the condition of complete additivity KPS 2, replacing it with a requirement of simple or finite additivity, as in FPS 2. We may, therefore, be tempted to conceive a probability space in de Finetti's sense as a species of structure with the same characteristic features as our finite probability spaces, except that the underlying sample space need not be finite. (I shall occasionally refer to a structure of this species as a 'probability space with finitely additive probability function'.) However, the matter is not quite so simple, for de Finetti would rather have us associate a sample space S with *every conceivable* function from $\mathcal{P}(S)$ to [0,1] that satisfies FPS 1 and FPS 2.[14]

4.3 Chance setups

In my reading, Galileo's analysis of dice games depends on the following two conditions: (i) the purportedly known facility of single throws (with one die) and (ii) the postulated immunity of the facility of each kind of throw against the outcome of other throws. But, one is bound to ask, what on earth is *facility*? What physical feature of actual events is designated by this word? A moment's reflection will persuade us that the question, though seemingly natural, is not well posed, for the facilities we are talking about are ascribed to *possible*, not to *actual*, events. (Any actual throw rolls a definite number and lacks every facility for displaying another one.) Philosophers of probability have sidestepped the question, by reducing facility to frequency; or they have dismissed it offhand, by denying that probability functions measure a physical property; or, finally, as in the recent rich flowering of "propensity" theories, they have offered an answer to some suitably rephrased version of it. In Sections 4.4 to 4.6 I shall discuss these three alternatives. But before doing so, I propose to reflect for a while on the naïve conception we all seem to have of the comparative ease with which diverse, mutually exclusive events can happen. For some such conception apparently guided the first steps of the theory of probability in the analysis of games of chance and its subsequent extension to physics.

Students of public opinion investigate common notions by framing questionnaires and getting people to answer them. If, as is often the case, the questionnaire is steeped in unanalyzed commonplaces, the answers are likely to be boring and unilluminating. On the other hand, if the interviews include critical discussions of the concepts in question—as in the notorious polls conducted by Socrates in the Athenian agora—most interviewees will be deterred, and the few that remain may have to be disqualified as unrepresentative of the common run of mankind. I think, therefore, that in the study of common notions there is still room for the tried armchair method of philosophy, namely, to "search into myself"[15] and submit the findings to the reader, who may accept them as enlightening, reject them, or improve on them, as she or he sees fit.

I begin with a very general and almost trivial remark: Among all the physical objects—things, processes, events—that I may single out for attention, only those that are ostensibly on the verge of extinction, or are in the nature of instantaneous events (such as transitions, e.g., from sound to silence), are grasped by me as not having a future. Otherwise, every object has its prospects. Moreover, those that, as I have just noted, do not have them arise only in a setting of objects that do. The blatant fact that momentary and vanishing objects are everywhere surrounded and supported by continuants has not dissuaded some philosophers from trying to "reduce" the latter to a mass of fleeting, strictly actual "data," leavened by our subjective memories and expectations. But I am writing here about things as they show up in my life, not as they are made out to be by headstrong fanciful men. Of course, a particular object of observation will be seen as having a wealth of prospects only inasmuch as it is grasped as an instance of a universal. (EXAMPLES: The white billiard ball is heading for a collision with the red one inasmuch as they are both *solid bodies* moving relatively to each other with this or that *velocity*; the rectangular thing in the middle of my room offers me the prospect of a night's rest inasmuch as it is a well-appointed *bed*; the mountain I see through the window can yield several million tons of high-grade copper inasmuch as the chemical composition of its rocks is *such-and-such*.) Bare particulars, cut off from all classes and relations, do not reach prospectively—or retrospectively—beyond their instant givenness. But then, neither can they pose as objects for attention. For how could a particular ever happen or, as we say, "take place" without exemplifying some rudimentary but nonetheless universal neighborhood relations with other particulars? Falling under general concepts is the very mark of objectivity.

As the above examples suggest, most of the concepts by which physical objects are grasped in science and in everyday life ascribe to them prospects of some sort. Such prospects are part and parcel of what the objects are held to *be* when they are conceived in just that way. I may be wrong in thinking that the composition of the mountain I see through the window is such-and-such; however, if I am right, the mountain has now all that is necessary to yield the expected amount of copper if properly mined. The rectangular object in my room might not be my bed but an empty cardboard box substituted for it and covered with my bedspread by a malicious joker; however, if, as I think, it is indeed my bed, then it is now ready to hold me and hug me for a night's rest. I base my own subjective expectations on the prospects I ascribe to physical objects. But the prospects are themselves objective, neither more nor less so than any other attribute for which objectivity is claimed; for they are implicit in the very concepts by which observed particulars are grasped and understood as presentations of objects.

Prospects are uncertain—we are often told. It will be well to distinguish here between our subjective uncertainty, due to ignorance, and the objective uncertainty, or, as I shall hereafter call it, the *indeterminacy,* of prospects pertaining to things or situations. For example, I am fairly certain of my bed's disposition to hold me comfortably, and should I have any doubts, I can promptly dispel them, without having to lie on it, by pressing firmly with my fingers on judiciously chosen points of its upper surface. But the bed's prospects for holding me tonight do not depend only on its present makeup, but also on many external factors, including my own free will. Hence, even in the light of certain knowledge of the bed as such, its prospects are indeterminate. Consider now an object Ω that can reasonably be viewed as an isolated Newtonian system. Then, the prospects of Ω to which this description is relevant are fully determinate, within the margin of imprecision with which the description is applicable. But surely any such physical object Ω has actual and prospective features to which the said description is irrelevant and may run into circumstances in which the description would no longer be appropriate. For example, let Ω be the Solar System. By conceiving it as an isolated Newtonian system we are able to calculate the future relative motions of its component bodies, from their present positions and momenta. But from such data we cannot predict, e.g., whether there will be a nuclear war on planet Earth during the next quarter of a century. Moreover, according to current ideas, the system could be unexpectedly gobbled up or perturbed out of all recognition by a black hole lying in its path. Thus, Ω's prospects are determinate only up to a point. On the other hand, if Ω is a model of Quantum Theory, some of its prospects are inherently indeterminate, by virtue of its being just that. For example, if Ω is a single atom of ^{234}U it can disintegrate at any moment, and current physics gives odds of 15 to 1 that it will do so within less than a million years;[16] but it might also remain forever intact.

It will perhaps be objected that the examples of the bed and the Solar System, far from clarifying the distinction I proposed to make between uncertainty and indeterminacy, can only breed confusion. The indeterminacy I ascribe to their prospects issues only from the fact that the concepts under which, in the foregoing discussion, those objects were required to fall make allowance for our very imperfect knowledge of them and of the world they belong to. But—it will be argued—if we subsumed the objects in question under concepts commensurate to their full present circumstances, there would be little or no room left for prospective indeterminacies. As to the last example, it will no doubt be granted that it displays an instance of genuine indeterminacy, provided that the quantum theory of radioactive decay is substantially right. But then—it will be said—quantum indeterminism may still prove to be no more than a temporary asylum of ignorance, which will in the end give way to a "complete"—and therefore, presumably, deterministic—description of phenomena.

Hopes for a restoration of determinism in fundamental physics are dwindling steadily, and I shall not quarrel here with those who cherish them. My case for the indeterminacy of objective prospects is based on the general conditions under which we have access to physical objects, not on the acceptance of a particular theory or family of theories. Physical objects, such as we encounter in our lives, can only be grasped by concepts that, on any given occasion of their employment, provide for the imperfect determination and endless determinability of the objects involved. The notion that the same objects which our concepts underdetermine are "in themselves" *omnimode determinata*—wholly determined in every conceivable respect—is a theologically motivated, epistemically idle fancy of modern philosophy. As there is no rational or empirical evidence for it, it cannot be destroyed by argument. But exercises like the following may contribute to dissolve it.

I lift my eyes and see a fat fleecy cloud against the blue sky. I reckon it is no less than 300 or more than 30,000 feet above the ground. A pilot or a meteorologist could no doubt estimate its height correctly within much narrower bounds. A fairly precise value can be measured with an altimeter carried by aircraft to the cloud or by triangulation with optical instruments from the ground. However, it would be nonsense to assume that the cloud has now a definite height, representable by a real number to which our increasingly better estimates and measurements converge. For—as you will have noticed if you have ever flown into one—there is no way of establishing where a cloud begins. Even if we fixed by international agreement a minimum density below which a concentration of atmospheric water could not be said to constitute a cloud, this would, in any given case, yield only approximate, ever fluctuating boundaries.

Of course, nobody would dream of building metaphysical determinism on the analogy of clouds. Indeed the determinist will presumably regard 'cloud' as an incurably vague everyday term that has no place in a perfect language, suited for the exact description of reality. The trouble with such a view is that most things, as currently understood, from galaxies right down to atomic nuclei, somehow partake of cloudiness. Besides, there is no telling which, if any, of the terms employed in ordinary and scientific English could be retained in the idiom of Utopia. Terms that are not inherently vague have, as a rule, been discarded or drastically reconceived as scientific thinking has grown more discriminating and articulate.

A few examples will make my meaning clearer. The table before me can be usefully thought of as a rigid body with a definite length and width when it is a matter of ascertaining whether it will go through a door or fit well in a room. But this approach breaks down if we demand precision to within one billionth of a meter. A polished metal surface may be regarded as a smooth two-dimensional manifold with a definite Gaussian curvature at each point. However, under a microscope it turns out to be rugged and porous, more like a quivering sponge. Newtonian astronomy conceived the Solar System as held together by the mutual gravitational pull of the Sun and planets (and their moons). Approximate calculations based on this concept yield beautifully accurate predictions, which have made it possible, for instance, to bring a single space probe, in a 12-year flight, close to Jupiter, Saturn, Uranus, and Neptune. It is admittedly impossible to determine the exact magnitudes and directions of the Newtonian forces acting on the planets at any given moment. But it hardly makes any sense even to desire to be apprised of them, say, by divine revelation, for, on a closer look, the very idea of gravitational attraction has been found wanting. Since the sixties, Einstein's chronogeometrodynamic theory has gained the upper hand over the other extant theories of gravity. In a model of this theory, Newtonian forces do not exist. Moreover, free fall, at least in the ideally pure case of a nonrotating, uncharged test particle, follows a spacetime geodesic and therefore requires no pull whatsoever. On the other hand, Einstein's theory is incompatible with quantum physics, and everybody expects it to be dislodged by a future quantum theory of gravity. In the meantime at least this is clear: a contemporary metaphysical determinist who wished to communicate with God about the exact constitution of the Solar System would not be able to ask Him the right questions or to understand His answers.

Under the dogma of thoroughgoing determination, future states of affairs must either be unambiguously fixed by the present constitution of things or else be totally disconnected from it. But once we are no longer prey to the dogma we may come to understand and even to accept the prephilosophical idea that the present is—as Leibniz was wont to say—"pregnant with the future" but keeps, all the same, its options open. This idea pervades the common understanding of social and natural situations of all kinds. In Aristotle's famous example, even if the navies are ready for tomorrow's sea battle, it is still undetermined whether it will be fought (*De Interpr.*, 9, 18^b24 ff.). The old Hippocratic school of medicine saw illness as progressing according to a characteristic pattern towards a κρίσις, a decision, at which it would be resolved whether the patient would live or die (Hippocrates, *On*

Affections, viii; *On Airs, Waters, Places*, xi). As the clouds gather, everything is set for the impending storm, yet it is still undecided where the lightning will strike. Of the several mutually exclusive developments which are known to exhaust the prospects in such cases, some may come more easily than others. Themistocles, who at Salamis headed a well-trained navy fighting for survival, had better prospects of victory than the Great King. A Greek physician would, under certain circumstances, prognosticate recovery—while still fearing that nature might give him the lie. A man pointing upwards with a long iron rod on a desert plain is more likely to be smitten by lightning than his companion who lies in a ditch.

The idea of quantifying and measuring the ease or facility attributable to alternative prospects is not clearly documented before 1600. However, as I suggested at the end of Section 4.1, some very old games of chance and their paraphernalia do seem designed either to ensure that all players will have equal chances or to give a small advantage to the one who holds the bank. Dice, lotteries, roulettes are all symmetric. The diverse items associated with the alternative outcomes of the game—the faces of the die, the balls in the urn, the little compartments on the wheel—must carry different labels or else there would be no way of telling one outcome from another. But no effort is spared—at least, among gentlemen—in preventing such labels from exerting any influence on the development of the game and in making the labelled items indistinguishable in every respect that might affect that development. Similar aims obviously guide the standard procedures for shuffling, cutting, and dealing cards.

Symmetric gambling devices furnished the prototype of the *chance setup*, the sort of physical object that a theory of quantified facilities or objective probabilities—or *chances*—is intended to grasp. In his *Logic of Statistical Inference*, Ian Hacking introduces the term 'chance setup' with studied vagueness:

> A *chance set-up* is a device or part of the world on which might be conducted one or more *trials*, experiments, or observations; each trial must have a unique *result* which is a member of a *class of possible results*.
>
> (Hacking 1965, p. 13)

I take this to mean that each trial must be associated with a definite set of recognizably distinct, mutually exclusive states of, or events on, the setup, which can be achieved through the trial. (The class of possible results must have at least two members, lest the chance setup be degenerate.) On the naïve view I have adopted in the present section, 'can' must here be taken in a strong sense: a result is possible only if it lies within the actual prospects of the part of the world in question when the trial begins. A trial on a chance setup, even if it is to be carried out only once, must be grasped as an instance of a type of trial, performable on a type of setup, and associated with a class of typified results. In a quantitative treatment of chance setups there is of course no need to ascribe the same facility or chance to every one of the prospective outcomes. However, only if the alternative prospects are equally easy is there an altogether unexceptionable way of metrifying their chances, namely, by assigning to each and all the reciprocal value of the total number of alternatives.[17] To my mind, this explains why probabilistic thinking began as a study of games played with symmetric gambling devices and why we still resort to them whenever we look for examples of universally agreed upon—and hence presumably objective—assessments of probability.[18]

But, as Jacques Bernoulli noted in the *Ars conjectandi*, although games of chance are deliberately contrived "so that all players will have the same expectation of gain" (1713, p. 223), most natural and social phenomena do not sport such nicely symmetric prospects. Bernoulli thought, however, that the law of large numbers demonstrated in the last chapter of his book would enable one to conjecture the actual distribution of chances among the possible outcomes of a chance setup by observing their relative frequencies in a long run of trials. Bernoulli's Theorem, as stated and proved by him, applies to a type of chance setup—hereafter referred to as a *Bernoullian setup*— that produces two kinds of results, labelled 'favorable' and 'unfavorable', with constant, though possibly unequal, chances q and $1 - q$. Then the chance that in a given run of n trials the observed proportion of favorable outcomes falls within $1/n$ of q will come arbitrarily close to 1 as n increases beyond all bounds (Bernoulli 1713, pp. 225, 226). Bernoulli does not explain exactly what sort of conjectures about the unknown parameter q can be grounded, in his view, on knowledge of the relative frequency of favorable outcomes in a given run of trials. In the light of the modern formulation of Bernoulli's Theorem we used in Section 4.2 (see note 6), it is clear that, for any positive real number ε, no matter how small, the chance—call it $q_n(\varepsilon)$—that the relative frequency of outcomes in n trials lies in the interval $[q-\varepsilon, q+\varepsilon]$ depends on q (although it converges, of course, for every conceivable value of q—and for every $\varepsilon > 0$—to the same limit 1 as n goes to infinity). Bernoulli's Theorem does not, therefore, yield a chance, converging to 1 in the long run, that the unknown parameter of a Bernoullian setup lies close to the recorded relative frequency of favorable outcomes in a series of trials on that setup.

Indeed, from the naïve objectivist standpoint adopted in this section, it does not even make sense to speak of such a chance, unless the Bernoullian setup in question can be sensibly viewed, in turn, as a chance product of another chance setup. Such would be the case if the Bernoullian setup were, for instance, a biased coin freshly minted by a machine that produces coins with different biases, in a random fashion. The assumption that every chance setup can—on the analogy of such a coin—be regarded as a chance product was a key ingredient of Thomas Bayes' approach to the problem of inferring chances from frequencies (a problem he stated, with classical clarity, as follows: "*Given* the number of times in which an unknown event has happened and failed: *Required* the chance that the probability of its happening in a single trial lies somewhere between any two degrees of probability that can be named"—Bayes 1763, p. 376; on the same page Bayes explains that "by *chance* I mean the same as probability"). That farfetched assumption raises, of course, the specter of an infinite regress. Bayes, however, cut short the regress and solved the problem for any asymmetric Bernoullian setup by assuming that the chance—prior to testing—that its characteristic parameter q lies in this or that interval is uniformly distributed over [0,1].[19]

Bernoulli's Theorem involves no such unwarranted assumptions and does not entail a solution to Bayes' problem; but it does provide a means of grading alternative conjectures regarding the value of q. Consider a finite set Q of such conjectures, $q = q_1, \ldots, q = q_k$. For every index i ($1 \leq i \leq k$) and every pair of positive real numbers ε and δ, there is a positive integer $N(i)$ such that, in the notation used above, $(q_i)_n(\varepsilon) > 1 - \delta$ whenever $n \geq N(i)$. Denote the largest of those k integers by N. Let F be the number of favorable outcomes recorded in an actual experiment E consisting of N trials on the setup under consideration. Then, the set $\{q_1, \ldots, q_k\}$ can be partitioned into a subclass $A = \{q_i \mid |q_i - F/N| \leq \varepsilon\}$ and its complement A'. If q belongs to A, the chances of obtaining F favorable outcomes in n trials are greater than $1 - \delta$, but they are equal to or less than δ if q belongs to A'. Hence, if δ is very small, the conjectures in A do seem preferable, by far, in the light of experiment E, to the conjectures in A' (unless, that is, one prefers those hypotheses under which it would be extremely difficult for what has actually happened to happen). By judiciously choosing Q, δ, and ε, one can ensure that A is small—and therefore instructive—but not empty.

In the course of his polemic against the very idea of chance or objective probability, de Finetti has acknowledged that "even from a subjectivistic point of view" it makes sense to attribute definite, though unknown, probabilities to alternative drawings from a lottery urn of unknown composition. However, according to him unknown probabilities cannot be meaningfully ascribed to the alternative throws of an untried coin of irregular appearance, or, generally speaking, to other such uncertain alternatives (de Finetti 1937, in Kyburg and Smokler 1980, p. 101). De Finetti's remarks suggest that having conceived chance setups on the analogy of lottery urns, objectivists have gone astray by overstretching and overtaxing this analogy. In an urn the alternatives are there for all to see and can literally be grasped with one's hands, but in other chance setups they enjoy at best a purely notional existence in a fictitious realm of possibilities. I shall explain de Finetti's philosophy of probability in Section 4.5, but I find it convenient to examine this particular question here, as it pertains to the conception of chance setups and their alternative prospects.

Consider a large urn filled with many balls of equal size and weight, an unknown number of which are white. We draw one ball at a time, note down its color, and replace it in the urn, which is thoroughly mixed before each drawing.

> If, after numerous drawings, the observed frequency of the white balls is f, why do we attribute a value close to f to the probability that the ball will be white in one of the drawings which is going to follow? It can be answered that after the observation of such a frequency we attribute a very large value to the probability that the number of white balls will come very close to the fraction f of the total, and further, by supposing this fraction to be ρ, we judge that the drawings are independent and have all the same probability $\mathbf{p} = \rho$.
> (De Finetti 1937, in Kyburg and Smokler 1980, p. 102)

According to de Finetti, the same argument cannot be extended to a game of heads or tails with an irregular coin, because "the corresponding terms which would allow analogous reasoning do not exist." Frankly, I am unable to see how the actual differences between these two cases can warrant the distinction that de Finetti draws between them. There is indeed a definite number of white balls inside the urn. But the ratio between this number and the total number of balls in the urn can be equated with the probability of drawing a white ball the next time only if it is tacitly assumed that each individual ball has the same chance of being drawn as any other.[20] The same assumption is of course implicit in the claim that the relative frequency of white balls observed after numerous drawings very probably comes close to the actual proportion of white balls in the urn. This claim rests on a straightforward application of Bernoulli's Theorem to a finite probability space $\langle S,\mathbf{p}\rangle$, where each element of the sample space S represents the drawing of a distinct individual ball from our urn, and the probability function $\mathbf{p}$ assigns to every singleton in $\mathcal{P}(S)$ the same number in the interval $[0,1]$, namely, the reciprocal value of the total number of balls. Under these conditions, if the ratio of white balls to all balls happens to be h, the probability that the relative frequency of white balls in a series of n drawings will differ from h by less than some small number ε will be very close to 1 if n is large enough. The urn may not be set apart from the coin merely because the ratio of white balls to all balls in it is, as de Finetti observes, "an objective fact which can be directly verified" (Kyburg and Smokler 1980, p. 101). After all, so is the ratio of the head side to all sides of the coin (viz., 1/2). For probabilistic thinking, the significant—though not terribly important—difference between the two examples proposed by de Finetti is that, in the case of the urn, both prima facie discernible kinds of drawing, namely, white and nonwhite, can be analyzed into a definite—even though unknown—number of *equiprobable* outcomes; whereas in the case of the coin, the two ultimate alternatives, heads and tails, are presumably not equiprobable. This makes it well-nigh impossible to assign a definite chance to each possible outcome by direct inspection of the coin and of the circumstances in which the game is played. All the same, the tossings of the coin can be regarded as trials on a chance setup represented by a finite probability space $\langle\{\mathsf{H},\mathsf{T}\},\mathbf{p}\rangle$, where the sample space $\{\mathsf{H},\mathsf{T}\}$ consists of the two kinds of outcome, *heads* and *tails*, and $\mathbf{p}$ is an unknown probability function whose values at $\{\mathsf{H}\}$ and $\{\mathsf{T}\}$ can be conjectured, in agreement with Bernoulli's Theorem, as I explained above. It is, of course, impossible to determine $\mathbf{p}$ in that way with full precision and perfect certainty. In this respect, however, the probability function $\mathbf{p}$ is no more deficient than the other real-valued functions by which we represent purportedly objective quantities.

That de Finetti's distinction between the symmetric lottery and the asymmetric game of heads or tails is unreasonable—despite its apparent plausibility—will perhaps more readily be seen if we take a look at a third game, which is, so to speak, intermediate between the other two, viz., an asymmetric lottery. Fill the urn with balls of different weights and sizes, some of which are more easily drawn than the others, either because they rise above the rest when the content of the urn is mixed, or because the person in charge of the drawings finds them more pleasant to touch or to hold, or for any reason you like. In this case, the definite, though unknown, proportion of white balls in the urn cannot be inferred—within known bounds of uncertainty and imprecision—from the observed relative frequency of white balls in a long series of drawings. Is it therefore meaningless to assign a definite but unknown value to the chance that the next ball will be white? I am unable to accept this conclusion. Asymmetry can be introduced gradually in a perfectly symmetric lottery by continuous variation of the initially uniform mechanical properties of the balls. Yet the estimation of unknown probabilities allowed by de Finetti in the case of his urn surely does not become meaningless at the slightest departure from symmetry. In fact, if the asymmetry is small—which is all one can demand from real lotteries—we are still entitled to base an estimate of the actual composition of the urn on the observed relative frequencies of the several kinds of drawing. But our reasoning evidently proceeds in a direction opposite to that indicated by de Finetti in the quoted passage: the recorded frequency of white balls yields an estimate of the probability that the next outcome will be white, from which in turn, if the game is almost symmetric, one obtains an estimate of the actual ratio of white balls to all balls in the urn.

The type of reasoning just sketched lies at the heart of the most familiar and pervasive use of probability thinking in contemporary life: the method of random sampling from a population. Let P be any collection of objects that we wish to study but which it would be difficult or impossible to examine one by one. A *sample* s is drawn from the *population* P by picking out a certain number n of its elements. If each element of P can turn up in s at most once, the sample is a subset of the population and we speak of *sampling without replacement.* But if every element of P is liable to be chosen repeatedly, s is an element of the Cartesian product P^n, i.e., a list or n-tuple of population elements, and we speak of *sampling with replacement.* The theory of probability entitles us to infer features of the population P from homologous features of the sample s, if the sample is picked *at random,* i.e., if at any stage of the selection process every element of the population that is available for choice has an equal chance of being chosen.[21] The population is thus regarded, in effect, as the content of an imaginary urn from which the sample is drawn by some admissible lottery procedure. Of the various idealizations ordinarily involved in sampling, this assumption of randomness is perhaps the mildest. (Statisticians often practice sampling without replacement but subject the results to the more convenient mathematics of sampling with replacement; they are also wont to handle a finite and changing population as if it were infinite and fixed.) However, unless per impossibile all members of the population have exactly the same chances of being included in the sample, the best inference from sample to population conforms to the paradigm of a *nearly* symmetric lottery.

The mathematical theory of chance setups began its thenceforth uninterrupted public career in 1654, in letters exchanged between Pierre Fermat and Blaise Pascal about problems in gambling raised by the Chevalier de Méré.[22] The attention the theory received from such formidable mathematicians as Huygens, the Bernoullis, de Moivre, Euler, Lagrange, Laplace, Poisson, Gauss, etc., was motivated in part, no doubt, by its intrinsic beauty; but also, we may assume, by the applications it soon found to such matters of general interest as the size and composition of human populations and the life expectancy of individuals.

In 1662 the London merchant John Graunt published his *Observations upon the Bills of Mortality*, a demographic study based on the parish registries of christenings and burials kept in London and its outskirts since 1603. The parish bills, as they were called, recorded the number of persons of either sex who had been christened or buried each year, and also the cause of death of those buried (including such entries as 'affrighted', 'cold and cough', 'executed and prest to death', 'grief', 'rising to the lights', 'worms', and, of course, the Plague), but not their age. From such data, plus a rough estimate of the ratio of yearly births to child-bearing women,[23] the tax borne by the several parts of England and Wales, and the information supplied by Richard Newcourt's map of London of 1658, Graunt drew inferences concerning the population of the city and the entire realm, the number of fighting men (males aged between 16 and 56) in and near London, the comparative unhealthiness of town and country, and the Age of the World,[24] among other matters. He also managed to construct the first life table.[25] In this extraordinary work I find no clear indication that the author conceived the human person in her habitat as a chance setup with a quantifiable proclivity, dependent on age, to die within the next year (or the next decade). But passages such as these may easily have suggested it:

> Considering, that it is esteemed an even lay, whether any man lives ten years longer, I supposed it was the same, that one of any ten might die within a year.
>
> (Graunt 1676, p. 81)

> If in any other Country more than seven of the 100 live beyond 70, such country is to be esteemed more healthful than this our City.
>
> (Graunt 1676, p. 81)

> The *Lunaticks* are also but few, *viz.* 158 in 229250, though I fear many more than are set down in our *Bills*, few being entred for such, but those who die in *Bedlam*; and there all seem to dye of their *Lunacy*. [. . .] So that, this *Casualty* being so uncertain, I shall not force my self to make any inference from the numbers and proportions we find in our Bills concerning it; only I dare ensure any man at this present, well in his Wits, for one in a thousand, that he shall not dye a *Lunatick* in *Bedlam* within these seven years, because I find not above one in about one thousand five hundred have done so.
>
> (Graunt 1676, p. 31)

Graunt was the first to note that there are more boys born than girls, in a stable proportion he roughly set at 14/13. He did not link the stability of this and other statistical ratios to any "Laws of Chance"; but the suggestion must have been inevitable to someone mindful of the stability of the long-run relative frequencies of alternative outcomes in dicing (as recorded in *la lunga osservazione* mentioned by Galileo), on the one hand, and of the randomness inherent in the process of human fecundation (as disclosed by Leeuwenhoek's discovery of spermatozoa, communicated to the Royal Society of London in 1677), on the other. It would seem that the connection was already established in John Arbuthnot's mind when he wrote in the preface to his translation of Huygens' *De ratiociniis in aleae ludo*, published in 1692, that "it is odds, if a woman be with child, but it shall be a boy, and if you would know the just odds, you must consider the proportion in the Bills that the males bear to females" (quoted by Hacking 1975, pp. 166–67). However, the same Arbuthnot published 18 years later a notorious "Argument for divine providence taken from the constant regularity observed in the births of both sexes" (1710), where he contended that if the said proportion were due to chance alone it would fluctuate much more wildly than it had done according to the London Bills for the 82 years from 1629 to 1710. Commenting on this argument, Nicholas Bernoulli, a nephew of Jacques, wrote in 1713 to Pierre Rémond de Montmort that the recorded stability was not amazing (and did not portend a miracle), for there was "a great probability that the numbers of males and females might fall within even narrower bounds than those observed" (Montmort 1723, p. 388). To prove it, he assumed that there are 14,000 children born each year in London, of which 7,200 would be boys and 6,800 girls if the ratio of males to females did not deviate from the recorded mean of 18/17. He bid his correspondent imagine a set of 14,000 dice, with 18 white and 17 black faces each, and went on to show that the odds are 226 to 1 that in a sequence of 82 throws of all 14,000 dice, the number of white faces will not exceed 7,633 or fall beneath 7,037 in more than 11 throws (11 being the number of years between 1629 and 1710 in which the number of male births was not contained within such bounds). The young mathematician did not explain how to build a symmetric die with 35 faces. (Since a regular pentekaitriakontahedron cannot exist, the problem can be solved only by deftly loading an irregular one.) But it is clear that he saw in the process by which a child's sex is determined at its conception a close analogue of just such a gambling device.

Is there a common feature of natural processes or situations which warrants their being regarded as chance setups? The men who developed the probability calculus and its applications from the 17th to the 19th century were, for the most part, convinced determinists. As Jacques Bernoulli pointedly expressed it in the *Ars conjectandi*, right before the proof of his law of large numbers:

> That anything should be uncertain or indeterminate in itself, according to its own nature, we can so little understand as that God could have at once created it and not created it. For everything that God has created, He has also determined, by the very act of creating it.
>
> (Bernoulli 1713, p. 227)

Full determinacy extends not only to all things past and present, which "precisely because they are or have been exclude the possibility of not existing or not having existed" (Bernoulli 1713, p. 210), but also to all things to come, which will certainly come about as they will "because of divine foreknowledge and predestination."

> For if that which lies in the future will not happen with certainty, one cannot understand why the infinite glory of omniscience and omnipotence should be attributed to the Supreme Creator.
>
> (Bernoulli 1713, p. 211)

As there is no indeterminacy in things, the uncertainties underlying any application of the doctrine of chances can only be subjective. Thus, according to Bernoulli, a die thrown with a definite velocity from a definite position above the gambling table cannot fall otherwise than how it falls. Likewise, tomorrow's weather cannot be different from what it will be, given the present state of the atmosphere, involving a "definite quantity, distribution, motion, direction, and speed of the winds, vapors, and clouds, and definite mechanical laws by which all these move together" (Bernoulli 1713, p. 212). Nevertheless, we regard future throws of dice and upcoming weather as contingent, because we do not know well enough the present circumstances that make them necessary.

> Thus, contingency depends chiefly on our knowledge, insofar as we do not perceive any reason contrary to a thing's not being or not coming to be, although due to a proximate cause still unknown to us, that thing necessarily is or must be.
>
> (Bernoulli 1713, p. 213)

Or, as d'Alembert put it in a text written about 1750: "Strictly speaking, there is no chance, but there is its equivalent: the ignorance in which we are of the true causes of events."[26]

However, mere ignorance is not—and never was held to be—a sufficient condition for the fruitful application of the doctrine of chances to a fragment of reality. Not every piece of matter whose inner workings we do not happen to know can therefore be treated as a chance setup. Henri Poincaré, working still within the classical program of determinism in physics but no longer obeisant to a deterministic ontotheology, gave a more precise characterization of potential candidates for a probabilistic approach. They must involve physical processes in which a small variation of the initial conditions leads to a considerably different outcome.

> A very slight cause, which escapes us, determines a considerable effect which we can not fail seeing, and then we say this effect is due to chance. If we could know exactly the laws of nature and the situation of the universe at the initial instant, we should be able to predict exactly the situation of this same universe at a subsequent instant. But even when the natural laws should have no further secret for us, we could know the initial situation only *approximately*. If that permits us to foresee the subsequent situation *with the same degree of approximation*, this is all we require, we say the phenomenon has been predicted, that it is ruled by laws. But this is not always the case; it may happen that slight differences in the initial conditions produce very great ones in the final phenomena; a slight error in the former would make an enormous error in the latter. Prediction becomes impossible and we have the fortuitous phenomenon.
>
> (Poincaré, SM, pp. 68–69; Halsted translation, pp. 397–98)

Consider, for instance, the game of roulette as it is ordinarily played in casinos. A very slight change in the momentum initially impressed on the ball is apt to lead to a completely different outcome. In such cases, if the process in question is deterministic, the relation between the diverse possible initial conditions and the alternative outcomes can be represented by a mapping of the set of the former onto that of the latter, but even if, as is habitual in physics, both sets are endowed with a topology, the said mapping is not, or cannot be usefully thought of as being, continuous at any point of its domain.

To give a better feeling of what this implies I shall propose an idealized deterministic representation of what happens when a die is thrown on a table, on which it bounces several times until it comes to rest upon one of its six faces. If we neglect air resistance, the die can be regarded as a rigid body falling or jumping in a uniform gravitational field. It can therefore be suitably represented by a point in a 12-dimensional phase space that describes a curve satisfying Hamilton's equations—for the appropriate Hamiltonian. (As a rigid body has six degrees of freedom, its dynamical state is fully described by six position and six momentum coordinates.) This representation breaks down as soon as the die hits the table, for it will then suffer deformations incompatible with its supposed rigidity. However, the die will again behave as a rigid body when it leaves the table after bouncing from it. Hence the motion of the die from the moment it is thrown until it comes to rest upon one of its faces can be represented by a smooth curve in phase space with finitely many discon-tinuities (one for each bounce). Let D stand for the set of points in $\mathbf{R}^{12}$ representing the diverse possible initial conditions of the throw, with the standard subset topology inherited from $\mathbf{R}^{12}$ (U is open in D only if U is the intersection of D with an open set in $\mathbf{R}^{12}$.) Let S stand for the set of alternative outcomes {1,2,3,4,5,6}, endowed with the discrete topology. For the sake of the argument, let us assume (I) that the six position and six momentum coordinates of the die at the time it hits the table and begins to undergo deformation determine its position and momentum coordinates when it returns to a constant shape after bouncing. (In the light of current physics, this assumption is hardly tenable, inasmuch as the solid state is subject to quantum theory.) There is then a mapping f of D onto S, assigning to each possible initial state of the die one and only one of the six outcomes. Let us assume, moreover, (II) that the slightest variation in the die's initial state changes the outcome of the throw. Then f is nowhere continuous. Consequently, every possible initial state x is densely surrounded in D by alternative initial states leading to outcomes different from $f(x)$. In other words, the inverse image $f^{-1}(\omega)$ of each outcome $\omega \in S$ has an empty interior and consists exclusively of boundary points. The thorough mixing in D of the six sets of possible initial conditions leading to different outcomes makes a travesty of the deterministic bonds represented by the mapping f.

This result depends, indeed, on the strongly idealized assumption (II), which, it will be said, reflects only our own incapacity for distinguishing between the slight changes of initial conditions that are sufficient to overturn the outcome and other, even slighter such changes that would leave the outcome invariant. Now, our estimation that the slightest change in the initial conditions of a deterministic process will bring about a significant change in the outcome may well be justified only insofar as we refuse to consider the effect of variations too small to be detected by means of our instruments. If such is the case in our particular example, the emergence of chance in the midst of determinism, conjured up by the dramatic streamlining of the mathematical model as a consequence of assumption (II), does no more than express our ignorance.[27] But in this and in other analogous instances it might also happen that, when improved instruments will enable us to consider hitherto neglected differences, we shall be driven to change the very concepts by which the process in question, its initial conditions, and deterministic laws are now grasped. In such cases, the infestation of determinism by chance, in the manner I have sought to illustrate with the example of dice throwing, turns out to be a constitutive feature of the physical situation *as objectified by those concepts.* Thus, chance setups—as explicated by Poincaré—will be found *objectively* in any domain of reality structured by a deterministic physical theory *T* wherever noticeable changes in expected effects result from unnoticeable changes in their determining conditions, provided that *T* is liable to be dislodged and replaced by a different theory if such unnoticeable changes become noticeable. Because all physical theories are liable to this and every domain of reality structured by a deterministic theory contains processes in which considerable changes ensue from inconsiderable ones, we must conclude that chance setups pertain to the realm of objectivity disclosed by physics, even if we make no allowance for indeterministic theories or assume that they will be eventually dislodged and replaced by deterministic ones.[28]

4.4 Probability as a limiting frequency

The frequentist philosophy of probability, independently initiated in the 1840s by A. A. Cournot (1843), R. L. Ellis (1849), J. F. Fries (1842), and J. S. Mill (1843), adopted by George Boole (1854) and systematically developed in John Venn's *The Logic of Chance* (1866; 3d. ed., 1888), reached maturity towards 1930 in two variant versions expounded in the treatises of Richard von Mises (1931, 1964; cf. von Mises 1928, 1957) and Hans Reichenbach (1935, 1949). 19th-century frequentism reacted against the prevailing tendency—nourished, as we have seen, by faith in determinism—to view probability as an expression of subjective uncertainty, a measure of expectations arising from ignorance. Boole stated with characteristic clarity what he thought was wrong with this view:

> The rules which we employ in life-assurance, and in the other statistical applications of the theory of probabilities, are altogether independent of the *mental* phenomena of expectation. They are founded on the assumption that the future will bear a resemblance to the past; that under the same circumstances the same event will tend to recur with a definite numerical frequency; not upon any attempt to submit to calculation the strength of human hopes and fears.
>
> (Boole 1854, pp. 244–45)

The frequentists did not take issue with determinism. They noted that the actual uses of probability in scientific inquiry and practical decision making involve large families of events or states of affairs among which some feature turns up or fails to turn up, for whatever reasons, seemingly erratically, yet recurs with a more or less stable frequency in the long run. Thus, for example, different causes can prevent the delivery of a letter. The number of undeliverable letters returned by a particular postman on any particular day may fluctuate widely. Nevertheless, the percentage of undeliverable letters posted in Paris in Laplace's time remained approximately the same from year to year. With similar examples in view, Venn proclaimed the "distinctive characteristic" of "the science of Probability" to consist in this:

> That the occasional attributes [e.g., to be undeliverable] as distinguished from the permanent [to be posted in Paris], are found on an extended examination to tend to exist *in a definite proportion of the whole number of cases.* We cannot tell in any given instance whether they will be found or not, but as we go on examining more cases we find a growing uniformity. We find that the proportion of instances in which they are found to instances in which they are wanting, is gradually subject to less and less comparative variation, and approaches continually towards some apparently fixed value.
>
> (Venn 1888, p. 11)

Considerations such as these inspired the frequentist definition of the probability of finding a feature *A* among objects of a class *B* as the limit of the relative frequency of *A*'s in an infinite sequence of *B*'s.

As I am unable to dwell here on the history of frequentism, I shall concentrate my attention on its mature formulation by von Mises. But before proceeding any further, it is worth noting that the frequentist philosophy makes the naïve understanding of probability stand on its head. In the naïve view, we anticipate that the relative frequencies of the alternative outcomes in a long run of future trials on a chance setup will almost inevitably fall within some narrow interval *because* we can estimate their chances in a single trial on the basis of our grasp of physical symmetries. Where symmetries are lacking or unknown, we can still reach for the likeliest conjecture concerning the distribution of chances among the several possible outcomes of a single trial by studying their relative frequencies in a long run of past trials (inasmuch as we can tell what hypothesis or what family of closely related hypotheses about the single case would, if true, make the observed frequencies almost inevitable). Thus, our reasoning goes from a *single* trial to a *single* run of (finitely) many trials, or vice versa. The trials must, of course, be conceptually grasped in order to be subjected to inference. But chances are attributed to the particular instances of the appropriate concept (e.g., *one* throw of a die; *one* series of throws of a die). The known or conjectured distribution of chances is understood to belong to the individual trial—or to the individual finite run of trials—*qua* individual, in its impending realization in the physical world; not *qua* term—or segment—of a fancied infinite sequence. Single-case probabilities do indeed entail consequences concerning the limits of relative frequencies in such ideal sequences, and it is often convenient to refer to those limits in mathematical proofs and calculations. But naïve probabilistic thought begins and ends with the single case, even where a consideration of its repeatability and actual repetition is required to set that thought in motion. In stark contrast with the naïve view, the frequentist philosopher will typically

> regard the statement about the probability of the single case, not as having a meaning of its own, but as representing an elliptic mode of speech. In order to acquire meaning, the statement must be translated into a statement about a frequency in a sequence of repeated occurrences. The statement concerning the probability of the single case thus is given a *fictitious meaning*, constructed by a *transfer of meaning from the general to the particular case.*
>
> (Reichenbach 1949, pp. 376–77)

According to von Mises, the theory of probability is the mathematical theory of a certain type of phenomenon, just as the Newtonian theory of gravity is the mathematical theory of planetary motions and free fall. The phenomena to which the theory of probability is applicable are "mass phenomena" in which "either the same event repeats itself again and again, or a great number of uniform elements are involved at the same time" (von Mises 1957, p. 11). To illustrate this general idea, von Mises mentions three examples: (i) throws of a pair of dice; (ii) insurance taken by many different people against a particular kind of risk; and (iii) "the random motion of colloidal particles which can be observed with the ultramicroscope" (von Mises 1957, p. 10). Experience has shown that in all cases such as these "the relative frequencies of certain attributes become more and more stable as the number of observations is increased" (von Mises 1957, p. 12). According to von Mises, this feature of mass phenomena suggested the development and warrants the application of the concept of probability. He approvingly quotes the following passage from the introduction to Poisson's *Recherches sur la probabilité des jugements en matière criminelle et en matière civile* (1837), in which the said feature is very clearly described:

> In many different fields, empirical phenomena appear to obey a certain general law, which can be called the Law of Large Numbers. This law states that the ratios of numbers derived from the observation of a very large number of similar events remain practically constant, provided that these events are governed partly by constant factors and partly by variable factors whose variations are irregular and do not cause a systematic change in a definite direction. Certain values of these relations are characteristic of each given kind of event. With the increase in length of the series of observations the ratios derived from such observations come nearer and nearer to these characteristic constants. They could be expected to reproduce them exactly if it were possible to make series of observations of an infinite length.
>
> (Poisson 1837, quoted by von Mises 1957, p. 104–5)

Poisson also employed the expression 'Law of Large Numbers' to designate the mathematical theorem that I have called the weak law of large numbers, which had been proved by Jacques Bernoulli for sample spaces with two elements and was generalized by Poisson himself. Von Mises insisted that this theorem ought to be carefully distinguished from the empirical Law of Large Numbers described in the foregoing quotation.

The mathematical structure proposed by von Mises for representing mass phenomena is the von Mises *collective*. Its conception was guided by the following three ideas:

(1) Mass phenomena are given through long, open-ended series of observations, classifiable into different types. A series of this sort can be ideally represented by an infinite sequence of terms drawn from a fixed set of labels, corresponding to the several types to which the successive observations belong. E.g., an open-ended series of coin tossings can be represented by an infinite sequence of **H**'s and **T**'s. A *collective* is just such a sequence.

(2) The relative frequency of observations of a particular type fluctuates unpredictably, but as the number of observations increases it tends to fall within an ever narrower neighborhood of some definite ratio. This is the empirical law of large numbers. To represent this property of mass phenomena, von Mises assumed that the relative frequency of each label occurring in a collective converges to a limit, the *probability* of that label.

(3) The record of the first n observations of a mass phenomenon can give no hint as to the outcome of the $(n + 1)$th observation. It is, indeed, impossible to devise a successful gambling system for betting for or against a particular outcome at certain points of the series on the basis of one's knowledge of the preceding outcomes. Thus, a gambler who, following some

such system, bets on heads only at the k_1th, k_2th, k_3th, . . . throws of a coin and otherwise refrains from playing will in the long run fare no better than if he bets on heads in all throws. Von Mises regards this "principle of the impossibility of a gambling system" as a fundamental natural law of mass phenomena, comparable to the principles of thermodynamics (which posit the impossibility of perpetual motion of the first and the second kind). In order to represent it, von Mises postulated that any subsequence K' extracted from a collective K by deleting terms according to any rule suitable for use as a gambling system is also a collective in which the relative frequencies of all labels converge to the same limits as in K. The exact nature of the rules admissible for constructing such a subsequence K' from a given collective K has long been controversial. Now, as Alonzo Church (1940) aptly noted, a gambling system surely ought to be governed by an algorithm that tells the gambler when and how to bet. Following Church's Thesis of 1936—to which no counterexample has yet been discovered—we take it that every conceivable algorithm corresponds to a recursive or, equivalently, to a Turing computable function.[29] This condition, which is obviously necessary, will here be treated as sufficient for the rules in question. In fact, however, it is perhaps too liberal, and several alternative refinements have been proposed to narrow it down. (Martin-Löf 1966; Schnorr 1970a; Fine 1973.)[30]

To translate these ideas into a concise definition of the collective I must introduce some new terms. Let L denote our set of labels, or *label space.* We shall assume that L is finite or at most countable. Let σ be a sequence in L, i.e., a mapping of the set $\mathbf{Z}^+$ of positive integers into L. I write σ_n for $\sigma(n)$, the value of σ at $n \in \mathbf{Z}^+$. If σ is given, there is given also, for each positive integer n, a function f_n from L to the closed interval $[0,1]$, defined as follows: for every $\lambda \in L$, $f_n(\lambda)$ equals $1/n$ times the number of occurrences of λ in the n-tuples $\langle \sigma_1, \ldots, \sigma_n \rangle$. Thus, f_n assigns to each label λ the relative frequency of λ's among the first n terms of the sequence σ. f_n can clearly be regarded as the value at n of a sequence f that maps $\mathbf{Z}^+$ into the set $\mathbf{F}$ of functions from L to $[0,1]$. If ζ denotes a sequence of positive integers, i.e., a mapping of $\mathbf{Z}^+$ into $\mathbf{Z}^+$, then $f \circ \zeta$ is again a sequence in $\mathbf{F}$. If ζ is order preserving (i.e., if $\zeta(m) < \zeta(n)$ if and only if $m < n$), we say that $f \circ \zeta$ is a *subsequence* of f. An order-preserving sequence ζ of positive integers is fully determined by the characteristic function of its range, i.e., by the function $X_\zeta: \mathbf{Z}^+ \to \{1,0\}$ such that for any $n \in \mathbf{Z}^+$, $X_\zeta(n) = 1$ if n lies on the range of ζ and $X_\zeta(n) = 0$ if n does not lie on the range of ζ. We shall say that an order-preserving sequence ζ is a *place selection* if the characteristic function of the range of ζ is Turing computable. If ζ is a place selection, the subsequence $f \circ \zeta$ is obtained by place selection from the sequence f.

Let us now return to the sequence $\sigma: \mathbf{Z}^+ \to L$ and its associated sequence

$f_1, f_2, \ldots$ of relative frequency functions on L. The sequence σ is a *collective* in the sense of von Mises if and only if:

VMC1. The sequence $f_1, f_2, \ldots$ converges pointwise on L to a function $\mathbf{p}$: $L \to [0,1]$. (In other words, for every label λ there is a real number $\mathbf{p}(\lambda) \in [0,1]$ that satisfies the following condition: for every positive real number ε there is a positive integer N such that $|f_n(\lambda) - \mathbf{p}(\lambda)| < \varepsilon$ whenever $n > N$.)

VMC2. $\sum_{\lambda \in L} \mathbf{p}(\lambda) = 1$.[31]

VMC 3. Every subsequence $f \circ \zeta(1), f \circ \zeta(2), \ldots$ obtained by place selection from $f_1, f_2, \ldots$ converges pointwise to $\mathbf{p}$ on L.

If σ is a collective with label space L, and $\lambda \in L$, $\mathbf{p}(\lambda)$ is called by von Mises the *probability* of λ. If σ is not a collective but satisfies VMC 1 and VMC 2, I shall say that it is a *semicollective.* In that case, $\mathbf{p}(\lambda)$ is called by von Mises the *chance* of λ. (As far as I know, this usage is peculiar to him.) Since $\mathbf{p}$ is defined on L, not on its power set $\mathcal{P}(L)$, the triple $\langle L, \mathcal{P}(L), \mathbf{p}\rangle$ is *not* a probability space with sample space L and probability function $\mathbf{p}$. But then, neither is 'probability' a registered trademark of Kolmogorov & Co. Moreover, as we shall see, the von Mises probability function $\mathbf{p}: L \to [0,1]$ generates a function on $\mathcal{P}(L)$ that complies with the axioms FPS 1 and FPS 2 of Section 4.2 (but not necessarily with the Kolmogorov axioms KPS 1 and KPS 2).

For von Mises, the task of the probability calculus is to construct new collectives from given collectives and to compute the $\mathbf{p}$-functions of the former from those of the latter. This is achieved by the single or combined application of four basic operations on collectives that I shall now describe. Simple proofs of the several claims I shall make can be found in von Mises 1964, pp. 15–39. Unless otherwise noted, σ stands for a collective with label space L and von Mises probability function $\mathbf{p}$.

1. Selection. If ζ is a place selection, then, by VMC 3, $\sigma \circ \zeta$ is also a collective with the same label space and probability function as σ. The new collective is said to be formed by *selection* from the first.

2. Mixing. Let L' denote a partition of L and let σ' be the sequence in L' defined as follows: for each $n \in \mathbf{Z}^+$, $\sigma'_n = A \in L'$ if and only if $\sigma_n \in A$. Then σ' is a collective formed by *mixing.* L' is its label space. Its probability function $\mathbf{p}'$ is such that for any $A \in L'$, $\mathbf{p}'(A) = \sum_{\lambda \in L} \mathbf{p}(\lambda)$. This fact has an important implication. If Γ is any subset of L and Γ' is its complement in L, the set $L_\Gamma = \{\Gamma, \Gamma'\}$ is a partition of L, so we can form by mixing a collective with label set L_Γ. Let $\mathbf{p}_\Gamma$ denote the probability function thus defined on L_Γ. Now let Γ range

over $\mathcal{P}(L)$ and define a function $\mathbf{p}^*$ from $\mathcal{P}(L)$ to [0,1] by $\mathbf{p}^*(\Gamma) = \mathbf{p}_\Gamma(\Gamma)$. Then $\mathbf{p}^*$ satisfies the probability axioms FPS 1 and FPS 2, and $\langle L, \mathcal{P}(L), \mathbf{p}^*\rangle$ is a probability space with finitely additive probability function.[32] This result holds also if σ is a semicollective. Because $\mathbf{p}^*$ is uniquely determined by $\mathbf{p}$, I shall hereafter imitate von Mises (1964) and denote both functions by the same letter $\mathbf{p}$.

3. PARTITION. Let A be a finite or countable subset of L such that $\mathbf{p}(A) > 0$. Let σ_A be the sequence formed by deleting all terms of σ which do not belong to A and suitably renumbering the remaining terms. Then σ_A is a collective with label space A obtained from σ by *partition* (von Mises' term). We denote its probability function by $\mathbf{p}_A$. If B is a subset of A, $\mathbf{p}_A(B) = \mathbf{p}(B)/\mathbf{p}(A)$. Now let Z be an arbitrary subset of L and put $B = A \cap Z$. Then, clearly, $\mathbf{p}_A(B) = \mathbf{p}(A \cap Z)/\mathbf{p}(A)$ is none other than the conditional probability of Z given A, usually designated by $\mathbf{p}(Z|A)$.

4. COMBINING. Let κ and κ' be two collectives formed by selection from σ. Then the sequence of pairs (κ_1,κ'_1), (κ_2,κ'_2), ... is a collective with label space L^2 and probability function $\mathbf{p}_2$ defined by $\mathbf{p}_2(\lambda,\mu) = \mathbf{p}(\lambda)\mathbf{p}(\mu)$. Under certain conditions this result may hold even if κ and κ' are arbitrary collectives with different label spaces and probability functions (see von Mises 1964, pp. 30–31).

Von Mises (1964, Chapter II) develops a theory of collectives with uncountable label space, presumably in order to bring his ideas in line with the standard practice of statisticians and to provide a proper conceptual foundation for it. To someone acquainted with Kolmogorov's approach this theory looks curiously unwieldy. I shall not go into it here, but I cannot help wondering whether it is not an alien growth in the body of von Mises' frequentism. For surely, if σ is a sequence, the range of σ—i.e., the set of its terms—is at most countably infinite. (In effect, it is counted by σ itself, possibly with repetitions.) Thus, even when a collective is drawn from an uncountable label space L, the labels that actually contribute to the collective's characteristic frequencies and "probabilities" constitute at most a countable subset of L.

The defining properties of collectives, as expressed by VMC 1, "the axiom of convergence," and VMC 3, "the axiom of randomness," are meant to capture the peculiar combination of regularity and unpredictability characteristic of mass phenomena. Some mathematicians—accustomed to having their convergent sequences given by a law of construction from whose very form the existence of the limit can be inferred and its value computed—questioned the compatibility of those two axioms and demanded a proof of consistency.[33] Now, one cannot simply prove the consistency of von Mises' concept of the collective by producing a particular instance of it, for any rule employed for identifying its successive terms can also serve as the basis of a gambling system and therefore cannot yield a genuine collective.[34] Abraham Wald (1937) found, however, a nonconstructive consistency proof subject to the condition that the place selections mentioned in the axiom of randomness be drawn from a countable set. This condition is automatically fulfilled if, following Church (1940), we require the place selections to be computable.

Tougher objections have been raised with regard to the possibility of finding an application for the theory of collectives, a physical model of axioms VMC 1, 2, and 3. The perceived strength of these objections has led to the gradual abandonment of frequentism after it enjoyed much acceptance about half a century ago. Since the objections concern the relationship between the mathematical core of a purported physical theory and the domain of reality for which it was intended, they are of particular interest to us.

It ought to be clear that a real-text, in Ludwig's sense, can never be read in such a way that it provides an example of a collective. We may, indeed, single out a list of days, classified according to the maximum temperature recorded on each at a given meteorological station; or a list of persons buried over a series of years in a certain country, classified by the stated cause of death; or a list of dice throws, classified by the number of points made each time. But any such list will be finite. It cannot therefore be a collective but is at most a segment of a collective whose label space contains all the classes of items occurring in the list (or the labels by which we designate those classes). For the list to be grasped as such a segment, infinitely many additional terms must be supplied *by postulate.* The practice of enriching a real-text with hypothetical objects or attributes in order to make it intelligible is, of course, common to all fields of physics. Typically, however, the postulated entities are supposed to exist and to contribute by their active presence to the observed composition and configuration of the real-text under study. But when it comes to reading a given list of events as a segment of a collective, the required supplementary events need not be real. It is enough to assume that such-and-such events *would be* forthcoming if, per impossibile, the list in question were extended to infinity. Moreover, this assumption must not specify the particular terms of the extended list but only (i) the set of labels under which they must all be classifiable and (ii) the convergence of the relative frequencies of the various labels to definite limits in agreement with the VMC axioms. The recorded composition of the real-text imposes only a minimal restriction on (i), viz., that the said set of labels must contain all the labels exemplified in the observed list—though it may indeed contain many more. Moreover, the order and frequency of the observed items does not in any way constrain (ii). Let $\alpha_1, \ldots, \alpha_n$, be a list of terms drawn from a set K and let $\beta_1, \beta_2, \ldots$ be a collective with label set L and probability function $\mathbf{p}$. Put $\omega_i = \alpha_i$ for every positive integer $i \leq n$, and $\omega_i = \beta_{i-n}$ for $i > n$. Then $\omega_1, \omega_2, \ldots$ is a collective with label set $K \cup L$ and probability function $\mathbf{p}'$ such that $\mathbf{p}'$ agrees with $\mathbf{p}$ on L and is identically 0 on $K \backslash L$. Hence, no matter how large we choose n, the composition of the initial segment $\alpha_1, \ldots, \alpha_n$, is totally irrelevant to the distribution of probabilities in the collective $\omega_1, \omega_2, \ldots$ (The same can be said if the collective $\omega_1, \omega_2, \ldots$ is formed by interspersing the α_i's among the β_i's.) Thus, any record whatsoever of so-called mass phenomena can be embedded in an arbitrarily designed collective; but, for this very reason, such an embedding will not throw light on those phenomena.

It is true, indeed, that similar difficulties lurk in any empirical application of the mathematical concept of a limit. Thus, for example, the assignment of an instantaneous velocity to a moving body is based on measurements of its average velocity over short times. But the latter can remain stable or change slowly within a narrow interval while the instantaneous velocity fluctuates wildly over even shorter times. In assigning an instantaneous velocity we assume as a matter of course that it varies as little and as smoothly as is compatible with the measured averages—unless, indeed, there are theoretical grounds for assuming a stronger variation. Analogously, the hypothetical collectives into which statistical data are embedded are built, according to von Mises, on the "silent assumption" that *"in certain known fields of application of probability theory* (games of chance, physics, biology, insurance, etc.) *the frequency limits are approached comparatively rapidly* (the rate of approach being different for different problems)" (von Mises 1964, p. 108). Therefore, the assigned probabilities should lie close to the relative frequencies observed in long, though finite, series of observations. *"This assumption has nothing to do with the axioms of the probability calculus* [. . .] and is not explained by any results of theoretical statistics," for the latter, in fact, depend upon it (1964, p. 108). But "on the other hand, the whole body of our experience in applications of probability theory seems to prove that rapid convergence indeed prevails—at least in such domains—as a physical fact, confirmed by an enormous number of observations. [. . .] In such domains, and only in them, statistics can be used as a tool of research" (pp. 109, 110).

Von Mises understood with rare lucidity the unfastidious pragmatism of science. However, it is not clear to me that his theory of collectives adequately or even conveniently expresses the tacit assumptions underlying statistical practice. Let us take another look at the above example of velocity assignments. The basic assumption here is that moving bodies have a position in a metric space that changes continuously (and differentiably) in time. If this is not assumed, the concept of instantaneous velocity does not make any sense. But if, on whatever empirical or transempirical grounds, the assumption is made—as indeed it was in both classical and relativistic mechanics—then each body has a definite velocity at every instant, and it is only sensible to calculate it by curve fitting from the best available data about the body's average velocities. (Surely it would be perverse to make allowance for undetectable velocity fluctuations if there is no known reason for expecting them.) Such a method of calculation is pragmatic indeed; but its task is set by our intellectual grasp of the phenomena of motion. If we now direct again our attention to the "known fields of application of probability theory" we find a similar situation there. In order to experience the rapid convergence of relative frequencies in certain ordered collections of data one must first have gathered the data into such collections. The "silent assumption" voiced by von Mises rests, therefore, on other more basic assumptions governing the constitution of the real-texts on which the concept of probability is brought to bear. Can the theory of collectives account for the usual way of grouping statistical data so that they display quick convergence to limiting frequencies (as the theories of space and time justify—and indeed necessitate—our reading of average velocities as approximations to instantaneous velocities)? In my judgment, the answer to this question is no. In order to single out, label, and order a mass of data for probabilistic analysis it is not sufficient to grasp them under the concept of a collective, i.e., to view them as terms in an infinite sequence satisfying axioms VMC 1, 2, and 3. This conclusion follows from our earlier proof that any given finite segment of a collective is irrelevant to the collective's overall structure—and vice versa. But the following example will perhaps make it clearer, and at the same time suggest what other concept is at work in the actual construction of applications of probability theory.

Consider dice in the guise of well-balanced cubes made of light, very hard plastic, on whose faces the familiar points are marked by small gold studs of exactly the same shape and weight. Suppose that when playing with one of these dice, the studs usually begin falling off before the 100th but not before the 20th throw, and that none ever remains in place after 2,000 throws. On the other hand, even in a much longer series of throws, the body of the die shows no sign of wear and tear except for the holes the falling studs leave behind. The throws of such dice can be grouped for statistical study in many ways. By throwing a single die and replacing it with a brand new one as soon as it sheds a stud, we can estimate the chance of making each number while all studs are in place. Or we can estimate such chances for a specific constellation of missing studs by recording the throws of several dice that lack precisely the designated studs until they lose another one. Or we can take a sample, say, of 1,500 new dice and count the times each is thrown until one, two, or three, etc., studs come off. Let us grant, for the sake of the argument, that in each sort of grouping the data are viewed as terms of a suitable collective. *Such a view does not command the selection and reading of the data.* It is not through it that one comes to see, for instance, that differently studded dice will display different biases, so that their throws should be embedded in different collectives. Suppose that in over 20,000 throws with the same die the relative frequency of 'six' becomes stabilized at or near q. Under the circumstances of our example, we may safely choose q, or a number very close to it, as the probability of 'six' in the collective representing the throws of any single die. Yet we would unhesitatingly refuse odds of $1 - q$ to q on 'six' when playing with a new die in which the face bearing that number is still loaded with six gold studs while the opposite face carries only one. This refusal evidently does not issue from any thought of limits in some infinite and therefore ideal sequence but, quite simply, from a grasp of the present prospects of the die at hand based on knowledge of its physical properties. The construction of the appropriate potential collective reflects the perceived potentialities of the single case.

Before closing this section on frequentism, I shall mention two more objections which, if sustained, would prove that the representation of chance events as terms in collectives or semicollectives has precious little to do with the actual meaning of probability in science. One of them was stated with unexcelled clarity by R. B. Braithwaite. He noted that "since the notion of limit is an ordinal notion, the [Limiting-Frequency view defended by von Mises and Reichenbach] makes the value of the probability depend upon the order in which the observations are grouped into the collective or semicollective $s_1, s_2, \ldots$." A different grouping would yield a different sequence of relative frequency functions $f_1, f_2, \ldots$, "with quite possibly a different limiting value."[35] The Limiting-Frequency view "does not, it is true, require that the observations should be ordered in time, but it does require that there should be some principle according to which [they] can be assigned successive numbers so that they can be ordered in a series." On the other hand, "any notion of order seems quite irrelevant to the scientific notion of probability in general, which is concerned with the significance of statements like '51% of children are born boys' in which there is no reference to any order whatever" (Braithwaite 1953, p. 125).

D. A. Gillies has remarked that it is wrong to believe that order is indifferent for the purposes of probability theory. For "we have to examine the order of the results obtained to see whether they exhibit the requisite randomness or not" (Gillies 1973, p. 127). Thus, if the outcomes of 30 coin tosses are given by

HTHTHTHTHTHTHTHTHTHTHTHTHTHTHT

I should think that the chance of heads was not the same in each toss. Gillies' remark does not entail, however, that the *notion* of order has any role to play in the *definition* of probability (as indeed it must according to frequentism). What the above example shows is only that, if a given succession of events is much less probable on one hypothesis—viz., randomness—than on another—viz., deliberate contrivance—one can invoke it to reject the former in favor of the latter, just as one would invoke for a like purpose any relevant occurrence not involving order.[36]

The other objection I wish to mention results from comparing the identity established by frequentism between probability and limiting frequency with the much weaker relation asserted by the strong law of large numbers. As a matter of fact, I find it difficult even to state the strong law in a sensible way using the frequentist vocabulary. In the naïve language of Section 4.3, the strong law simply says that if σ is an infinite sequence of trials on a chance setup and if the probability of outcome λ on each trial is $\mathbf{p}$, then the probability $\mathbf{P}$ that the relative frequency of λ's converges to the limit $\mathbf{p}$ is equal to 1. Thus, the strong law derives the inferred probability $\mathbf{P}$ that a certain sequence of trials displays a particular limiting frequency from the given probability $\mathbf{p}$ of a certain kind of outcome on each single trial of the sequence. For the frequentist, however, single trial probabilities are only a manner of speaking of limiting frequencies in the collective to which the trial belongs. To him, the given probability $\mathbf{p}$ is by definition identical with the limiting frequency of λ's in σ. The inferred probability $\mathbf{P}$ must also be identified with a limiting frequency, namely, the limit—in a sequence Σ of collectives such as σ in which the probability of λ is $\mathbf{p}$—of the relative frequency of collectives in which the limiting frequency of λ's is $\mathbf{p}$. Since every collective in which the probability of λ is $\mathbf{p}$ has, *by definition*, a limiting frequency $\mathbf{p}$ of λ's, it is not surprising that $\mathbf{P} = 1$. But this equation is not, like the strong law of large numbers, a deep and laboriously achieved mathematical result, but a facile consequence of the frequentist definition of probability.[37] We are reminded of similar situations in geometry, where nontrivial properties of a figure are trivially true of its degenerate instances. One cannot avoid the suggestion that the theory of collectives does not explicate the scientific concept of probability in its full generality, but only a special, degenerate case.[38]

4.5 Probability as prevision

A completely different understanding of probability is proposed in the writings of Bruno de Finetti (1931, 1937, 1972). It is embodied in didactic treatises by Leonard J. Savage (1954), Dennis V. Lindley (1965), and de Finetti himself (1970; English translation, 1974), and has found application in the epistemological ventures of Isaac Levi (1980), Brian Skyrms (1980, 1984), and others. Independently, Frank P. Ramsey had taken a similar, though somewhat less radical, approach in his essay "Truth and Probability," written in 1926 and posthumously published in 1931; but his early death prevented him from developing it.

De Finetti's basic tenet is uncompromisingly stated in the preface to his *Theory of Probability*:

PROBABILITY DOES NOT EXIST.

Like phlogiston, the cosmic ether, absolute space and time, and fairies and witches, probability, "if regarded as something endowed with some kind of objective existence" is a "misleading conception," for it is only a reification of our subjective uncertainty.[39] De Finetti denied the objective reality of probability as a physical quantity on the strength of the strict positivist criterion of cognitive meaning, to which he unswervingly and unqualifiedly subscribed:

> Every notion is only a word without meaning so long as it is not known how to verify practically any statement at all where this notion comes up.
>
> (De Finetti 1937, in Kyburg and Smokler 1980, p. 108)
>
> Statements have *objective* meaning if one can say, on the basis of a well-determined observation (which is at least conceptually possible), whether they are TRUE or FALSE.
>
> (De Finetti 1974, vol. 1, p. 6)

If probability is conceived as facility of occurrence, it evidently does not meet de Finetti's standard of objective meaning. Take again Galileo's example of a fair die. The statement that the probability of rolling an ace in the next throw is 1/6 will not be proved *true* or *false* by the outcome of that throw. Since the same statement applies in this example to *each* throw, it cannot be proved or disproved by *any* outcome. A sequence of 500 consecutive aces is indeed extremely improbable on Galileo's assumptions, but it is not therefore impossible. As a matter of fact, the probability of making an ace 500 times in a row with a fair die is exactly the same, namely, $(1/6)^{500}$, as the probability of obtaining any particular list of 500 numbers, chosen from among the six smallest positive integers, in the next 500 throws. Since some definite sequence or other is bound to occur if the die is thrown 500 times, the very low probability attached to the one—whichever it may be—that will eventually come true must of course be compatible with its occurrence.

The frequentists—some of whom did not hide their sympathy for positivism—thought they could bestow an objective meaning on probability statements in agreement with the positivist criterion by conceiving probability as an attribute of collections, not of single events. For them, probability was not the facility with which an event of a certain kind might occur when certain conditions are fulfilled but the relative frequency of events of that kind among instances of fulfillment of those conditions. But then, although relative frequency in a finite class of occurrences may well be observable but certainly cannot be equated with probability, relative frequency in an infinite class is nonsense, and statements concerning the *limit* of relative frequencies in an infinite sequence are compatible, as we have seen, with every conceivable composition of a finite segment of the sequence and can therefore be neither proved nor disproved by observations, which, of necessity, are confined to such finite segments. Hence, the frequentists were unable to furnish the notion of probability with an objective meaning in a manner acceptable to de Finetti—and to themselves.

If probability is not a measurable property of things or events, what do its numerical values stand for? De Finetti says time and again that they measure a person's *degree of belief* in the occurrence of uncertain—future or unknown—events.[40] Now, this looks prima facie like an ill-advised attempt to supplant a not directly observable but plausible physical quantity with an implausible psychical quantity which is no more observable. I, for one, know of many situations in which, to my mind, it is equally easy to achieve or to avoid a certain effect; but I can think of none in which I could positively believe two contradictory propositions with equal strength. Fortunately, however, the evaluation of probabilities according to de Finetti does not have to rely on a supposed awareness of metricizable or at least gradable states of mind, for he provides two alternative, presumably equivalent "operational definitions" of probability in purely behavioral terms. To explain them, I must first refer to the quantity that de Finetti calls 'prevision', of which, in his view, probability is only a special case.

Prevision (It. *previsione*, Fr. *prévision*), in de Finetti's peculiar sense, does not imply, as etymology would suggest, an ability to see the future. On the contrary, it necessarily involves a reference to some—past, present, or future—situation that we cannot see or foresee. "Prevision is not prediction!" (de Finetti 1974, vol. 1, p. 98; cf. pp. 70–71, 207). Having to act on matters that we are unable to ascertain, we resort to prevision in lieu of knowledge. According to de Finetti, a fact is either known to be true or false—and then its denial would be false or true—or it should be reputed possible, in which case its denial is just as possible, for "there are no degrees of possibility" (de Finetti 1974, vol. 1, p. 71).[41] Nevertheless, in the face of uncertainty "one feels, and You feel too, a more or less strong propensity to expect that certain alternatives rather than others will turn out to be true" (ibid.). Prevision, in de Finetti's sense, is the measure of such subjective propensities. 'You', with a capital Y (a typographical convention I shall follow only when quoting from him), designates here, as throughout his work, any arbitrary—but fixed—subject who feels them. Roughly speaking, your prevision of a certain advantage, contingent on a particular event E, is the sacrifice you are willing to make—the price you are ready to pay—while E is uncertain, in order to secure that advantage, when E becomes certain; or, alternatively, the benefit you would demand—the price you would charge—while E is uncertain, in exchange for your commitment to procure that advantage for someone else, when E becomes certain.

This rough description can be made definite if the advantage in question can be assigned a numerical value, say in units of "utility" or simply in money. We speak then of your prevision of an unknown quantity X which is equal to the said value if E obtains, and equal to 0 is E fails to obtain. While E is uncertain, your prevision of X can be readily evaluated if you are willing to play for the prize $c(X - \mathbf{P}(X))$, where $\mathbf{P}(X)$ is a real number to be fixed by you without regard to the identity of your opponent and c is a real number chosen by the latter after you have declared $\mathbf{P}(X)$, it being understood that if the prize turns out to be a positive number, you must disburse it, but you collect it from your opponent if it is negative. $\mathbf{P}(X)$ is then your prevision of X. Note that if $\inf X \leq \mathbf{P}(X) \leq \sup X$, your opponent may bet *for* the occurrence of E by choosing $c > 0$, or *against* it by choosing $c < 0$. On the other hand, whenever you set $\mathbf{P}(X) < 0 = \inf X$, or $\sup X < \mathbf{P}(X)$, you are bound to lose, no matter what happens, if your opponent chooses $c > 0$ in the former case and if he chooses $c < 0$ in the latter.

However, it is unreasonable to assume that you will always be ready to play on such terms for any advantage contingent on an unknown event. As Richard W. Miller (1987, pp. 331–33) pointedly observes, you may suffer from risk aversion. De Finetti has therefore devised a different method for eliciting your prevision of X, which is more likely to work under all circumstances. You are asked to name a quantity $\mathbf{P}(X)$, while E is uncertain, in exchange for a fee to be paid to you in any case when the success or failure of E becomes known. The fee is set at $F - k(X - \mathbf{P}(X))^2$, where F and k are positive real numbers chosen in advance so as to ensure that you will find it worthwhile both to accept the deal and to do your part, viz., to fix $\mathbf{P}(X)$ with great care. $\mathbf{P}(X)$ is then your prevision of X. Note that you secure the maximum fee F if you set $\mathbf{P}(X) = X$. However, since X is unknown while E remains uncertain, you may prefer to minimize your loss $k(X - \mathbf{P}(X))^2$ by choosing $\mathbf{P}(X)$ somewhere between inf X and sup X, although X, by definition, cannot take a value in this open interval but only at either one of its extremes. On the other hand, if you choose $\mathbf{P}(X) < \inf X$, you are bound to make a greater loss, whatever happens, than if you put $\mathbf{P}(X) = \inf X$; and if you choose $\mathbf{P}(X) > \sup X$, you are bound to make a greater loss, whatever happens, than if you put $\mathbf{P}(X) = \sup X$.[42]

Having come this far in the elucidation of prevision, it is an easy matter to define probability. Let χ_E denote the *indicator* or characteristic function of event E; that is to say, $\chi_E = 1$ if E occurs and $\chi_E = 0$ if E fails to occur. The probability of event E, for you, is simply $\mathbf{P}(\chi_E)$, your prevision of the unknown quantity χ_E. De Finetti denotes the indicator of an event E by the same letter that names the event itself and therefore writes $\mathbf{P}(E)$ for what I have just called $\mathbf{P}(\chi_E)$. The **P** that generally stands for 'prevision' can be read, in this special case, as 'probability'. We are now in a position to understand de Finetti's concise characterization of probability.

> The probability $\mathbf{P}(E)$ that You attribute to an event E is therefore the certain gain p which You judge equivalent to a unit gain conditional on the occurrence of E: in order to express it in a dimensionally correct way, it is preferable to take pS equivalent to S conditional on E, where S is any amount whatsoever, one Lira or one million, \$20 or £75.
>
> (De Finetti 1974, vol. 1, p. 75)

Let X be, as above, an unknown quantity contingent on event E, and let E designate also the indicator of that event. Then, obviously, $X = E \sup X$. It can be easily verified that, in either method for evaluating the prevision of X, $\mathbf{P}(X) = \mathbf{P}(E) \sup X$. In other words, $\mathbf{P}(X)$ agrees with the so-called mathematical expectation of the unknown quantity X. Indeed, de Finetti himself says that 'prevision' is just a better word for 'mathematical expectation' (*speranza matematica* in Italian, that is, literally 'mathematical hope'!). However, this may be misleading, for 'expectation' is usually *defined* in terms of 'probability', whereas in de Finetti's theory 'probability' is introduced as a specification of 'prevision', which is taken as a primitive.[43]

A comparable inversion of the habitual procedure can be observed in de

Finetti's treatment of conditional probability. The probability $\mathbf{P}(E \mid H)$ of the event E conditional on the (hypothetical) event H is usually defined by the quotient $\mathbf{P}(E \,\&\, H)/\mathbf{P}(H)$. (Thus, $\mathbf{P}(E \mid H)$ is not defined if $\mathbf{P}(H) = 0$.)[44] But de Finetti's understanding of probability as prevision enables him to replace this conventional definition, which he deems devoid of "substantive meaning" (de Finetti 1972, p. 81), with a naturally motivated one (which also avoids the requirement that $\mathbf{P}(H) \neq 0$): The probability $\mathbf{P}(E \mid H)$ of the event E conditional on the event H is the probability that you attribute to E if you think that your present information will be enriched with one and only one additional item, namely, that H obtains (de Finetti 1974, vol. 1, p. 134). The habitual definition can then be inferred as a theorem from the conditions of coherence discussed below, which lead from probability as prevision to the calculus of probability.

The definition of the probability of an event E for a person Y as Y's prevision of E's indicator has several interesting consequences. In the first place, all probabilities are *single-case* probabilities, assigned to one or the other *individual event.* You may, of course, attribute a probability to an aggregate of such events (e.g., to the fact that among the next n automobile collisions in Colorado, m will involve a drunken driver). You may also, if you feel like it, attribute the *same* probability to *any* event that you regard as belonging to a certain class (e.g., to the death within the first year of life of any newborn Nicaraguan child). But you are by no means constrained to do so. Nor should probability be construed as pertaining to the individual event qua member of some class.

In the second place, assignments are confined to *unknown* events, inasmuch as "an evaluation of probability only makes sense when and as long as an individual does not know the result of the envisaged event" (de Finetti 1937, in Kyburg and Smokler 1980, p. 107). The reason for this is simple enough: prevision is our answer to uncertainty; as soon as the success or failure of an event becomes known, its indicator takes a definite value, 1 or 0, and is no longer a matter of prevision. De Finetti's implicit claim that probability evaluations of known events are nonsense presumably does not apply to truth-functions of unknown events, which, of course, may be known to be true—if tautological—or false—if contradictory—even while their components remain uncertain. And yet, if probability is just a form of prevision, one must indeed ask oneself what meaning can be attached to the prevision of a blatant tautology.[45] Similar questions can be raised with respect to the probability $\mathbf{P}(E \mid \text{not-}E)$ of the event E conditional on its failure, and the probability $\mathbf{P}(E \mid E)$ of the event E conditional on itself. Mathematicians avoid such difficulties by adopting clever conventions. But one may wonder whether this way out is open to de Finetti, who has condemned the practice of tinkering with concepts for the sake of mathematical elegance or convenience (de Finetti 1972, p. 89; cf. de Finetti 1974, vol. 1, pp. 119, 140, and vol. 2, pp. 261ff., 279, 339).[46]

In the third place, *every* unknown event can be assigned a probability. One need not suppose that the event whose probability is being evaluated issued or will issue from a chance setup, conceived in some sense as a physical model of the theory of probability. In sharp contrast with all purportedly objective concepts of probability, which are confined to some specific domains of reality, de Finetti's subjective concept is equally relevant to all fields of life. This gives a semblance of truth to his allegation that quantitative probability, thus understood, only refines and makes precise the ordinary commonsense notion of the probable that we traced back, in Section 4.1, to Cicero's Latin translation of πιθανός. The man in the street will give or take odds on all sorts of events, not only on the numbers, say, that may be obtained in craps or roulette with carefully built contraptions that are the classical paradigm of chance setups, but also on the outcome of football games and presidential elections, which lack a definite physical basis for the assignment of probabilities. There is only one limitation to the scope of betting, and it is that bets must be decidable: when the game is over, there ought to be no doubt as to who won and who lost. De Finetti emphasizes that the same restriction applies to the domain of probability. Here is his definition of *random* (as in 'random event', 'random function', etc.):

> Let us make clear the meaning that we give to 'random': it is simply that of 'not known' (for You), and consequently 'uncertain' (for You), but *well-determined* in itself. [. . .] To say that it is *well-determined* means that it is *unequivocally individuated.* To explain this in a more concrete fashion: it must be specified in such a way that a possible bet (or insurance) based upon it can be decided without question.
>
> (De Finetti 1974, vol. 1, p. 28)

For the proper handling of probability as a real-valued function, the well-determined events that are to be its arguments should form together a well-determined domain of definition. At first sight it does not seem possible that de Finetti's approach can meet this requirement. His subjective probability function would have to be defined on the aggregate of everything that might be the case, and it is all too obvious that there is no such aggregate. Uncertain events can only be identified by description, and we do not possess a stable all-encompassing system of concepts by which to describe anything we may wish to single out as a particular event. The best we have for this purpose, the theories of mathematical physics, are each restricted to a specific domain of reality and are not always mutually consistent where their domains overlap. De Finetti must have been rather insensitive to the role of concepts in the articulation of facts, or else he would not have found in Hume's writings "the highest peak that has been reached by philosophy" (Kyburg and Smokler 1980, p. 115) nor spoken so patronizingly of "poor Kant" (de Finetti 1974, vol. 2, p. 201; cf. vol. 1, p. 22). However, since he maintains that "probability is the result of an evaluation" and that "it has no meaning until the evaluation has been made" (de Finetti 1974, vol. 1, p. 145), he will not "pretend that [the probability] **P** could actually be imagined as determined, by any individual, for *all* events (among which those mentioned or thought of during the whole existence of the human race only constitute an infinitesimal fraction, even though an immense number)." On the contrary, he invites us to assume "at each moment, and in every case," that **P** is "defined or known for all (and only) the random quantities (or, in particular, events) belonging to some completely arbitrary set χ" (de Finetti 1974, vol. 1, p. 84).[47]

For de Finetti a probability is a real number assigned to an event by a person, at her own discretion. The information available to this person may indeed inspire or guide her choice of that number in some obscure fashion that de Finetti never quite succeeds in explaining, but it does not in any way determine it. On the other hand, there are practical reasons that impose constraints on a person's probability assignments, if she actually heeds those reasons. De Finetti believed that someone who did not heed them would behave incoherently, which is why he calls them *conditions of coherence.*[48] I have already explained the two methods proposed by de Finetti for eliciting a person's previsions. Each method involves a different condition of coherence, but the constraints imposed on previsions—and thus on probabilities—by either condition are exactly the same. For simplicity's sake let us concentrate on probabilities. The unknown quantity X, conditional on event E, mentioned in the presentation of both methods, will therefore be equated with the indicator χ_E of E. Following de Finetti, I designate χ_E by E. It takes the value $\sup E = 1$ if E occurs and the value $\inf E = 0$ if E fails to occur.

Suppose now that your probability assignments are evaluated by the method of betting described above (de Finetti's first method of eliciting previsions). In this case, the condition of coherence takes the following shape: You must assign a probability $\mathbf{P}(E)$ to each event E under consideration in such a way that it is not certain that you will suffer a loss on every possible outcome. Or, as de Finetti expresses it:

> It is assumed that You do not wish to lay down bets which will with *certainty* result in a loss for You. A set of your previsions is therefore

> said to be *coherent* if among the combinations of bets which You have committed yourself to accepting there are none for which the gains are *all uniformly negative.*
>
> (De Finetti 1974, vol. 1, p. 87)

When I explained de Finetti's first method for eliciting previsions I noted that the condition of coherence just stated imposes an upper and a lower bound on their range. For an arbitrary event E, this constraint reads as follows:

$$0 \leq \mathbf{P}(E) \leq 1 \tag{1}$$

Now consider two mutually incompatible events E_1 and E_2. The condition of coherence implies that

$$\mathbf{P}(E_1 \vee E_2) = \mathbf{P}(E_1) + \mathbf{P}(E_2) \tag{2}$$

(where '$E_1 \vee E_2$' designates the event that occurs when either E_1 or E_2 obtains and fails to occur when neither E_1 nor E_2 obtain). To prove it, put $E = E_1 \vee E_2$, and let c, c_1, and c_2 be the coefficients chosen by your opponent when betting on E, E_1, and E_2, respectively. If your probability assignments do not satisfy eqn. (2), we have that $\mathbf{P}(E) - \mathbf{P}(E_1) - \mathbf{P}(E_2) = r$ for some real number $r \neq 0$. You will suffer a loss, no matter what happens, if your opponent puts $c_1 = c_2 = -c$, and chooses $c > 0$ if $r > 0$ and $c < 0$ if $r < 0$.[49] It is clear also that if E' denotes the negation or complement of an event E, the condition of coherence entails that

$$\mathbf{P}(E \vee E') = 1 \tag{3}$$

For, by (1) and (2), $\mathbf{P}(E) + \mathbf{P}(E') = \mathbf{P}(E \vee E') \leq 1$. Hence, if you set $\mathbf{P}(E \vee E') < 1$, your opponent can make you suffer a loss at any event by choosing the same factor c when betting on E, E', and $E \vee E'$. (E will occur or will fail to occur; in either case you pay $2c(1 - \mathbf{P}(E \vee E'))$.)

If your probability assignments are elicited by de Finetti's second method—that is, through your participation, in the manner explained earlier, in setting the penalty to be deducted from the fee offered to you for doing just this—the condition of coherence is as follows: You must assign a probability $\mathbf{P}(E)$ to each event E under consideration in such a way that the said penalty is not greater, after every possible outcome, than it would be if you chose a different assignment of probabilities. In de Finetti's words:

> It is assumed that You do not have a preference for a given penalty if You have the option of another one which is *certainly* smaller. Your set of previsions is therefore said to be *coherent* if there is no other possible choice which would certainly lead to a uniform reduction in your penalty.
>
> (De Finetti 1974, vol. 1, p. 88)

In note 42, I proved that this condition of coherence entails the inequality (1). It can be readily seen that it also implies (3). For, generally speaking, if T designates the indicator of a tautology such as $E \vee E'$, your fee for estimating $\mathbf{P}(T)$ will in any event be subject to a penalty proportional to $(1-\mathbf{P}(T))^2$. Such a penalty is minimized by setting $\mathbf{P}(T) = 1$. The derivation of (2) from the second condition of coherence is more involved and will not be given here.[50]

The requirements (1), (2), and (3) that previsions must meet in order to satisfy de Finetti's conditions of coherence agree precisely with the properties that define a probability function on a finite probability space (Section 4.2). The inequality (1) gives the postulated range of such functions, and equations (2) and (3) are equivalent to the axioms FPS 2 and FPS 1, respectively. Without this result—independently reached by F. P. Ramsey and de Finetti in the 1920s—the subjectivist philosophy of probability would not be viable. De Finetti argued persuasively that personal probability, as elicited by his methods, does indeed quantify the prescientific concept of the probable (described in Section 4.1). But this alone could not have gained him a respectful hearing. He had to show that personal probabilities, when constrained by the conditions of coherence, provide a model of the structure created by the mathematical theory of probability. For this structure—through its multiple applications—embodies the meaning of 'probability' in contemporary science and philosophy.

There is still one difficulty I ought to mention here. The conditions of coherence do not furnish a straightforward justification for the Kolmogorov axiom KPS 2, that is, the requirement of complete additivity usually prescribed for probability functions on infinite sample spaces. E. W. Adams (1962) has shown how, with some ingenuity, KPS 2 can indeed be vindicated by the conditions of coherence. But de Finetti himself rejects KPS 2, which, in his opinion, has been adopted only on despicable grounds of mathematical expediency. I have already mentioned one of the disadvantages of complete additivity, viz., that under the usual set-theoretical axioms it rules out the assignment of a probability to every subset of an infinite set. This necessitates the restriction of probability functions to some proper part of an infinite set's power set. According to de Finetti, such a restriction would be philosophically unjustifiable. He also points out that complete additivity would preclude a uniform distribution of probabilities over a countably infinite sample space. Consider, for example, a lottery with one ticket for every natural number. If the lottery is fair, each ticket has the same probability of being drawn, viz., 0. But then, if complete additivity is required, the probability of drawing any ticket at all must also be 0.[51]

I should emphasize that—contrary to what de Finetti's words might suggest—we are under no logical compulsion or moral obligation to obey the conditions of coherence. In certain circumstances—e.g., if you happen to be a benevolent grandmother playing chess with a five-year-old grandson, or a female professor wary of the envy of your untalented male colleagues—you may prefer to lose at any event, or you may aim for a less-than-optimal fee. You may also refuse to play or to estimate penalties on de Finetti's terms. However, there are persons, both single and incorporated, whose avowed business it is to make just the sort of deals that de Finetti describes, and who manifestly accept the conditions of coherence. Surely it was not by accident that the probability calculus was initially developed in answer to their needs. I do not refer only to the gamblers who proposed problems to Galileo and to Pascal and Fermat, but also to the Dutch Republic, whose business of selling annuities prompted some pioneering work by Hudde and de Witt (Hacking 1975, Chapter 13). For the like of them, the calculus provides in effect what de Finetti has called a *logic of uncertainty*, that is, a set of rules for controlling the coherence—in the stated sense—of probability assignments and for computing further probabilities, coherent with those already assigned.

If probability exists only in the eye of the beholder, by virtue of his subjective previsions, it makes no sense to speak of 'unknown probabilities' or to advance hypotheses as to their probable value. This conclusion flies in the face of current scientific practice. We are all acquainted with hypotheses to the effect that, with less than, say, .05 probability of error, the probability of contracting illness *A* under circumstances *B* is such-and-such. According to de Finetti this manner of speech is meaningless. For Jill, a statistician, the probability that Jack, a chain-smoker, will develop lung cancer within the next 10 years has whatever value she thinks fit to ascribe to it, and that value is known to Jill and to anyone else to whom she reports it, without any error or uncertainty. We may, indeed, conjecture—and also bet on—how much higher Jill's true estimate is than her declared value; for instance, if she is in the pay of the tobacco lobby. But the uncertainty concerns here a well-determined, though to us unknown, fact in Jill's life, not a pretendedly stochastic property of the interaction of Jack's lungs with cigarette smoke.

On the other hand, de Finetti professes to show that the familiar talk of 'unknown probabilities' and their conjectural values, though philosophically inept, is somehow justifiable as a *façon de parler* about uncertain events, inasmuch as it can be accurately translated into the subjectivist idiom at least in one important family of cases. The paradigm of this family is a game of heads or tails with an untried, presumably unfair coin. One normally assumes that, no matter what the coin's peculiar bias might be, it will not change easily, so that the unknown probability of making heads on any trial can be viewed as constant. Thus the outcome of the earlier throws will exert no influence whatsoever on the chance of heads in subsequent throws, and the game can be regarded as a series of independent trials on a Bernoullian setup (hereafter, *Bernoulli trials*). Given the outcome of a long series of throws, one may therefore use Bernoulli's Theorem to estimate the likelihood of any particular hypothesis concerning the unknown probability of heads. As the number of throws increases beyond all bounds, the likelihood becomes vanishingly small that the probability of heads does not lie arbitrarily close to their observed relative frequency.[52] De Finetti dismisses this whole approach. He contends that if the throws were all mutually independent, it would be impossible to learn from experience, i.e., to improve one's estimate of the probability of heads by trying the coin. By definition, the throws are thus independent if and only if, for any positive n, the unconditional probability of heads on the $(n + 1)$th throw is equal to the probability of heads on the $(n + 1)$th throw conditional on any particular outcome of the previous n throws. But, if such were the case, your prevision of heads ought to remain unchanged, no matter how the coin behaved; whereas in real life, after obtaining 800 heads in 1,000 consecutive throws, you would surely reject the usual assignment of $\mathbf{P}(\text{heads}) = .5$. This argument will not impress objectivists, for it rests squarely on the identification of the probability of heads with your personal estimate of it. For an objectivist, it is only the latter that will change with experience, as you reach for the unknown probability, which may well comply with the requirement of independence and be the same on every throw, regardless of the outcome of past throws. But de Finetti is not trying here to refute objectivism, which, in his view, still has to make its case plausible. His aim is rather to show that we can safely dispense with the objectivist reading of the game of heads or tails in terms of unknown probabilities, and yet reap the fruits of all the work done on this important model. For this purpose, it is not necessary to assume that the successive throws of the coin are mutually independent but only that they are exchangeable, in the sense I shall now define.

Let E_i stand, as usual, for an individual event (e.g., 'heads on the ith throw') and also for its indicator. $\mathbf{P}(Q)$ denotes the probability of the event designated or the fact described by Q. The events $E_1, \ldots, E_n$ $(n \geq 1)$ are said to be *exchangeable* if and only if, for every list $\langle e_1, \ldots, e_n \rangle$ of zeroes and ones and every permutation σ of $\{1, \ldots, n\}$,

$$\mathbf{P}(E_1 = e_1, \ldots, E_n = e_n) = \mathbf{P}(E_1 = e_{\sigma 1}, \ldots, E_n = e_{\sigma n}) \tag{4}$$

An infinite sequence of individual events $E_1, E_2, \ldots$ is exchangeable if and only if all finite subsequences drawn from it are exchangeable. Exchangeability is strictly weaker than mutual independence (i.e., the latter entails it but is not entailed by it) but is nevertheless a rather stringent condition. Thus, for example, the sequence $E_1, E_2, \ldots$ is exchangeable if and only if, for every fixed positive integer n, the probability $\mathbf{P}(E_{i(1)} = 1, \ldots, E_{i(n)} = 1)$ of obtaining n successes in n trials is the same for each list of n events from the sequence (de Finetti 1937, in Kyburg and Smokler 1980, p. 81). If the throws of our coin are exchangeable, the probability of obtaining k heads in a series of n throws (for any positive integers k and n) is the same for every list of places that the k heads might take in the series and therefore depends on n and k alone (see eqn. (8)). The concept of exchangeability, as I have defined it, makes good sense and is important also from an objectivist standpoint (see, e.g., Dawid 1985). On de Finetti's subjectivist understanding of probability, a finite set or an infinite sequence of individual events can, of course, be said to be exchangeable only with respect to—and by virtue of—someone's previsions.[53]

According to de Finetti, in the statistical analysis of the game of heads or tails—or of any comparable situation—the burden usually placed on the assumption of mutual independence can be carried in full by the weaker assumption of exchangeability. This conclusion follows from his celebrated Representation Theorem, as I shall now try to explain.[54] Let $E_1, E_2, \ldots$ be the indicators of a sequence of exchangeable events. Let S be the set of all mappings of the positive integers into $\{0,1\}$ (all infinite sequences of zeroes and ones). S gathers then all the alternative values that the said sequence of indicators can take. A probability function $\mathbf{P}$ on the power set[55] $\mathcal{P}(S)$ will be said to be *Bernoullian*, with parameter q, if and only if, for all positive integers h and k, $\mathbf{P}(E_h = 1) = \mathbf{P}(E_k = 1) = q$, and $\mathbf{P}(E_1 = 1, \ldots, E_k = 1) = q^k$. A Bernoullian function $\mathbf{P}$ is completely determined by its parameter q and may therefore be designated by $\mathbf{P}_q$. Obviously, $\mathbf{P}_q(E_i = 0) = (1 - q)$ for each $i = 1, 2, \ldots$, and

$$\mathbf{P}_q(E_1 = 1, \ldots, E_k = 1, E_{k+1} = 0, \ldots, E_n = 0) = q^k(1 - q)^{n-k} \tag{5}$$

for all positive integers n and $k \leq n$. Since (5) holds also for every permutation of the indices $\{1, \ldots, n\}$, it is clear that

$$\mathbf{P}_q\Big(\sum_{i=1}^{n} E_i = k\Big) = \frac{n!}{k!\,(n-k)!}\, q^k (1-q)^{n-k} \tag{6}$$

De Finetti's Representation Theorem says that, if $E_1, E_2, \ldots$ and S are as stated above, *any* probability function **P** on $\mathcal{P}(S)$ is equal to a so-called mixture of Bernoullian functions.[56] More accurately, what the Representation Theorem says is that for every such probability function **P**, there is a normed distribution function F on the real line, concentrated on [0,1],[57] such that, for any positive integers n and $k \leq n$,

$$\mathbf{P}(E_1 = 1, \ldots E_k = 1, E_{k+1} = 0, \ldots E_n = 0) = \int_0^1 q^k(1-q)^{n-k}d\mathsf{F}(q) \quad (7)$$

and

$$\mathbf{P}\left(\sum_{i=1}^{n} E_i = k\right) = \frac{n!}{k!(n-k)!}\int_0^1 q^k(1-q)^{n-k}d\mathsf{F}(q) \quad (8)$$

Moreover,

$$\mathsf{F}(x) = \lim_{n\to\infty}\mathbf{P}(n^{-1}\sum_{i=1}^{n} E_i \leq x) \quad (9)$$

That is to say, for any real number x, the probability—as evaluated by **P**—that the relative frequency of successes ($E_i = 1$) among the first n events in the sequence is not greater than x converges to the limit $\mathsf{F}(x)$ as n grows beyond all bounds.

Exchangeability evidently entails that eqns. (7)–(9) also hold if any list of n distinct integers is substituted for the indices $1, \ldots, k, k+1, \ldots, n$, on the left-hand side of (7), or if any list of n distinct indicators $E_{i_1}, \ldots, E_{i_n}$ is substituted for the summands on the left-hand sides of (8) and (9). Bearing this in mind as we compare (7) and (8) with (5) and (6), we see at once that

$$\mathbf{P} = \int_0^1 \mathbf{P}_q d\mathsf{F}(q) \quad (10)$$

The arbitrary probability function **P** is therefore equal to a mixture of Bernoullian functions whose parameters take all admissible values. As I indicated in note 57, the normed distribution F determines a unique function M on the Borel sets of [0,1], which has all the properties of a probability function (in Kolmogorov's sense).[58] For any real number x in [0,1], $\mathsf{F}(x) = \mathsf{M}([0,x]) = \mathsf{M}(\{q \mid q \leq x\})$. For any set D in the domain of M,

$$\mathsf{M}(D) = \int_D d\mathsf{F}(q).$$

Let us now go back to the objectivist analysis of the game of heads or tails with an untried coin. The successive tosses of the coin are conceived as Bernoulli trials with unknown parameter q. Let E_i be the indicator of heads on the ith toss. The probability of n heads in n tosses is the same, viz., q_n, no matter what tosses are chosen for consideration. Thus, for the objectivist, the tosses form a sequence of exchangeable events. While q remains unknown, the probability P assigned by him to the outcome of one or more tosses will be a mixture of Bernoullian functions weighted by the probabilities he ascribes to the several admissible values of the parameter q. Suppose, for example, that he knows that the coin has been picked at random from a pile of 100 coins of known bias, 70 of which yield heads with probability r and 30 of which yield heads with probability s. Then, $P = .7P_r + .3P_s$. On the other hand, if the Bernoullian parameter q lies with the same probability in any interval of equal length (in [0,1]), $\mathbf{P} = \int_0^1 \mathbf{P}_q d\mathsf{F}(q)$. In the most general case, the probability that q lies in a particular interval will be given by a normed distribution F, concentrated on [0,1], such that $\mathsf{F}(x)$ is the probability that $q \leq x$ (and therefore $\mathsf{F}(v) - \mathsf{F}(u)$ is the probability that $u < q \leq v$). Thus the objectivist's probability assignment to the outcomes of the game of heads or tails is well represented by eqn. (10). This is formally justified by the Representation Theorem: Any coherent assignment of probabilities to exchangeable events will take this shape. Hence, all the mathematical consequences of the objectivist analysis are valid and stand at the subjectivist's disposal. But, according to de Finetti, the objectivist reading of (10) is sheer nonsense. In his view, no meaning can be attached to the notion of an unknown Bernoulli parameter q, or to a "statistical hypothesis" F regarding the distribution of objective chances among the possible alternative values of q. Objectivist talk is for him a metaphysically pompous, though mathematically innocuous, manner of speaking about subjective previsions of exchangeable events.

The objectivist, however, sees the Representation Theorem in a different perspective, not as a *reduction* of the right-hand side of (10) to the *arbitrary* assignment of probabilities to exchangeable events named on the left-hand side, but as an *analysis* of the latter as a combination of Bernoullian functions weighted by a distribution of probabilities over the possible values of the characteristic parameter. Any meaning implicit in the less articulate side of the equation is shared and explicated by the other side. Hence, by merely grasping a class of events as exchangeable you are being led to conceive them after the classical statistical model of Bernoulli trials of unknown probability.[59]

The Representation Theorem bestows an objective meaning on this model if, but only if, exchangeability is an objective property of events. (Then, but only then, will objectivity flow, so to speak, through the sign of equality, from the left to the right of (10).) But this is precisely what de Finetti denies. Events are exchangeable if their probabilities satisfy certain equations, and to him probabilities are no more than previsions. However, while one can easily figure out certain sets of previsions under which some events turn out to be exchangeable, one can hardly base exchangeability on prevision where no definite previsions have been made. How, for example, can you assign to every series of n tosses with an untried coin exactly the same chances of totalling, say, k heads before even attempting to evaluate those chances? How, indeed, if not on the strength of your perception of the objective conditions of the game? Of course, de Finetti acknowledges that subjective previsions depend—in some unspecified way—on available information. But should your probability assignments be determined or constrained, at least in some cases, by your grasp of facts, then, to the extent of that determination or constraint, they would lay claim to objectivity. Thus, the following passage by de Finetti would indicate that he was not altogether steadfast in his rejection of objective probability:

> It is *intuitively obvious* that drawings from an urn with unknown composition are exchangeable (eg. an urn containing an unknown number of black and white balls, with the standard method of drawing with replacement). The same applies to tosses of a possibly asymmetric coin, and, more generally, to all those cases which are commonly referred to as 'repeated trials with a constant but unknown probability of success'. It is less obvious, but none the less *true*, as we shall see, that we still have exchangeability in the case of drawings without replacement, or with double replacement.
>
> (De Finetti 1974, vol. 2, pp. 211–12; my italics)

Can a statement be true, and obvious, if it is not objective? Indeed, de Finetti might here be putting forward a purportedly objective claim concerning our subjective makeup, viz., that in making decisions under risk about events that—to an uncommitted observer—fall under the given description, everyone will treat those events as exchangeable. But would such a claim be warranted? Is it "intuitively obvious" that everyone will act in this way? Can it be "true" that such a behavior will always be forthcoming even in those cases in which de Finetti grants that this is not quite so obvious? Rather than attributing to de Finetti such an implausibly facile psychological generalization, I prefer to read the above passage as saying that, whenever an objective situation is grasped under one of the stated descriptions, then, by virtue of that very grasp, any coherent assignment of probabilities to the relevant events issuing from that situation must meet the condition of exchangeability. This reading respects de Finetti's interpretation of probability as prevision and yet extracts from him the concession that in the cases mentioned, the ascription of exchangeability is necessitated by—and is therefore apt to disclose—our judgment of the real prospects of the situation.

It is a corollary of de Finetti's Representation Theorem that on *every* assignment of probabilities to an exchangeable sequence of events, the probability of success on the $(n + 1)$th event, conditional on the relative frequency of successes in the preceding n events being k, converges to k as n increases beyond all bounds (de Finetti 1937, in Kyburg and Smokler 1980, pp. 102ff.; de Finetti 1974, vol. 2, pp. 218ff.; Savage 1954, pp. 54–55).[60] This important result is the basis of de Finetti's claim that his "subjectivistic theory solves the problem of induction completely in the case of exchangeability" (de Finetti 1937, in Kyburg and Smokler 1980, p. 105). According to him, it explains why different statisticians sharing the same large pool of data form similar previsions and, generally, how "we learn from experience." Not, indeed, because we correct our previsions in the light of incoming information. "Nothing can oblige one to replace one's initial opinion [i.e., prevision], nor can there be any justification for such a substitution" (de Finetti 1974, vol. 2, p. 211). But as more and more events become known, our initial previsions *conditional* on such events must replace our unconditional initial previsions.[61] Therefore, wherever we have to do with exchangeable events, "a rich enough experience leads us always to consider as probable future frequencies and distributions close to those which have been observed" (de Finetti 1937, in Kyburg and Smokler 1980, p. 102). Note, however, that all such "learning" rests on the learner's grasping the future events as belonging to the same exchangeable sequence as the events observed. Prevision can benefit from information only if linked to it by concepts that provide a scheme for describing the realm of the former in the same terms employed for conveying the latter. By thus creatively projecting the past onto the future, our understanding extends the domain of objectivity. Note further that capricious or unwise projections are dearly paid for. Imagine, for instance, a bookie who believes that American presidential elections are exchangeable with the tosses of a silver dollar. What would have been the cost for him of laying odds on a Democratic victory in 1988 based on the outcome of 10,000 throws of such a coin? Learning from experience in de Finetti's way will not teach one which individual events are exchangeable with which, nor what schemes of description are ripe for revision.

For a sick man it is cold comfort to hear that the harsh medical treatment he must undergo has a limiting frequency of success of .95. What worries him is his individual probability of survival. De Finetti assures him that it makes good sense to inquire about it, provided that he expects no more than a subjective prevision in reply. But then, why ask the doctor, and not rely on one's own feelings? Indeed, if both doctor and patient regard the latter's case as exchangeable, their previsions, enlightened by medical statistics, will, if coherent, tend to agree with each other—and with the relative frequency of successes observed among such cases. But exchangeability is in turn a matter of prevision, and one on which statistics will throw no light. For in order to tell what statistics are relevant to a given case one must first decide with what other cases it is exchangeable. It may well be that subjectivism is the only sensible approach to medical probabilities. Even if we view a surgical operation as a trial on a chance setup, in the objectivist sense, its circumstances often are so peculiar that the distribution of chances is anyone's guess. But this is not ground enough for banishing real chances from human discourse, or from the world as described by it. I, for one, am reluctant to admit that if I hold 999,999 out of 1,000,000 tickets in a lottery in which lots will be drawn in the regular way, I shall not be closer *in point of fact* to winning the prize than if I held just one ticket. To measure and compute this kind of factual closeness we need a concept of objective probability.

4.6 Probability as a physical propensity

In April 1957, Karl R. Popper, at the time the foremost philosopher of science in Europe, communicated to a gathering of scientists and philosophers in Bristol that he no longer endorsed the frequentist philosophy of probability put forward in his *Logik der Forschung* (1935), for he now considered probability to be a physical property, a disposition or *propensity* of experimental arrangements.[62] In Popper's words:

> Every experimental arrangement is *liable to produce*, if we repeat the experiment very often, a sequence with frequencies which depend

> upon this particular experimental arrangement. These virtual frequencies may be called probabilities. But since probabilities turn out to depend upon the experimental arrangement, they may be looked upon as *properties of this arrangement. They characterize the disposition, or the propensity,* of the experimental arrangement to give rise to certain characteristic frequencies *when the experiment is often repeated.*
>
> (Popper 1957, p. 67)

In the discussion that followed, W. B. Gallie remarked that, although Popper could not actually explain the notion of probability merely by using the word 'propensity', "he may have given us the beginning of a philosophical exposition" of this notion, by emphasizing that "probabilities are bona fide elements of reality—that they just do occur and they are as real properties of processes as are ordinary properties, although they are only displayed through a long sequence of events" (Körner 1957, p. 81). R. B. Braithwaite recalled that Popper's approach had been anticipated by C. S. Peirce when he wrote in 1910 that "the statement that the probability, that if a die be thrown from a dice box it will turn up a number divisible by three, is one third [. . .] means that the die has a certain 'would be'; and to say that a die has a 'would be' is to say that it has a property, quite analogous to any habit that a man might have" (Peirce, CP, 2.664; cf. 8.225f.).[63]

Popper has explained that he was moved to revise his views on probability as he reflected about Quantum Mechanics. Specifically, it was the consideration of the notorious two-slit thought-experiment which ultimately led him to see probabilities as propensities.

> It convinced me that probabilities must be 'physically real'—that they must be physical propensities, abstract relational properties of the physical situation, like Newtonian forces, and 'real', not only in the sense that they could influence the experimental results, but also in the sense that they could, under certain circumstances (coherence), interfere, i.e. interact, with one another.
>
> (Popper 1959b, p. 28)

But Popper's decisive argument against frequentism can be illustrated with the simplest and oldest games of chance.

> Let us assume that we have a loaded die, and that we have satisfied ourselves, after long sequences of experiments, that the probability of getting a six with this loaded die very nearly equals 1/4. Now consider a sequence *b*, say, consisting of throws with this loaded die,

> but including a few throws (two, or perhaps three) with a homogeneous and symmetrical die. Clearly, we shall have to say, with respect to each of these few throws with this fair die, that the probability of a six is 1/6 rather than 1/4, in spite of the fact that these throws are, according to our assumptions, *members of a sequence* of throws with the statistical frequency 1/4.
>
> (Popper 1959b, pp. 31–32)[64]

The objection has no force if the sequences whose convergent relative frequencies define probabilities are given, not extensionally, by listing, but intensionally, by a set of generating conditions. This, says Popper, "only states explicitly an assumption which most frequency theorists (myself included) have always taken for granted" but which involves in effect a switch "from the frequency interpretation to the propensity interpretation" (Popper 1983, p. 355). Under the said assumption, the probability distribution must be seen as a property of the generating conditions, which transmit it to every single case generated under their sway, and *therefore* to every sequence of such cases.

While most scientists continue to profess—or, at any rate, to pay lip service to—some often nebulous version of frequentism, Popper's understanding of objective probability as a propensity has been generally taken up by philosophers who will not subscribe to de Finetti's drastic subjectivism. They have, of course, developed several variants of it. Variations may concern what probability is a propensity *of* (viz., a physical object—e.g., a die—or a process or procedure—e.g., dice throwing); what it is a propensity *for* (viz., for producing an event with certain features, or an infinite sequence of events with such features); and what logical or ontological category it belongs to (is propensity a kind of disposition or tendency, or does it constitute a category sui generis?). To illustrate these differences and to motivate a choice among them, I shall now briefly examine the conceptions of probability as propensity put forward by Bas van Fraassen (1979; 1980, Chapter 6), Ronald N. Giere (1973, 1976a), D. H. Mellor (1971), and Isaac Levi (1967, 1977a, 1980).

On the face of it, van Fraassen's "modal frequency interpretation" of probability could be taken as a faithful elucidation of Popper's meaning. "In slogan formulation," the modal frequency interpretation says that the probability of an event "equals the relative frequency with which it would occur, were a suitably designed experiment performed often enough under suitable conditions" (van Fraassen 1980, p. 194). The latter subjunctive conditional is clearly equivalent to a statement in the indicative mood describing a factual property of the arrangement for that "suitably designed experiment," namely, its disposition to produce the event in question with the said relative frequency in the long run. This is precisely what probability statements are supposed to say, according to the first quotation from Popper given above.[65] On the other hand, one may well wonder whether frequentism, as intended by its creators, could in the end amount to anything very different from van Fraassen's slogan formulation. For surely, when von Mises equated the probability p of a certain event A with the limiting relative frequency of A's in a suitable collective K, he did not imply that the infinite sequence K must exist *in rerum natura* lest the statement that A has probability p be false or only vacuously true. He knew full well that no gambling device will conserve its original probability distribution after being used, say, a trillion times, and that men will not be around forever to restore it or replace it. It would appear, therefore, that in his mind the distribution could only refer to the limiting frequencies which the alternative outcomes *would* approach if, contrary to fact, the game *were* to continue indefinitely under constant conditions. Of course, his positivistic commitment to extensionality precluded the explicit use of the potential and subjunctive moods. But in the transition from formal syntax to live talk his if-then statements must have implicitly regained their ordinary force.

By equating probabilities with limiting frequencies, frequentism goes beyond the implications of the mathematical theory of probability. In the thus strengthened theory, the subtle link between single-case probabilities and probabilities of relative frequencies, laboriously established in the mathematical laws of large numbers, becomes a trivial corollary. (If the single-case probability of A is *equated* axiomatically with the limiting relative frequency of A's it is no wonder that this equation should hold with probability 1—it is indeed not just practically but logically certain!) Models of the frequentist theory of probability satisfy, of course, the weaker standard theory, just as a straight line, say, satisfies the equation of a conic. This remark applies not only to the earlier forms of frequentism but also to van Fraassen's postpositivist version of it. For here, again, the equality between probability and limiting relative frequency is necessarily true, not just overwhelmingly probable, as in the original mathematical theory.

In contrast to von Mises, van Fraassen does not include this equality among the axioms of probability, which he takes unchanged from Kolmogorov. It follows, however, from his definition of the intended scientific models of probability theory, the *good families of ideal experiments*. Such a *good family* (van Fraassen 1979, pp. 370–71; 1980, p. 193) is essentially a collection of infinite sequences of items drawn from a non-empty set of "possible outcomes," partitioned into countably many types, in ways that may differ from one sequence to another (and also for a given sequence).[66] Let σ_ρ be such a sequence associated with a partition G_ρ. We say that a class of outcomes is *relevant* to σ_ρ if it belongs to the Borel field generated by G_ρ.[67] By condition

(iii) of van Fraassen's definition, in every sequence belonging to a good family the relative frequency of outcomes of a given relevant class converges to a limit. By condition (v), the relative frequency of outcomes of a class that is relevant to two sequences of the family converges in both of them to the same limit. Van Fraassen has no difficulty in showing that $\langle K,F,P\rangle$ is a probability space if (α) K is the set of possible outcomes of a good family of ideal experiments (meeting, besides the conditions already mentioned, two additional requirements that we need not state here), (β) F is the union of the Borel fields of relevant classes of outcomes in K (relevant, that is, to one or more sequences of outcomes in the family), and (γ) P is the function that assigns to every element of F its limiting frequency in the family of sequences to which it is relevant. He also shows that if $\langle K,F,P\rangle$ is an arbitrary probability space, there is a maximal good family E relative to which K, F, and P satisfy the foregoing descriptions. Such a good family E is constructed as follows: for every countable measurable partition G of K, included in F, E contains every infinite sequence σ of elements of K in which the relative frequency of any $A \in G$ converges to a limit equal to $P(A)$. Thus, a natural one-to-one correspondence is established between the probability spaces and van Fraassen's good families of ideal experiments. Of course, the good family E assigned by this correspondence to a given probability space with sample space K is only a proper subset of the uncountable collection $\mathbf{G} \times K^{\infty}$ of all infinite sequences of elements of K, indexed by the set $\mathbf{G}$ of countable measurable partitions of K included in F. To prove that such a good family E is not empty, van Fraassen invokes the strong law of large numbers.[68] Think now of a simple application of the theory of probability, involving some series S of past and future physical events that we choose to regard as trials on one or more Bernoullian setups with the same parameter q. As usual, the possible outcomes are classified into favorable (+) or unfavorable (–). According to the modal frequency interpretation this manner of regarding S consists in conceiving it as a fragment—in effect, a random selection (van Fraassen 1979, p. 366)—of an infinite sequence σ belonging to a good family E of ideal experiments.[69] Let $\langle K,F,P\rangle$ be the probability space associated with E. Without any significant loss of generality, we may take E to be the maximal good family associated with $\langle K,F,P\rangle$. Then, in the modal frequency view, the said manner of regarding the series S commits us to the hypothesis that as S is indefinitely extended the proportion of positive outcomes will converge to the limit $P(+) = q$. The probability that we might be wrong in accepting or in rejecting this hypothesis can be assessed by the standard methods of statistical testing, thanks to the assumption—very aptly introduced by van Fraassen—that if S is a part of σ, it is a *randomly* selected part.

The modal frequency interpretation makes, of course, no allowance for the possibility that S might be properly seen as a series of Bernoulli trials with parameter $P(+)$ and yet appear, when indefinitely extended, to be part of a sequence in which the proportion of favorable outcomes does not converge to $P(+)$. Now, in the straightforward traditional understanding of probability this possibility is, no doubt, extremely unlikely—such sequences constitute at most a countable subset of the uncountable set of all sequences of trials of Bernoulli trials with parameter $P(+)$ in which we have chosen to locate the ideal extension of S to infinity—but it remains open, all the same. It is true, indeed, that in actual practice the statistical evidence that the proportion of outcomes of a certain kind is not converging to the probability assigned to that kind is sufficient ground for rejecting the conjectured probability distribution. Nobody in his right mind will continue to assume that a coin is fair and is being tossed without cheating if tails have turned up 1,000 times in a row. This might seem to favor the modal frequency view of the matter. However, such statistical evidence can only show that it is extremely *improbable* that the observed series of outcomes constitutes a random selection from an infinite sequence of trials on a chance setup featuring the distribution in question; and the likelihood of this hypothesis given the evidence is not significantly smaller if such an infinite sequence is required to meet van Fraassen's criteria for belonging to a good family of ideal experiments, instead of being just *any* sequence of Bernoulli trials with the said distribution. Thus, the rejection of a statistical hypothesis on the strength of evidence such as that described above is no less justifiable from the traditional than from the modal frequency standpoint. On the other hand, there is a powerful reason for preferring the former to the latter, which I shall now explain.

Once again, let us countenance the hypothesis—call it H—that a certain series S of past and future events consists of independent trials on chance setups modelled by the probability space $\langle K,F,P\rangle$. Note that the past and known and the future or unknown parts of S each bears an essential, yet an essentially different, relation to H. H must be assessed in the light of the former but is proposed only with a view to the latter. Aside from the guidance it may afford to our expectations regarding the unknown, the assignment of probabilities to known events is utterly pointless. Therefore, H must presuppose a link between the known and the unknown terms of S. The existence of such a link is, of course, implicit in the assumption, basic to H, that each term of S was or will be the outcome of a physical process arranged—by man or by nature—to produce certain results (collected in K and classified by F) with certain probabilities (given by the function P). This assumption, however, does not entail that if S is extended to infinity the relative frequency of each kind of outcome *must* converge to the probability assigned by P to that kind, but only that it will do so *almost always*.[70] And I cannot think of any empirical grounds which might warrant a strengthening of the assumption to meet the requirements of modal frequentism.

If Popper's view of probability as propensity is adequately explicated by modal frequentism, it will, of course, be liable to the same objection. But is it? Any doubts one could entertain on this point were apparently dispelled by Popper when he wrote that "the propensity interpretation" asserts *precisely* that "we have to visualize the [generating] conditions [of statistical data] as endowed with a tendency, or disposition, or propensity, to produce sequences with frequencies equal to the probabilities" (Popper 1983, p. 356; formerly printed in Popper 1959b). But this seemingly unequivocal statement must now be read—and suitably qualified—in the light of Popper's criticism of von Mises (Popper 1983, pp. 367ff.), which turns on essentially the same objection levelled here against van Fraassen. In the course of it, he stresses that the strong law of large numbers "does not (and cannot) establish that *all* sequences converge, only that *almost all* converge" (Popper 1983, p. 372). More interestingly, he rejects a neofrequentist proposal to *define* the probability of an event x as the frequency of events of the type x within *almost all* sequences of an infinite set of (random or randomlike) sequences.[71] The proposal, he says, is "utterly unsatisfactory" and "it really means putting the cart before the horse," because the probability measure on sequences in a sample space mentioned in the strong law of large numbers and relative to which the italicized expression 'almost all' must be understood depends on the distribution of probabilities for single events of that space and cannot therefore be employed in the definition of single-event probability. Therefore, he concludes, "a satisfactory interpretation will not be one which explains [the probability of an event] $p(x)$ in terms of frequencies" (Popper 1983, pp. 382–84).

This conviction is shared by the authors to whom we now turn. According to Ronald Giere, "the intuitive idea behind the propensity interpretation of probability is that the probability distributions associated with stochastic systems are distributions of causal tendencies *not reducible to relative frequencies whether actual or possible*" (1976a, p. 327; my italics). The philosopher's task is to explicate this intuition in a clear and precise way. In the pàper from which the quotation is taken, Giere gives truth conditions for statements of propensity in terms of measures over sets of possible worlds. I cannot enter here into the minutiae of Giere's formal semantics, and unfortunately—as is usually the case with such constructions—an abridged presentation would be both inaccurate and unintelligible. On the other hand, the paper contains several remarks of a general philosophical nature that are of considerable interest to us. Giere recalls the familiar classification of physical systems into deterministic, stochastic, and chaotic systems. A system is *deterministic* if its exact final state is determined uniquely by the initial state. An indeterministic system is therefore liable to evolve from a given initial state to one of—finitely, countably, or uncountably—many final states. An indeterministic system is said to be *stochastic* (or *probabilistic*) if each initial state determines a probability distribution over the possible final states. Otherwise, the indeterministic system is *chaotic*. A stochastic *trial* is a particular instance of the process by which the stochastic system evolves from an initial state to one of its possible final states; in other words, it is what—following Hacking—I have called a trial on a chance setup. To the question "Metaphysically speaking, then, what are physical propensities?" Giere replies as follows:

> They are weights over physically possible states of stochastic trials—weights that generate a probability distribution over sets of states. The function u [i.e., the measure over a set of possible worlds] provides a formal and rather shallow analysis of this distribution. But is it not just the task of a propensity interpretation to explain what these weights are? No, because it cannot be done. We are faced with a new metaphysical category.
>
> (Giere 1976a, p. 332)

Thus, in the end, the formal semantics turns out to be only an attractive modern dress for the classical notion of quantified facility we met in Galileo's paper on dice. This is indeed a newcomer among 'metaphysical categories'; there are only a few unclear indications of its presence before 1600 and its current pervasive use set in much later. But an archeologist of consciousness could trace its roots to another, more primitive notion that is quite possibly shared by every human being; namely, the thought that a particular situation may or may not, under the circumstances, lead to a particular result. Thus, any Third World subsistence farmer is apt to think that an overcast sky may or may not bring rain today (whereas it is obvious to him that it cannot rain while the sky is clear), that the family hen may or may not lay an egg tomorrow (whereas it is out of the question that the family dog might do so). Only someone possessed of *furor analyticus* could dream of reducing such an elementary thought to still simpler components.[72] There is but an easy step from grasping such alternative results as genuinely open possibilities to sizing up their prospects as 'greater' or 'smaller'. The subsequent assignment of numbers required creativeness and a will to idealize that is perhaps distinctively modern. However, at a time when natural philosophers were intent on substituting precise numerical valuations for all rough estimates of size, this momentous step must have seemed quite reasonable, especially if it involved assigning equal fractions, adding up to 1, to mutually exclusive and exhaustive alternatives which, in view of some perceived symmetry, were judged to be equally possible. (As I suggested in Section 4.1, some such judgment may have inspired the much earlier invention of symmetric gambling devices.) The further transition from rational to real numbers took place unawares, following the mainstream of quantitative thinking in science, before anyone noticed the gaping abyss which had thus been crossed.

Giere rightly stresses that a propensity distribution is to be regarded as a property of a particular setup (stochastic system, "experimental arrangement") at the particular time for which its prospects are being estimated. "Thus when a probability value is assigned to an outcome, that value attaches directly to the particular trial—no reference to any set of trials is necessary" (Giere 1976b, p. 71). But, of course, a given distribution can be attributed to the single case here and now only insofar as the latter is grasped under a suitable universal, e.g., as a model of a reliable probabilistic physical theory or as an instance of a class from which a fair statistical sample has been taken. Giere himself somehow implies this when he grants, right after the preceding quotation, that "any direct *test* of a propensity hypothesis will require frequency data on a number of trials with identical initial conditions." Evidently, this requirement can be consistently complied with in the actual conduct of scientific inquiry only if it refers to the *perceived* equality of such initial conditions as are *understood* to be relevant to the case in point. However, as Giere fittingly notes, the distribution of probabilities over a long run of independent trials follows from the propensity distributions attached to the individual trials and should properly be seen as the propensity distribution that pertains to the single case consisting of all those many trials.

Much as I appreciate Giere's overall approach to probability, I must take exception to what he says on the assignment of propensity distributions to mechanical devices employed in standard games of chance. As our disagreement on this point is of some philosophical importance, I shall quote him at length.

> What, finally, can be said about macrophenomena which are, to a very good approximation, deterministic systems throughout, though the relevant variables may be unknown and practically uncontrollable? This class contains the mechanisms used in many classical games of chance, e.g., roulette and dice. Due to the operation of many uncontrollable variables, series of trials of such mechanisms are often experimentally indistinguishable from series that would be generated by a sequence of genuinely indeterministic individual trials. This empirical fact justifies our using probability theory in such cases, and even makes it natural to apply

> probability language to individual trials. But it must be realized that this is only a convenient way of talking and that *the implied physical probabilities really do not exist.* To forget this is to invite conceptual confusion.
>
> (Giere 1973, p. 481; my italics)

To deny the existence of physical probability in the cases for which this concept was introduced in the first place, and to which we still go back whenever we need to make it plausible, will seriously undermine its standing, even if, with Giere, we vigorously uphold its reality in the controversial domain of quantum theory. For all we know, the latter could well be just "a convenient way of talking" justified by "empirical fact." It seems to me, however, that Giere's words do in turn betoken a conceptual confusion, viz., between the piecemeal grasp of physical phenomena laboriously achieved in real life through suitably blurred applications of physical theories and the purportedly adequate global pictures of this or that aspect or "level" of Nature which flash out from such theories in the Sunday fantasies of some scientists and philosophers. This is the sort of confusion you would be in if, at a time when nobody is able to write, let alone to solve, Hamilton's equations for the standard process of shuffling cards (with the usual variability as to the number of times the cards are mixed), you referred to this process as a realization of Classical Mechanics, governed by the said unwritten equations. Giere's examples, dice and roulette, involving as they do fairly rigid motions, are indeed susceptible of mechanical description. But such a description must rely on blurred data, i.e., initial positions and momenta represented by a small region, not just a point, of phase space; and the intervention of an unpredictable live croupier and the discontinuity of the motion every time the ball or the dice jump ensure that the admissible trajectories issuing from each such region end up in *every* possible outcome of the game. Under the circumstances, the blurred predicates which go into the application of Classical Mechanics to such gambling devices characterize them in effect as stochastic systems. To insist that the "true"—though to us unknown—initial state in a particular trial must nevertheless be represented in phase space by a point, on a trajectory leading to a unique outcome, so that the system is "really" deterministic, is to indulge in a Utopian dream of scientific prowess. "Truth" and "reality" can be realistically attributed to the statements of Classical Mechanics—or, indeed, of any physical theory—only in the context and with the qualifications (e.g., blurring) of its successful applications to concrete states of affairs.

In *The Matter of Chance* (1971), the British philosopher D. H. Mellor takes a view of probability as propensity which is broadly similar to Giere's.

Although this deep and eloquent book was published before Giere's papers, I have chosen to deal with the latter first because Mellor's work contains some refinements of its own that make it somewhat less representative of the general tendency. I shall now touch upon some of them. Mellor accepts, on the whole, the subjectivist account of probability statements as expressing a person's prevision or—as he calls it—"partial belief" concerning uncertain events; but he maintains that some probability statements "can be made true by things having a certain dispositional property" which, "following Popper," he calls *propensity* (Mellor 1971, p. 1). In his terminology, however, 'propensity' stands for something other than objective probability or *chance*. Propensities belong to *things*; chances pertain to *trials* carried out on those things under the appropriate circumstances. A prevision of a particular outcome of some such trial will then be 'true' (in the sense of the foregoing quotation), not indeed if that outcome eventually occurs, but if the prevision agrees with its chance. As Mellor neatly puts it:

> The relation between the propensity and personalist theories is this. According to the latter the making of a probability statement expresses the speaker's "partial belief" in whatever he thereby ascribes probability to, say that a coin *a* will land heads when tossed. Knowledge of the coin's propensity on the present theory is what in suitable circumstances makes reasonable the having of some particular partial belief in the outcome of the toss. The chance of the coin falling heads when tossed is then the measure of that reasonable partial belief.
>
> (Mellor 1971, p. 2)

The chance ascribed to an outcome of a trial constitutes "an objective constraint on the partial belief reasonably held on its occurrence." It must therefore be a feature of the trial itself. However, "the pertinent feature of the trial cannot just be the chance of the result that actually occurs but the chance distribution over all possible results" (Mellor 1971, p. 59). The chance distribution is to be regarded, in turn, as the display of a dispositional property—a "propensity"—ascribed to a more permanent entity—a "chance set-up" (Mellor 1971, p. 67). Although Mellor knowingly picks these terms from Popper and Hacking, he gives them a somewhat different connotation. Thus, in the familiar example of coin tossing,

> the propensity must not be attributed to the whole assembly of coin, tossing device and environment that is only present when the coin is actually being tossed. To do that is to remove completely the point

> of ascribing a disposition as something that is present whether or not it is being displayed. It is to confound propensity with the chance distribution that displays it and hence indeed to make 'propensity' no more than a new name for chance.
>
> (Mellor 1971, p. 75)[73]

To single out within the transient situation of the trial some more permanent entity to bear the propensity is somehow conventional—as Mellor readily acknowledges. But the freedom exercised here is just one more instance of the creativeness that comes into play whenever a fragment of life's flux is articulated and understood as an objective state of affairs. Mellor's distinction between the thing to which the propensity belongs and the experimental arrangement in which it is tried and displayed has been regarded by some as a backward step. But it seems particularly well-tailored to fit the concepts of Quantum Mechanics. In Dirac's formulation, the state of a quantum system is represented by a vector $|\psi\rangle$ in a complex vector space endowed with an inner product. The theory provides an algorithm for calculating the chance that a measurement of a particular physical quantity on the said system will yield a particular value from among a spectrum of admissible values of that quantity. The algorithm associates with every measurable physical quantity—with every *observable*—a self-adjoint linear operator on the space of states. The admissible values of the quantity are the eigenvalues of this operator.[74] It is sufficient for our purpose to consider only the simple case of an operator Λ which has a finite spectrum with a single eigenvector for each eigenvalue. Then, if, say, $|\beta\rangle$ is the eigenvector corresponding to a particular eigenvalue b (i.e., if $\Lambda|\beta\rangle = b|\beta\rangle$), the probability that a measurement of the physical quantity associated with Λ on a system in state $|\psi\rangle$ yields the result b is equal to the squared modulus of the inner product of $|\beta\rangle$ and $|\psi\rangle$ (i.e., $\text{prob}(b) = |\langle\beta|\psi\rangle|^2$). The state vector $|\psi\rangle$ may therefore be said to represent a propensity of the quantum system which is displayed in diverse chance distributions depending on the observable that is being measured and hence on the type of experimental arrangement into which the system is brought.

The talk of chance distributions *displaying* propensities may sound awkward, for such displays are nowhere to be seen (as, e.g., we see the hands of a clock display the time). Mellor admits that chance distributions are "inferred indirectly and inconclusively from frequencies and from other properties related to them by statistical laws." But a like condition is shared by numerous physical properties which, nevertheless, bear witness to other, deeper properties. Indeed, the chance distributions "need no more be visible or tangible as features of single trials than is the passing of an X-ray" (Mellor

1971, p. 81). A tougher ground for resistance to Mellor's doctrine is the supposition—common among philosophers of science—that a genuine physical quantity must take (and hence, in principle at least, be susceptible of being attributed) an exact real value in each case of its occurrence. Now, as Mellor appositely notes,

> Precise values of chances, and hence of propensities, are notoriously not ascribable even in principle. On any recognised theory of testing statistical hypotheses [. . .] all that can ever be shown is that a chance lies in an interval of values. Even this assertion will be subject to a finite probability of falsehood. Not only can precise values of chance not be ascribed, it cannot even be shown conclusively that a value lies in any interval less than [0,1]. This all stems not from errors of measurement but from the very nature of chance.
>
> (Mellor 1971, p. 101)

Here Mellor plays the devil's advocate, making the statistician's predicament look worse than it is in fact. You can certainly ascribe a precise value to the chance of an event on the strength of a standard statistical test, provided that you are content to have no more than probability 0 of being right. But the same is true of *any* ascription of a definite number to a real-valued physical quantity on the basis of observations. Thus, the most that can be said after a long series of measurements of, say, the acceleration of gravity at a given place on the Earth is that its numerical value lies in such-and-such a (hopefully narrow) interval with such-and-such a (satisfactorily high) probability. The reason for this is, of course, that every experimental measurement of this sort—as they are currently and, I dare say, aptly understood—involves an element of chance, so that the best that a researcher can expect to obtain by an optimal arrangement is a random sample from an ideal population of possible results normally distributed about the so-called true value. Apart from the inevitable dispersion of observed values—called 'operational imprecision' by Mellor—many physical quantities suffer from another, so to speak constitutive, form of imprecision, which he terms 'conceptual'. A familiar instance of this is temperature, which due to "fluctuating molecular velocities" can never be ascribed a precise value. Mellor also mentions "lengths, pressures, electric currents, and the magnitudes of fields, electromagnetic, gravitational etc. connected with particles in indeterminate motion" (Mellor 1971, p. 103). He apparently thinks that chance and propensity also belong to this group. Now, such constitutive imprecision is the mark of quantities which have no application at purportedly deeper, more elementary levels of description. However, as quantum physics becomes more and more entrenched it is increasingly difficult to think of a level of description so deep that the concept of chance would not be applicable to it. It seems to me, therefore, that conceptual imprecision, in Mellor's sense, is not inherent in the concept of chance but is imported into it in specific uses from the conceptually imprecise quantities for which a chance distribution is being given.

"In a deterministic world there would be no propensities"—says Mellor. Therefore, by contraposition, "if propensities are ever displayed, determinism is false" (Mellor 1971, pp. 156, 151). The inference is sound, but its import remains in doubt. It depends, of course, on what one means by 'world'. Whatever its original acceptation,[75] in contemporary philosophical prose the word usually does the work of two Greek nouns, namely, τὸ πᾶν, meaning the All in which we live and move and have our being, and ὁ κόσμος, meaning the order fragmentarily and tentatively discerned in it and put together, through intelligent observation and reflection, by philosophy and science. It is tacitly assumed that 'world' can play this double role in one breath. It certainly did so in the discourse of theologians and metaphysicians who claimed they had unveiled the true order of the All. But in the company of a fallibilist epistemology such an assumption is wholly out of place. From our present standpoint, 'world' as κόσμος should, indeed, go beyond the fragments of order hitherto uncovered by accepted scientific theories and refer to the global system which scientific research is reaching for; but such a reference can only succeed insofar as the intended referent lies effectively in the purview of ongoing research, being truly alive in the thoughts that guide it and generally coherent with what is assumed to be known. On this condition, 'world' as κόσμος cannot also refer to the All, since there is as yet no articulate, viable, working scientific research project which aims at embedding in a single structure the various phenomena of gravity, electricity, poetry, vegetation, morality, mathematics, etc. Now, if in the above quotation from Mellor 'world' is taken to mean the natural order as disclosed by current physics, then surely propensities are being displayed in it, in Mellor's sense. Thus, to mention but one example, we have that *either* the experimental data from the great particle accelerators is construed as evidence of genuine chance distributions *or else*, as far as we can tell, these costly structures, intertwined with the lifework of a sizable community, are just meaningless scrap. On the other hand, if in the said quotation 'world' designates the All, we cannot ascertain whether it is deterministic or not. Indeed it is not even certain that a clear sense can in that case be attached to this very question. But then, by Mellor's valid inference, neither can we know whether any propensities are actually on display or whether all the purportedly objective chance distributions that are talked about in science are not just the expression of a misunderstanding. The identification of the two meanings of 'world', championed by scientific "realism," thus winds up in pointless agnosticism.

A direct link between objective and subjective probability is also a key feature of the elucidation of chance offered by Isaac Levi in 1977 (in papers reprinted in Levi 1984) and further elaborated in his book *The Enterprise of Knowledge: An Essay on Knowledge, Credal Probability and Chance* (1980). Levi, however, excludes objective probabilities from the category of dispositions in which they had been placed by Popper, Mellor, and Giere. While conceding that the received semantics of dispositional terms left much to be desired, these authors apparently found comfort in the thought that scientific discourse was teeming with such terms, so that epistemology would not have to face a new difficulty due to the interpretation of physical probabilities as propensities. But Levi took it upon himself to clarify the meaning of dispositions and came up with an analysis that simply would not fit probabilities. So he had to produce a separate theory of chance predicates. Nevertheless, there are on his view some important analogies between dispositional and chance predicates which may explain why Popper and his followers tried to put them all into the same bag.

Statements of disposition, such as 'copper is a good conductor of electricity' or Molière's celebrated 'opium has a somnific power (*vertus dormitiva*)', point so to speak in two directions. On the one hand, they remind one of the phenomena which they imply would set in if certain conditions were fulfilled (e.g., if a copper wire were inserted in an electric circuit or if a dose of opium were administered to a wailing child). But they also allude to a supposedly hidden structure of the objects to which the disposition is attributed; a structure that would, if made perspicuous, account for the onset of those phenomena under such conditions. Both references are implicit in the ordinary meaning of disposition predicates. I shall denote them by α and β, respectively. β is what makes disposition predicates such a source of annoyance to philosophers bent on banning obscurity from human discourse.[76] They have sought to eliminate it by reading disposition predicates as brief reminders that certain conditional statements are true. However, as the conditionals licensed by a statement of disposition use the subjunctive mood and generally sport contrary-to-fact protases, not much clarity is achieved by this ploy. In Levi's view "any effort to explicate disposition predicates in terms of the counterfactuals they support appears to put the cart before the horse" (Levi 1980, p. 248). His own analysis of disposition predicates—issuing from a proposal by Levi and Morgenbesser (1964)—is therefore focussed on β. By implicitly referring to an unknown or poorly understood property of things a statement of disposition portends an explanation of observed regularities and thereby provides a temporary warrant for predictions in the shape of subjunctive conditionals. A disposition predicate thus contains a promissory note to be cashed by future research.[77] When the ground of a disposition is made perspicuous by a satisfactory theory, the corresponding predicate can be taken as an abbreviation of an adequate description of that ground. Levi is well aware that standards of perspicuity depend on research programs and the state of inquiry. "Predicates that are deficient relative to some program for explanation may, nonetheless, be used in stopgap explanations pending further inquiry; this latter will either render those predicates acceptable for purposes of explanation or will replace them with predicates that are acceptable" (Levi 1980, p. 238). Levi's "conceptual analysis" of disposition predicates does not proffer to specify truth conditions for any statement containing such a predicate. According to Levi, such semantics is unnecessary. All that is required in order to understand the use of a disposition predicate is a set of instructions for evaluating hypotheses about the test behavior of objects of which the predicate is held to be true (Levi 1980, p. 243).

Disposition predicates play of course an essential role in talk about physical probabilities. To conceive a physical system as a chance setup implies ascribing to it a disposition to wind up in one of—finitely, infinitely, even uncountably—many specific states ("outcomes") when it is subjected to certain conditions ("trials"). But such an ascription says nothing about the distribution of chances among the several states. The disposition of a fair coin to make heads or tails when fairly tossed is not conceptually different from that of a biased coin. The task of distinguishing between such cases falls upon chance predicates. What sort of information do they convey that cannot be captured by disposition predicates? In Levi's eyes all attempts to explicate chance predicates in terms of limits of fictive sequences or measures over fictive worlds are "obscurantist diversions" (Levi 1984, p. 186). For him, the whole point of introducing chance predicates into our language is to have a means of describing conditions whose known presence would warrant the adoption of definite previsions or, as he calls them, "credal[78] states" regarding prospective events (Levi 1980, p. 261). For reasons carefully spelled out by him, the real-valued function that expresses a chance distribution ought to match the previsions warranted by the conditions to which that distribution is attached. Therefore, the said function must satisfy the standard requirements of a probability function (or the previsions it matches would be incoherent—see Section 4.5).

But if chance is parasitic on prevision in this way, why need chance predicates be used at all? Why not follow de Finetti's advice and dispense with them altogether? Levi's reply is embodied in a simple example. Suppose a man *X* has before him two coins, *a* and *b*, which he believes to be physically similar in all respects except that one came off the mint just before the other. Suppose further that his prevision for the hypothesis that *a* will land heads *r* times in *n* tosses conditional on its being tossed *n* times at time *t* is $\binom{n}{r}(.5)^r$, while his prevision for the hypothesis that *b* will land heads *r* times in *n* tosses conditional on its being tossed *n* times at time *t* is $\binom{n}{r}(.9)^r(.1)^{n-r}$. This set of previsions could not be taxed with incoherence, yet under the circumstances—according to Levi—not even de Finetti would say that they make sense. *X*'s discrepant previsions would be wholly unwarranted unless there was, in his view, some significant difference in the characteristics of *a* and *b*.

> That is not to say that *X* should be in a position to offer an explanatorily adequate characterization of the difference between the coins; but he should be committed to the view that there is a difference in traits. The coin *a* has some property *C* such that given knowledge that an object has *C*, ceteris paribus, *X*'s credal state for hypotheses specifying relative frequencies of heads on n tosses should be as specified above. Similarly, *b* has some property *C´* knowledge of the presence of which licenses a credal state of the sort attributed to hypotheses about *b*'s behavior. One way of putting it is to say that coin *a* is unbiassed whereas coin *b* is heavily biassed in favor of heads. Another way to put it is to specify the explicit simple chance predicates which are true of *a* and of *b* concerning outcomes of *n* tosses.
>
> (Levi 1980, p. 269)

Thus, though closely bound to the previsions or "credal states" of persons, statements of chance do not describe such states. They bear truth values and make claims about objects which purportedly hold good independently of the subjective states of knowing subjects. The latter can accept or reject chance statements or suspend judgment or even adopt credal states about them. Descriptions employing chance predicates—e.g., a description of the differences between the two coins in our example purely in terms of their respective chances of falling heads or tails—are admittedly deficient. But their deficiency is not to be remedied "by restricting credal judgments to hypotheses about test behavior and forbidding the acceptance of chance hypotheses into evidence or the assignment of credal probabilities to them," as de Finetti prescribed.

> The defects in chance predicates are to be removed not by eliminating chance statements from the language [. . .] but by integrating chance predicates into theories through inquiry as is

> attempted in genetics, statistical mechanics, and quantum mechanics in differing ways.
>
> (Levi 1980, p. 269)

So, according to Levi, chance predicates, like disposition predicates, provide a knowingly inadequate characterization of the objects to which they apply. They draw on the credit of future inquiry. There is, however, an important difference between both kinds of predicate that we ought to keep in sight. A disposition term may, by integration into a physical theory, become shorthand for a complex expression in which no disposition predicates occur. For instance, in Einstein's theory of gravity weight and attraction are supplanted by the standing geometry of spacetime. But chance predicates cannot hold the place of explanations from which all idea of chance is absent. Statements of chance cannot be inferred from nonchance premises. For chance predicates to be integrated into a theory of mathematical physics the latter's structure must include a probability space. Thus, unless we gratuitously assume that all chance statements are false or nonsensical and are ultimately dispensable, we must acknowledge chance to be an ineradicable feature of things—at least as we have to understand them from our radically contingent vantage point. Moreover, in a chancy environment, the dispositions of chance setups to yield alternative outcomes are also irreducible and cannot be described by nondispositional terms.

At first sight it may seem that Levi's view of the role of credal states in our grasp of chances will not fit well with our findings about the advent of quantitative probability (Sections 4.1, 4.3). We traced back to antiquity the use of 'probable' (and its Greek equivalent πιθανός) as an epithet for plausible, reliable prospects. Like so many other adjectives, 'probable'—in this sense—admitted degrees of comparison. But it was not associated with numerical values. Metricized credal states do not show up in Western Europe until the 13th century or later. The oldest documents concern betting rates on designated outcomes of games of chance. The first that are fully explicit—viz., Galileo's paper on dice, the Fermat-Pascal correspondence, Huygens' *De ratiociniis in aleae ludo*— base recommended betting rates on calculations of physical facility. The calculations proceed from the assignment of identical numerical values to alternatives that are judged equally easy because of a perceived symmetry in the physical situation. These facts suggest that far from being parasitic on credal states, the novel concept of physical facility or chance prompted—and indeed made possible—the quantitative formulation of such states (based on the rule that what is easy in itself—*in re*—is probable in our judgment—*in mente*). It is true that after the new arithmetical idiom for credal states became familiar, its use was extended as a matter of course to situations (common, e.g., in weather forecasting or in medical practice) that give little or no grounds for a grasp of chances. But such use should be reputed metaphoric, if not downright deceptive.

Further reflection on the matter will however bring us closer to Levi's view. To my mind, there is no question but that the quantitative concept of chance, as found in Galileo or in Huygens, is an intellectual creation of their time (broadened, if you wish, to include the 16th century). The concept was well attuned to the then dawning idea that—as Galileo put it—the Book of Nature is written in mathematical language. Still one may wonder why the earlier applications of the new concept always presupposed a uniform distribution of chances among elementary alternatives. The pervasiveness of this feature cannot be explained on physical or logical grounds. There was, on the other hand, a good epistemic reason for it: only on the supposition of uniformity could the initial distribution premised in any calculation of chances be readily and securely determined, viz., by counting all the ultimate alternatives and taking the reciprocal value of the total to be the chance of each. But then this procedure may, not unnaturally, have occurred to someone familiar with the premodern, nonmetric version of credal probability, thus sparking off the emergence of the quantitative concept of chance. Even if he had never thought of assigning numbers to his credal states he must have been able to say when he was in a state of indifference concerning the prospects of two or more mutually exclusive events. Thus, without impairing the originality of the modern concept of chance, we can admit that qualitative credal states acted the midwife at its birth.

But now that the concept of chance has been available for several centuries, prompting the metricization of credal states, does our grasp of chances sometimes come about by the introspection of such states? Their mediation is not needed if the chance distribution rests on a physical theory such as Statistical Mechanics or Quantum Mechanics. But credal hunches no doubt have a role to play in the development of such theories. And they continually intervene in everyday statistical practice. Much of it consists in drawing conclusions about a large population from the observation of a small sample chosen from it. If the sample is picked at random (i.e., in such a way that any member of the population has the same chance of being selected), the mathematical theory of probability enables one to determine the probability that the proportion ω of elements in the sample that exhibit a certain property lies within a given distance of the proportion π of elements with that property in the whole population—or, alternatively, the distance from π within which ω lies with a given probability.[79] The probability in question is objective if, but only if, the sample is random (or very nearly so). Not always is an established physical theory available to guarantee the fairness of a viable sampling procedure. (For instance, the people on whom a new vaccine will be tested cannot be picked out by a lottery in which everyone has a ticket.) Therefore, the original design and subsequent improvement of sampling methods owe much to the statistician's tact, i.e., to his credal judgments. On the other hand, the very success of the statistical predictions based on a particular method lends color to the hypothesis that it meets the condition of randomness.

4.7 Ideal chances

In the view advanced in Sections 4.3 and 4.6, the abstract structure of a probability space takes on a physical meaning when the probability function is understood as a measure of the facility of occurrence of certain definite types of event in suitably specified physical situations—or, if you wish, as a measure of the propensity of such situations to yield such events. The objection is often voiced that 'propensity' is just another name for probability. It could hardly be otherwise if the scientific notion of probability was, from the start, conceived in just this way. However, because there have been proposals for conceiving it differently, the characterization of physical probability as propensity or facility, though tautologous, is not redundant. The idea of propensity has also been called obscure, presumably because it is tantamount to physical probability. Now, physical probability is certainly not one of those notions—like 'reality', say, or 'time'—that have been with us for so long and are so deeply entrenched in human discourse that we cannot think of other, clearer notions in terms of which to elucidate them. But if it is a genuine intellectual invention, it has to be acknowledged as a primitive concept, notwithstanding its fairly recent date of emergence. The ordinary prescientific idea of the ease or difficulty with which an uncertain prospect can come about is presumably its direct ancestor, but there is an unmistakable "jump" from one to the other.

For brevity's sake, I shall speak—with de Finetti—of the *prevision* of an event when I mean its personal or subjective probability, and—with Hacking—of its *chance* when I wish to refer to its physical or objective probability. The term 'probability' can then be reserved for the abstract quantity characterized by the Kolmogorov axioms—with KPS 2 (complete additivity) or with FPS 2 substituted for it. When dealing specifically with the latter option we talk of 'finitely additive probability'. There is an important difference in the manner in which chances and previsions relate to probability. As we noted in Section 2.6, a quantitative predicate can be properly understood only in the context of a definite quantitative structure. The structure underlying the chance ascribed to an event *E* is, of course, the probability space associated with the chance setup from which *E* is apt to issue. You can refer to the chance of *E* only insofar as you view it as a possible outcome of a trial on a chance setup; and in order to grasp a physical system or situation as a chance setup you must bring it under the concept of a probability space. Thus, 'chance' is, if you wish, no more than applied probability; or—as I would rather put it—mathematical probability is abstract chance, i.e., what remains of this concept created for grasping physical states of affairs when we choose to forget its original reference to physical facility or propensity. On the other hand, a person's previsions (of events) exemplify (finitely additive) probabilities per accidens, in case those previsions happen to be coherent. As this circumstance cannot alter the meaning of 'prevision', the quantitative structure on which this meaning depends cannot be supplied by the mathematical theory of probability but must be found elsewhere. And indeed, whether those previsions are measured by the person's betting rates or by de Finetti's method of self-assessed penalties, the structure in question is not far to seek. In either case previsions are referred to the scale of multiples of a monetary unit and defined as a ratio between two such multiples.[80] Because previsions as such have their seat outside the mathematical theory of probability, there are requirements—e.g., finite additivity—that the latter must meet in order to be serviceable in calculations of the former. But no such exogenous prescriptions restrict the mathematical physicist's creativeness in the solution of problems and the grasping of situations by means of novel ideas of probability qua chance.

If a particular, contingent event is certain, neither chance nor prevision can be predicated of it. Because both concepts have this in common, it has seemed plausible to replace the former with the latter. But subjective prevision keeps clear of the subjectively certain, i.e., the known, while objective chance is excluded from the objectively certain, i.e., the determinate. Let *E* denote an outcome of heads on the 15th toss of a given coin in a game being played between Jack and Jill. After that toss has been made and the coin has come to rest, there are no chancy prospects for *E*. Either heads have turned up, and then *E* has more than just a good chance of occurring; or tails have turned up, and then *E* stands no chance at all.[81] Ascribing chances—even a chance of 1—to a determinate contingent event, qua determinate, involves, I dare say, a category mistake. And yet a good many of our chance estimates plainly concern events we know to have occurred, though we are still ignorant of their result. How are we to understand this? Must we concede that such estimates are not genuine *chance* estimates but rather subjective previsions? But their standing among our judgments and beliefs is not essentially different from that of chance estimates concerning future events. Shall we conclude that chance is just an intellectual mirage, that there is no such quantity in the nature of things? I see no reason for adopting this counsel of despair. Contingent determinate events qua determinate do not sport this quantity. But neither is it ascribed to them qua determinate. Suppose we lay bets on a card that has been placed face down in front of us. We can, indeed, grasp it as such-and-such a card, say, the Queen of Spades, but only if we turn it over. While we refrain from doing so, we grasp the card under a different concept, viz., as the first card dealt from a well-shuffled standard pack. Thus, we do not think of its denomination as a determinate but unknown fact but as the possible outcome of a known chance trial. It is therefore no less sensible for us to lay bets on that denomination now that the card is what it is and has no chance of being different, than it was before the pack was shuffled. (Indeed, it may be more sensible, for we can now be satisfied that the shuffling was properly done, so that we are in fact confronted with the outcome of a genuine chance trial.) Here, as elsewhere, we judge the unknown effect by its known provenance. If we can regard it as the result of a deterministic process, we calculate it. If it is the outcome of a trial on a chance setup, we figure out its chances. Conditional chances become relevant when more information is available. Thus, if in our card-guessing game three more cards are drawn from the pack and placed face up on the table, we shall base our betting rates on the chance that the first card might turn out to be such-and-such, conditional on the next three being what they are; or equivalently—if we would rather keep our causal chains in order—on the chance that such-and-such a card might occur among the first four, conditional on the other three being what they are.

I should say that the foregoing example typifies most everyday applications of chance predicates to determinate contingent events. But I am not sure that the use of chance distributions in science always follows this pattern, for such distributions are not always clearly linked to the interpretation of a definite physical system as a chance setup or a definite physical process as a stochastic process. Consider a physician who after reviewing a patient's symptoms and reading his blood count concludes that there is a .85 chance that he suffers from dengue fever. She would be hard put to say from what chance setup she sees him as issuing. Note that there are two stages in the physician's reasoning. One (which she finds ready-made in the medical literature) proceeds from empirical studies of human samples to a conclusion about the proportion of dengue cases among people displaying certain features. This is a standard statistical inference. The conclusion demonstrably has a definite probability of being right if it is based on a random sample. For the inference to be valid the sampling must involve some randomizing procedure, so that the resulting selection may properly be regarded as the outcome of a trial on a chance setup.[82] The second stage of the reasoning is carried out by the physician herself, and consists in estimating the patient's chance of having dengue fever from the statistically inferred frequency of that illness among people with the patient's symptoms and blood count. This requires that the individual patient be grasped by the physician as a random token of the population to which the medical statistics refer. But I cannot think of a physical process by which, in her view, the patient was actually drawn at random from that population. I suspect, therefore, that in such cases as this one, chance predicates are not used in the same way as in the game discussed in the preceding paragraph, or in the standard inference from a sample to a population.

I shall now make a proposal for understanding the scientific use of chance predicates in contexts in which no definite chance setup or stochastic process is in sight.

In order to make the patient into an object of scientific inquiry the physician must grasp him under a concept. Now, an object grasped under a concept C, without any further qualification, is just an *arbitrary* instance of C, on a par with any other member of C's extension. This levelling effect lies in the nature of ordinary concepts.[83] Ordinary generic concepts are internally articulated into more or less definite subordinate concepts (henceforth: subconcepts), viz., the concepts of the several species to which the members of their respective extensions can belong. If someone who is aware of the internal articulation of a generic concept C grasps an object under C, without any further qualification, she implicitly ascribes to it the possibility of being an instance of any one of C's subconcepts.

Many scientific concepts fall under this description, but others display additional refinements. Consider the concept of temperature. It is internally articulated into varieties represented—in the absolute, or Kelvin, scale—by the positive real numbers. But on grasping a physical state under this concept of temperature we do not ascribe to it the possibility of having a temperature of ρ kelvin, for some positive real number ρ, but rather that of having a Kelvin temperature in the interval $(\rho - \varepsilon(\rho), \rho + \varepsilon(\rho))$, for some positive real number ρ and a real-valued function ε, intrinsic to our grasp. We discussed such *blurred* concepts in Chapter 3.

In my view, the scientific application of chance concepts without the support of a definite physical chance setup depends on a different form of improvement in the internal structure of generic concepts. Suppose that the concept C not only is articulated into subconcepts but involves a probability distribution over them. We shall then say that C is a concept with intrinsic probability distribution, or a *p-concept*.[84] Less informally, let S_C denote a set of mutually exclusive subconcepts into which C is exhaustively articulated (S_C defines a partition on the extension of C); let $\mathcal{B}(S_C)$ be the Boolean algebra generated from S_C by the operations of negation and (finite or countable) conjunction and disjunction; let **p** be a mapping of $\mathcal{B}(S_C)$ into the interval [0,1]. C is a *p*-concept if $\langle S_C, \mathcal{B}(S_C), \mathbf{p}\rangle$ is a probability space.[85] It will be convenient to refer to such a concept as the *p*-concept $\langle C,\mathbf{p}\rangle$. On grasping an object as an arbitrary instance of the *p*-concept $\langle C,\mathbf{p}\rangle$, without any further qualification, one would ascribe to it a chance of lying in the extension of each $\zeta \in \mathcal{B}(S_C)$, equal to $\mathbf{p}(\zeta)$. On grasping the object under the *p*-concept $\langle C,\mathbf{p}\rangle$ with an additional specification $\sigma \in \mathcal{B}(S_C)$, one would ascribe to it a chance of lying in the extension of each $\zeta \in \mathcal{B}(S_C)$, equal to $\mathbf{p}(\zeta \mid \sigma)$ (the probability of ζ *conditional on* σ).

The application of this scheme to the reasoning of our physician is straightforward if we make a few idealizing assumptions. Suppose that current medicine has a *p*-concept of a human being, with a probability distribution over pathological states and symptoms. Suppose, moreover, that this distribution assigns to dengue fever a probability of .85, conditional on the presence of certain symptoms. Then, by the very act of grasping her patient as a human being with just those symptoms the physician would assign him a chance of .85 of having dengue fever. Of course, medical thinking only very exceptionally is so straightforward. But much of it follows this pattern, in subtler and more tentative ways.

Two questions must be faced before my proposal can be accepted as a plausible description of the use of chance predicates in the absence of a definite physical chance setup. Firstly, what exactly is being said or thought of a particular object when it is grasped as an instance of a *p*-concept? Secondly, what justifies the application to this particular object of a definite *p*-concept, endowed with a definite probability distribution? Let me deal first with the second question.

A *p*-concept can be formed by attaching a probability distribution to an ordinary concept already in existence (e.g., 'human being'), with an already established articulation by subconcepts. Instances of the *p*-concept can then be identified, without regard to the distribution, as instances of that ordinary concept. By random sampling from a large population of such instances one defines and refines the distribution. Suppose now that I grasp under the *p*-concept $\langle C,\mathbf{p}\rangle$ some arbitrary, not further specified individual instance A of the ordinary concept C. By virtue of this grasp I judge it practically certain that any large collection of individuals like A will display statistical ratios in agreement with the probability function **p**, as required by the laws of large numbers. The appropriateness of grasping A under $\langle C,\mathbf{p}\rangle$ stands and falls with this certainty, which, in turn, can only be vindicated by examining large collections of *C*'s and studying the statistical ratios displayed by them. This is typically the procedure by which the probability function **p** must have been associated with the ordinary concept *C* in the first place. Thus, if *A* is a *C* and I am satisfied, by statistical inference, that **p** should be attached to *C*, I am justified in grasping *A* under ⟨*C*,**p**⟩.

A *p*-concept can also be created as such, the concept *C* being linked from its inception to a probability function **p** that is part of its definition. Thus, in contemporary physics, the concept of a new elementary particle, the existence of which is entailed by some new theory, typically will include a specification of the nature (integral, half-odd-integral) of the particle's spin eigenvalues, which, in turn, determines the probability of finding a gas of such particles in each one of its possible alternative states.[86] Again, the use of such a *p*-concept ⟨*C*,**p**⟩ will often be vindicated by examining samples of its purported instances. But if the function **p** is determined by a theory, its ascription to an arbitrary instance of *C* will be supported by whatever evidence there is for the theory (which would then include, of course, any evidence obtained by sampling for the applicability of ⟨*C*,**p**⟩).

To appreciate the difficulty of the first question, let us consider once more the ascription of chance predicates to future or past outcomes of trials on a definite chance setup. Truth conditions for such ascriptions can be readily stated in terms of propensity or facility of occurrence. For example,

> 'The chance of obtaining (having obtained) a deuce on a standard throw of this die is (was) 1/6' is true if and only if the chance setup consisting of the die thrown in the standard way has a propensity of 1/6 to turn up a deuce.

Truth conditions of this kind are strictly unverifiable, but not thereby inadmissible in our postpositivist age. (In fact, they are no more unverifiable than the limiting relative frequencies that the positivists gleefully proclaimed as truth conditions for probability assignments.) But the intrinsic probability distribution of a *p*-concept cannot be thus simply anchored to a dynamical property—no matter how elusive—of a real physical system. I can think of only three ways of understanding a *p*-concept ⟨*C*,**p**⟩, each involving a different interpretation of the chances that the probability function **p** is supposed to measure. I shall now describe those three ways and then proceed to illustrate them with examples from the literature of classical statistical mechanics.

(1) **p** measures what I shall call *elpistic chances* (from ἔλπις = hope). The *p*-concept is used as a—hopefully provisional—surrogate for a still unavailable dynamic theory of the genesis of the objects of which it is predicated. It is expected that a tenable theory would conceive those objects as the outcomes

of trials on a chance setup. Someone blessed with this hope—and the matching faith—can accept truth conditions for the *p*-concept ⟨*C*,**p**⟩ tailored after the following schema:

> 'An instance *x* of *C* is a *B* with probability **p**(*B*)' is true if and only if the true theory of the genesis of *C*'s assigns to the chance process that produced *x* the propensity **p**(*B*) of yielding a *B*.

Note that, in this understanding, most *p*-concepts would have to be replaced by theories of such complexity that it is extremely unlikely that they would ever be forthcoming. Philosophers often seem to delight in epistemic promises that nobody is under any obligation to fulfill. But objective predications based on such promises are in effect a travesty of objectivity.

(2) **p** measures *cosmic chances*. The universe is regarded as a chance setup delivering *C*'s that are *B*'s, with propensity **p**(*B*). In this understanding, the promissory note implicit in the *p*-concept ⟨*C*,**p**⟩ need not be cashed by a future physical theory, for it is already backed by metaphysical gold. This may sound like a farfetched attempt to quantify Μοῖρα or Τύχη, but it makes good sense if the chance distribution can be plausibly based on definite cosmological considerations, however schematic and uninformative.[87]

(3) **p** measures *ideal chances*. We give up all pretense that the probability function **p** inherent in the *p*-concept ⟨*C*,**p**⟩ expresses the dynamic propensities of a given physical system, be it local or global. The notion of chance is transposed from the realm of causal to that of purely conceptual relations. Instead of a weighted spectrum of alternative *happenings* on a particular setup, **p** represents here a weighted spectrum of alternative *specifications* of a universal concept, without the least regard to the physical processes by which such specifications are brought about. Just as a Euclidian circle of radius *r* is *with necessity* a flat surface of area πr^2, so an instance *x* of the *p*-concept ⟨*C*,**p**⟩ is *with probability* **p**(*B*) a *B*. If **p**(*B*) is very close to 1, it is practically certain that *x* is a *B*. If **p**(*B*) is not close to 1 or 0, the meaning of the isolated statement that *x* is a *B* with probability **p**(*B*) will elude us, but its implications are clear enough: at suitable arguments, in apposite product spaces constructed from $\langle S_C, \mathcal{B}(S_C), \mathbf{p}\rangle$, the appropriate probability functions derived from **p** take values arbitrarily close to 1. The same holds, of course, for probability as propensity: middling propensities are only worth mentioning because of the practical certainties that can be inferred from them on the strength of the laws of large numbers. Hence, a *p*-concept involving ideal chances is no worse, on this count, than one involving cosmic or merely elpistic chances. It is clear, on the other hand, that *p*-concepts linked to ideal chances can be no less serviceable for the scientific enterprise of objectifying experienced particu-

lars by incorporating them into suitably structured domains of reality than, say, blurred concepts endowed with uniformities, or any other device of the creative understanding. The appropriateness of such epistemic contrivances should be judged by how they work in intellectual practice, not from sclerotic preconceptions concerning being and truth.[88]

J. C. Maxwell apparently had no qualms about resorting to ideal chances in physics. Right at the beginning of his first derivation of the Maxwell-Boltzmann distribution of velocities of the molecules in a gas he introduces a p-concept which, as far as I can see, must be understood in the above sense (3). Maxwell is trying to "demonstrate the laws of motion of an indefinite number of small, hard, and perfectly elastic spheres acting on one another only during impact" (Maxwell 1860 in Maxwell 1890, vol. I, p. 378), in order to compare the properties of such a system of bodies with those of gases. He proposes to find "the probability of the direction of the velocity [of two such spheres] after collision lying between given limits." The collision can only take place if "the line of motion of one of the balls passes the centre of the other at a distance less than the sum of their radii; that is, it must pass through a circle whose centre is that of the other ball, and radius (s) the sum of the radii of the balls. *Within this circle every position is equally probable* and therefore the probability of the distance from the center being between r and $r + dr$ is $2rdr/s^{2}$" (Maxwell 1890, vol. I, p. 379; my italics). It follows at once that the sought for probability does not depend on the angle between the original direction of the ball and its direction after impact, so that "all directions of rebound are equally likely." Now, the italicized assumption and Maxwell's chosen way of translating it into a probability measure on the said circle[89] are not based on the consideration of some actual, purportedly stochastic process by virtue of which the balls in question pass each other's center at this or that distance. Maxwell simply takes the general concept of two balls of the kind considered colliding with one another, states a necessary condition for such a collision to occur, lets this condition be articulated into the diverse ways in which it can be fulfilled and adopts a distribution of probabilities over such ways. Such a distribution does not display any real propensity of a particular ball to pass a particular ball at some distance or other. With apologies to Leibniz (see GP, vol. VII, p. 303), it may be said to express the chances of the several subconcepts of the general concept, as they vie with one another for the specification of its instances at the radical origination of things.

Maxwell's original derivation of the Maxwell-Boltzmann distribution was subsequently discarded, as both Maxwell and Boltzmann produced subtler and more satisfactory proofs from less restrictive assumptions. In their mature work, the momentary state of a mechanical gas model comprising N

molecules, each with f degrees of freedom, is represented by a point P in $2Nf$-dimensional phase space, while its dynamical evolution is represented by a path in phase space, viz., the unique solution through P of the Hamilton equations for the model. However, in any conceivable experimental situation, the measurable parameters will not determine a point in phase space but a continuum C of such points (e.g., a hypersurface of constant energy, or, rather, given the inevitable errors of measurement, a thin open neighborhood of such a hypersurface). The chance that the point P, representative of the actual situation, lay on a region $V \subset C$ was identified with the ratio of the volume of V with that of C. This identification defines a probability distribution over subconcepts of C, which may be understood as referring to strictly ideal chances. This way of understanding it is apparent in the work of Gibbs (1902). But Boltzmann sought to base the postulated distribution on the actual dynamical evolution of the system under consideration, which, he claimed, would typically take its representative point, in the course of infinite time, over the entire continuum C.[90] The time spent by the representative point on different regions of C would then be proportional to their respective volumes. The distribution in question follows, and can be regarded as a distribution of *cosmic* chances, if—but only if—the time at which the represented system is being considered is randomly selected from its total history. This property of randomness may indeed be granted to any time, including the present, if (i) time is homogeneous, (ii) the material systems under study exist from eternity, and (iii) the material conditions under which physical research is possible are not exceptional in nature. Boltzmann evidently held (i) and (ii) to be true. Surprisingly, he was not troubled by the obvious falsity of (iii).

Nor was he deterred by the incongruence of condition (ii) with the paradigmatic application of classical statistical mechanics to bottled samples of pressurized gas. On the other hand, Maxwell (1879) gave a clever reason for representing the actual state of a gas contained in a closed vessel by a *random* point in the region of phase space determined by its measured parameters: as the gas molecules, or some of them, encounter the vessel walls, the representative path of the system in phase space will shift incalculably from one admissible trajectory to another. Maxwell concludes that unless the vessel has a very special shape, "the system will sooner or later, after a sufficient number of encounters, pass through every phase consistent with the equation of energy" (Maxwell 1890, vol. II, p. 714). Since Maxwell did not have the faintest idea how the interaction of the gas molecules with the vessel walls might account for the said randomizing effect, the distribution, as vindicated by his argument, belongs to the class I have called *elpistic*.

5. Necessity

We saw in Chapter 1 that impersonal observation, as currently practised in the natural sciences, can supply information about the states and evolution of observed objects only if—or insofar as—the latter constitute necessary conditions of the recorded receiver states. It is therefore apparent that the natural sciences can thrive only in a setting strung together by necessary connections. I noted at the end of Section 2.1 that inferential explanation—at least in its strong deductive-nomological form—provides just that, as it imbeds the explained phenomena in a domain of objects linked by bonds of necessity. Thus the construction of such bonds does not merely cater to the cravings of Reason but is also required for collecting the empirical data that go into the fabrication of scientific facts. Our subsequent explorations—above all in Chapter 3—should have thrown some light on the intellectual means deployed for that purpose by physics. But nothing has yet been said about the kind of necessity that physics discloses in natural phenomena. The subject is notoriously thorny—so much so, that some philosophers eschew physical necessity altogether. I should not say that its condition has improved with the recent attempts to explicate necessity and the twin concept of possibility in terms of a plurality of worlds.[1] I shall briefly refer to such attempts by way of introduction to the informal review of ordinary forms of possibility and necessity that will occupy us in Section 5.1. Then I shall discuss the Greek discovery of necessity in geometry in Section 5.2, the natural laws in the guise of differential equations that are the backbone of physics in Section 5.3, and the uneasy coexistence between the type of necessary connection implied by such laws and the ordinary idea of cause and effect in Section 5.4.

5.1 Forms of necessity

The concepts of possibility and necessity, as applied to sentences or statements or to the truths and falsehoods ("propositions") conveyed by them, play the role of syntactic/semantic operators, which map sentences to sentences and propositions to propositions, just as negation maps a sentence to its denial or conjunction maps any pair of sentences to their joint assertion. In this role they are known as the modal operators—or the classical modal operators—and their close mutual relationship can be lucidly displayed with the symbols of modern logic. I choose the symbols ¬ (read: 'it is not the case that'), □ ('it is necessary that'), and ◊ ('it is possible that') to represent negation, necessity, and possibility, respectively, and I let the letter **p** stand for any sentence in the indicative mood (the sort of sentence you can use to make a statement, capable of truth or falsity). All philosophers agree that □**p** if and only if ¬◊¬**p** and that ◊**p** if and only if ¬□¬**p.** Following the fashion of formal semantics for the (nonmodal) predicate calculus, the polycosmist philosophers link the truth-conditions of ◊**p** and □**p** to the truth-values of **p** alone. Now, we may readily grant that it is possible that **p** if '**p**' is true[2] and that it is not necessary that **p** if '**p**' is false; but obviously being told that it is not the case that **p** will not teach us anything about the truth-value of '◊**p**', nor will knowledge that **p** is the case give us an indication as to the truth-value of '□**p**'. Thus, for someone trained to equate the world with the aggregate of everything that is the case, the truth-conditions for sentences governed by the classical modal operators must be found outside it—whence the need for many worlds so that modal talk will make sense in ours. In polycosmist semantics, '◊**p**' is true if and only if there is a world in which it is the case that **p**, and '□**p**' is true if and only if **p** is the case in every world.[3]

To the uncommitted observer this will seem a classic instance of philosophy getting mired in its own trash. Is the world no more than the aggregate of facts? Is understanding a sentence tantamount to knowing its truth-conditions?[4] Statements of possibility and necessity have been made and understood in every language—presumably since the dawn of man—without even the remotest hint of a plurality of worlds. Of course, the clever polycosmist will disclaim any intention to teach people how to use such ordinary modal operators as 'I can' or 'I have to', or to tell them what they are trying to say by means of them. His semantic rules define the modal operators contextually, i.e., they provide for every statement that contains a given modal operator another statement that does not contain it, yet allegedly always shares the truth-value of the former statement. Such definitions "in use" have a practical purpose when the definiens is more readily testable than the unaided definiendum. But the statements in question stand to each other precisely in the opposite relation. Since the definiens speaks of other worlds causally unconnected with our own, its truth-value must be taken on faith. However, it can be instantly computed from the known truth-value of the definiendum on the strength of the definition itself. For example, I know—of course, fallibly—that by pressing certain keys I can now type the

Greek letter sigma into the electronic manuscript I am writing. Hence, if I were to believe the polycosmist, I should rest assured that there is in another world a creature working on a device indistinguishable from a Macintosh computer whose life up to a certain time *t* is equal—as far as it and I and our respective friends can tell—to what has hitherto been my life, but which at time *t* does what I shall not do now, viz., to simultaneously press the keys command-shift-Q followed by S. When trying to figure out what I can or cannot do in a particular situation, so as to act accordingly, this sort of assurance leaves me cold. Polycosmist beliefs—or should we say conventions?—may favor the advancement of some theoretical projects,[5] but otherwise they are of no consequence. After all, what happens in *some* world will only matter to us if that world *can* be this one.

One of the polycosmist's chief worries is that rejection of his definitions will force us to take 'possible' or 'necessary' as primitive concepts. I say, what of it?[6] Although modal statements are often uncertain, this does not imply that their import is unclear. I find that our entire lives turn on our awareness of what is and what is not possible, no less than on our awareness of what is or has been actual. Indeed, much of our attention to the details of reality is prompted by our abiding quest for avenues—and barriers—to possibilities. A sense for the possible—and the impossible—is a major ingredient of the conscious performance of even the tiniest actions, e.g., bending the index finger or opening the mouth (faster? slower? more? less?—I sense it is up to me, within limits that I also sense). In fact, when no possibilities are sensed, one may not be said to be aware of *acting*. I dare say that this minute sense of what I can and cannot do right here and now, a sense that is sometimes hesitating but often cocksure, that is mostly hushful but never dormant, lies at the heart of my understanding of possibility and necessity. Plainly, it stands on its own no less than my sense of reality. Or rather, it is built into the latter. For if what I have just said is true, I sense my very being as a hub of possibilities. They mainly concern my worldly wheelings and dealings and are enabled and constrained by possibilities and necessities residing in the world.[7] Many of these I infer or surmise or accept on trust. Others, as the saying goes, I have never even dreamt of. But there are possibilities and necessities that I perceive in things no less immediately than their so-called sense qualities. For instance, as I press and roll a lump of wax between the palms of my hands I feel that it can take a different shape. However, if the wax were cold, I would feel no less clearly that it cannot, at least not by pressing it and rolling it with my hands. And, if I hugged it for a while, I would perceive, as it warms up, how it gradually acquires that possibility. No doubt it will be objected that what I perceive if the wax is not cold is that it *actually* changes its shape, not that it *can* assume another one. To this I reply that if, as I presume, we do *sense* change, and not just *infer* its occurrence from the sequential perception of different states of rest, we perceive in each change the possibility which it enacts much as we perceive our own possibilities in the course of action.[8]

The possibilities disclosed in such elementary experiences, or constructed from them in thought, may be termed *powers*—in good agreement with ordinary usage.[9] The corresponding impossibilities are often lived as powerlessness, for instance, when one watches the milk spilling from an overturned glass or tries in vain to stop a skidding automobile from crashing against another one. Our primary paradigm of necessity, brutal or coercive necessity, bemoaned by the poets and familiar to all, is rooted in such experiences. Philosophical attempts to reduce it to the purportedly more primitive notion of reality or existence seem to me extravagant, inasmuch as necessity thus perceived often constitutes an important criterion or indication of real existence. Is not our inability to abolish a fact a main ingredient of our awareness of its being factual? It is also worth noting how the effectiveness of power rests on the workings of necessity. I *can* unscrew a tight screw because the latter *must* yield to the torque transmitted by the screwdriver.

The common notions of possibility and necessity, as they show up in the ordinary usage of expressions that convey them, display features often ignored in philosophical discussions. Of course, this may be taken to mean simply that the philosophical concepts that are being discussed differ from their vulgar homonyms. I believe we ought to be aware of such differences and try to understand their motivation. Do they respond solely to a desire to trim off inconsistencies found in the common notions? Or did they arise in the course of extending these notions to a context for which they were not originally meant? I shall now review some peculiarities of the ordinary usage of modal expressions and speculate a little about the paths that may have led from it to the usage of philosophers.

Let me note, first of all, that in ordinary conversation we would never acknowledge a power or a possibility where there are no alternatives. You can easily fall while jogging on an uneven road, but not after you have already stumbled and are lying on the ground (unless, indeed, the ground itself lies at the top of a precipice from which you can go on falling). On the other hand, in philosophy whatever is known to be necessary is also said to be possible. This philosophical usage can trace its pedigree to Aristotle. However, as Jaakko Hintikka (1960) has pointed out, Aristotle often speaks on these matters in perfect agreement with our ordinary usage. For him, 'to be possible' involves a twofold possibility: if a thing may be, it may also not be.[10] Indeed, in a passage to which he subsequently refers as a definition (διορισμός) of possibility, he explicitly separates the necessary from the possible.

λέγω δ' ἐνδέχεσθαι καὶ τὸ ἐνδεχόμενον, οὗ μὴ ὄντος ἀναγκαίου τεθέντος δ' ὑπάρχειν, οὐδὲν ἔσται διὰ τοῦτ' ἀδύνατον.

I use the terms 'possibly' and 'the possible' of that which is not necessary but, being assumed, results in nothing impossible.

(*An. Pr.* I, 13, 32ᵃ18–20)

Aristotle admits that the necessary is sometimes said to be possible *by homonymy.*[11] But, strictly speaking, possibility and necessity are to be distinguished as two mutually exclusive and jointly exhaustive modes of being:

ἔσται ἄρα τὸ ἐνδεχόμενον οὐκ ἀναγκαῖον καὶ τὸ μὴ ἀναγκαῖον ἐνδεχόμενον.

That which is possible, then, will not be necessary; and that which is not necessary will be possible.

(*An. Pr.* I, 13, 32a28f.)

καὶ ἔστι δὴ ἀρχὴ ἴσως τὸ ἀναγκαῖον καὶ μὴ ἀναγκαῖον πάντων ἢ εἶναι ἢ μὴ εἶναι.

And, indeed, the necessary and not necessary are perhaps the principle of everything's either being or not being.

(*De Int.* 13, 23ᵃ18f.)

Now, if every possibility involves a plurality of options, there are no possibilities in what is already past. Aristotle did not hesitate to draw this conclusion:

οὐδεμία γὰρ δύναμις τοῦ γεγονέναι ἐστίν, ἀλλὰ τοῦ εἶναι ἢ ἔσεσθαι.

There is no power of having been, but only of being or going to be.

(*De Caelo* I, 12, 283ᵇ13)

Whence, of course, "what has already happened is necessary" (ἔχει γὰρ τὸ γεγονὸς ἀνάγκην—*Rhet.* III, 17, 1418ᵃ3-5; see also *Eth. Nich.* VI, 2, 1139ᵇ7–11). In this view, the past and the future do not just differ chronologically—in that they respectively precede and follow a particular moment of time, which happens to be now—but are two essentially distinct realms of being, one of which is being gradually transmuted into the other all the time. The limit between them—τὸ νῦν, the now—is the locus of fading options, where protean possibilities are ever taking a definite and definitive—necessary—shape.

Most philosophers find such metaphysics unpalatable, but it is certainly close to common sense. However, in everyday talk one often speaks of possibilities concerning the past. I do not mean just past possibilities, i.e., options which were once open but are now closed—for instance, the possibility of preventing World War I through diplomatic channels in July 1914. When such possibilities were in effect, they concerned the future and do not therefore constitute a problem for the stated view. But we also say quite commonly that it *is* possible—now—that such-and-such *was* the case—in the past. We may thus mention the possibility that X.Y.Z. murdered U.V.W., or that the Solar System resulted from a close encounter between two stars. Should we therefore conclude that the above views on modality and time do not, after all, agree with common sense? Questions about the contents of common sense are notoriously slippery, for any attempt to answer them naturally requires more decision and precision than one may wisely ascribe to it. However, it may be instructive to note that (i) we would not say that *it is possible* that X.Y.Z. murdered U.V.W. if we knew for certain that he did so, and (ii) we would regard the said modal statement as refuted if we learn, for instance, that X.Y.Z. was in New Zealand when U.V.W. was stabbed in Beirut. It appears, therefore, that when we talk of present possibilities with regard to past events all we mean to imply is that there are alternative ways in which our incomplete knowledge of such events may eventually be completed, not that there are any options open as to how they actually took place (or as to how their "counterparts" happen in different worlds).

By lifting our attention from particulars—be they individual events, processes, situations, states, or things—to their respective types, we can derive from the ordinary timebound notion of possibility as power a time-independent concept somewhat more akin to the possibility of philosophers. A natural way of doing it would be the following. Let us say that a type K of objects is possible if and only if there was or will be a time at which some object of type K is possible in the ordinary, timebound sense—a time, that is, at which the power exists to produce or not to produce such an object. Then, K is impossible, and the complementary type K′ is necessary, if and only if there never is a power to produce an instance of K. The possibility and necessity we have thus defined are not the timeless modalities of philosophy, but their ascription, if true, would not change with time. Evidently, a necessary type is actualized at all times and is, in this sense, eternal. But, of course, it does not follow that everything eternal is therefore necessary. Curiously, however, Aristotle draws just this conclusion:

τὸ γὰρ ἐξ ἀνάγκης καὶ ἀεὶ ἅμα· ὅ γὰρ εἶναι ἀνάγκη οὐχ οἷον τε μὴ εἶναι· ὥστ' εἰ ἔστιν ἐξ ἀνάγκης, ἀΐδιον ἐστι, καὶ εἰ ἀΐδιον, ἐξ ἀνάγκης.

> That which is *of necessity* is at the same time *always*. For what must necessarily be is incapable of not being. So that if something is of necessity, then it is eternal, and if eternal, then of necessity.
> (*De Gen. et. Corr.* II, 11, 337^b35–338^a2)

Now, if "capable of not being" (οἷον τε μὴ εἶναι) is understood as I have proposed, eternity does not entail necessity. A nuclear war is clearly possible even if in the end—viz., after mankind becomes extinct due to other causes—it turns out that there never was one. Aristotle's reasoning is valid, however, if a different concept of possibility is presupposed. The following Diodoran notion has been ascribed to him by Hintikka (1957, 1973) and others: A type *K* is possible if and only if there was or will be a time at which an instance of *K* is actual.[12] Suppose that *K* is, in this sense, not possible. Then, no instance of *K* is ever realized. Therefore, the complementary type *K´* is always actual. At the same time, since *K* is impossible, *K´* is necessary. In other words, a type *K´* is necessary if and only if it is actualized at all times. There are several passages in which Aristotle appears to endorse Diodoran modality. According to him, "it is evident that it cannot be true to say: 'this is possible, but it will not be'" (δυνατὸν μὲν τοδί, οὐκ ἔσται δέ—*Metaph.* IX, 4, 1047^b4–5). Also, "it is obviously impossible for that which is destructible not to be destroyed sometime" (ἀδύνατον φθαρτὸν ὄν μὴ φθαρῆναί ποτε—*De C.* I, 12, 283^a25). Other passages provide evidence to the contrary. The most remarkable among them concerns a cloak which it is possible to cut—or not to cut—but is never cut because it first wears out (*De Int.* 9, 19^a13ff.). Hintikka, however, does not think that this passage is at all relevant, because cutting the cloak is a particular event, and the concept under discussion applies to event types. Be that as it may, the issue is of no great consequence to us here, for the modalities of current philosophy are not Diodoran, and Diodoran modalities are blatantly at odds with ordinary usage. (Surely one would normally say of any given class of deployed nuclear missiles that it has posed a real danger, a possibility of violent death for millions of people, even if it grows obsolete and is decommissioned before ever being used.)

I surmise that to go from ordinary to philosophical modal discourse you do not have to climb from tokens to types, from particulars to universals, but that, as in other cases in which philosophers use common words in an odd way, you will find the missing link in medieval theology. The idea of God's power, or, rather, of God *as* power, is central to Judaic, Christian, and Islamic theology.[13] Such power is boundless, and therefore, in a sense, nothing should be impossible for Him. The theologian is required to explain exactly in what sense this is true. For surely God cannot destroy Himself or detract from His goodness. It will not do to say that God can do whatever is possible in the time-independent sense proposed above, i.e., whatever lies within anyone's or anything's capacity at any time and any place. God's omnipotence is more than just the sum total of all worldly powers.[14] On the other hand, we do not want a circular explanation that simply equates what is possible for God with what lies in His own power. To solve this conundrum, Aquinas resorts to the following concept of absolute possibility:

> Relinquit igitur quod Deus dicatur omnipotens, quia potest *omnia possibilia absolute*, quod est alter modus dicendi *possibile*. Dicitur autem aliquid possibile vel impossibile absolute, ex habitudine terminorum: possibile quidem, quia praedicatum non repugnat subiecto, ut Socratem sedere; impossibile vero absolute, quia praedicatum repugnat subiecto, ut hominem esse asinum.
>
> It remains to say that God is omnipotent because he can do everything that is absolutely possible. This is another sense of 'possible'. Something is said to be absolutely possible or impossible because of the disposition of terms; viz., possible, if the predicate is compatible with the subject, as in 'Socrates is seated'; but absolutely impossible, if the predicate is incompatible with the subject, as in 'the man is an ass'.
>
> (*Summa Theol.* 1, qu.25, a.3)

This is, of course, the concept of possibility that has prevailed in modern philosophy. According to Aquinas, it agrees with one of several meanings of 'possible' distinguished by Aristotle, viz., "the possible conceived without reference to a power" (δυνατά λεγόμενα οὐ κατὰ δύναμιν).[15] Next to the passage I have just quoted, and again, more clearly and at greater length, in the *Summa contra Gentiles*, 2.25, Aquinas shows why the word must be used in this particular acceptation to explain what is or is not possible for God. The gist of his argument is that since the object and effect of an active power is *ens factum*, a real product, and since no such power can operate on an ontologically deficient object, therefore, just as sight cannot see in the dark, God cannot do anything that goes against the very concept of being. But only one thing is contrary to the concept of being, namely, nonbeing ("nihil autem opponitur rationi entis, nisi non ens"). Hence, God cannot make that one and the same thing simultaneously is and is not ("hoc igitur Deus non potest, ut faciat simul unum et idem esse et non esse"). I do not see that apart from such ontotheological considerations, there is any reason for divorcing possibility from power.

From his logical understanding of modality Aquinas derives also the necessary fixity of the past—which not even God can undo.

> Deus non potest facere quod praeteritum non fuerit. Nam hoc etiam contradictionem includit: eiusdem namque necessitatis est aliquid esse dum est, et aliquid fuisse dum fuit.
>
> God cannot make that the past has not been. This too involves a contradiction: for indeed the same necessity pertains to something being while it is as to something having been while it has been.
>
> (*Summa c. Gent.* 2.25)

From our ordinary vantage point here and now the necessity of the past is, of course, so trivial that one does not care to mention it. What matters to us is the necessity of future events, or, more precisely, their necessitation by the present. The perceived success of the physical sciences in understanding it and bending it to our advantage is probably the chief source of the popular admiration for them. (And the popular contempt for philosophy would no doubt increase if more people knew that many of its practitioners—including some self-styled "realists"—dispute that any such necessitation exists). Our next—and final—task will be to examine the intellectual means employed by theoretical physics in the pursuit of that twofold aim. I shall maintain that they have consisted in grasping the seemingly blind fatality of events as a manifestation of conceptual relations. But we shall be able to see better how the necessity of natural processes can be epitomized by conceptual relations of a certain type if we first take a look at another, no less familiar, form of necessity.

When a speaker of English says that you *cannot* do a certain thing, he need not refer to an impossibility that constrains you with what I have been calling brutal necessity. For example, if you are told that in some parts of Antarctica you cannot sleep in the open air, this does not mean that you absolutely cannot do it, but that you cannot do it and still avoid freezing to death during the night. Should one rather say that you *can* but *may not* do it? However, ordinary usage stubbornly refuses to make a fast and clear distinction between 'can' and 'may'. Indeed, the former is the proper verb to use whenever a necessary connection is sensed—or assumed—between what is declared (conditionally) impossible and some result that one (absolutely) wants to preclude.

The same expression 'you cannot' is employed to convey the myriad impossibilities that entangle—and enable—our life in society. But the necessity here at play is plainly of another sort. And within this family, the conditional shades almost insensibly into the absolute. Consider a few examples:

(1) You cannot drive northwards on Fifth Avenue.

(2) You cannot read a Spanish text from right to left.

(3) You cannot obtain an Irish divorce.

(4) In most countries, you cannot sell real estate without a deed.

You can drive, of course, on a Sunday morning from the New York Public Library to Saint Patrick—you merely risk being stopped and fined by a policeman. Also, if you speak Spanish, you can manage, with a little exercise, to read fluently from right to left. Juan de la Cruz will sound like a Martian poet, but a poet all the same. However, his words will not make sense. Thus, whether the impossibility stated in (2) is absolute or conditional depends on what one means by reading. On the other hand, at the time of writing it is absolutely impossible to obtain a divorce in Ireland. An Irishman can indeed leave his Irish wife and never see her again. He can also divorce her in Mexico and marry a third party in Madagascar. But under Irish law they will still be married till death them do part. Example (4) illustrates the same idea, but there is an interesting twist to it, due to the mingling of physical with legal necessity. In exchange for a lump sum I can surrender my home in Puerto Rico to an illegal alien who does not wish to sign a legal document. But he will not own it. So he cannot sell it (without fraud). And he obviously cannot take it to another country where our verbal contract might be acknowledged as a valid sale.

Ever since Greek intellectuals pondered the contrast between φύσις and νόμος—nature and convention—there has been a philosophical tendency to lump into the latter category all forms of social necessity—linguistic, legal, ethical, etc.—and to oppose them to the former. Such polarity is evidently simplistic. In an attempt to defuse it, some 5th century Athenian coined the oxymoron ὁ νόμος τῆς φύσεως[16]—nature's convention—subsequently rendered in Latin as *lex naturae,* the law of nature. The phrase is still very much alive in papal encyclicals and epistemological tracts. We cannot delve here into its diverse and often conflicting connotations. We have to consider only a particular form of socially induced necessity which I believe can throw some light on the handling of natural necessity in physics; viz., the thoroughly conventional necessity that restricts the admissible moves in a game of strategy.

In every game there are things that a player can never do, or cannot do in certain situations. In sports, however, the impossibility is conditional: the inadmissible move—e.g., a forward pass in rugby football—cannot be undone, so if such a move is performed—and seen by the umpire—the culprit is penalized. But in pure games of strategy, such as chess or tic-tac-toe, impossible moves are absolutely impossible: you cannot make your queen jump over a bishop and still pretend that you are playing chess. The moves are best thought of as a succession of ideal elements, symbolized by certain physical events. Chess moves are represented by motions of the pieces on the chessboard, but not every physically possible motion of a piece represents a move in the game. At any given stage of the game, the person whose turn it is to play can choose her move from a fixed set characteristic of the position that was reached in the preceding move. (In some positions, she may be restricted to a single move.) The chosen move leads to another position, in which the rival player can in turn choose from a fixed set a move leading to another position, etc. Thus, the alternative courses that a game can follow from any given point are adequately conceived as paths in an oriented graph,[17] whose vertices are chess positions and whose edges are chess moves. Such a graph is a subgraph of what I shall call the graph of chess, an oriented graph that contains a vertex representing the initial position of the game and all the paths that issue from that position and can be followed by successively playing legal moves.[18] Let a complete path be one that begins at the initial position and ends in checkmate or in a legal draw. Every conceivable game of chess is represented by some initial segment of a complete path, and almost every complete path represents a legal game of chess. (The rare exceptions are explained towards the end of note 18). Each time two persons sit down to play chess, every game thus represented in the graph of chess is, in principle, possible: not, indeed, because in some world or other, counterparts of either player really play that game, but because *hic et nunc*, in this very world, the one and only in which both players are alive, their free submission to the rules of chess enables all those games. As their particular game progresses, more and more of the alternatives initially open become closed. Eventually, a position may be reached where, say, Black checkmates in four moves, i.e., a position such that by judiciously choosing his move on his next three turns Black can constrain White into a path ending with checkmate after White's third turn. As this case clearly shows, the possibilities and necessities in question are to be understood in the ordinary sense: they concern the players' actual powers at a particular juncture in their lives. Yet the graph of chess furnishes a universal scheme which covers many such junctures and furnishes an exact idea of the alternatives available at each one of them. The graph even provides the means of calculating the utility of the various possible moves for the purpose of winning the game. Thereby, the players' conduct of the game becomes understandable and—up to a point—predictable. Intelligibility and predictability follow here from the free agreement between two intelligent creatures to play a game of strategy they both conceive in the same way. But why shouldn't inanimate things likewise fall of themselves—or by the Demiurge's will—under a conceptual pattern spanning their evolution in time so that their past enables and constrains their future? Mathematical physics was originally rooted in just this belief. Whether it can long survive without it is indeed a moot question.

5.2 Geometry

Games of strategy have been around for a long time. Obviously they cannot be played without some form of awareness of their underlying structure. But a mathematical theory of games was not developed until fairly recently. The deliberate investigation of mathematical structures as a source of necessity and possibility did not begin with the study of games but with the study of geometry.

We tend to think of geometry as the science of space. Kant, who viewed it this way, saw space, in turn, as the "condition of possibility" of our perception of things; but it is perhaps more appropriate to think of it as a primary enabling and constraining condition of action—of our own action on things and of their reaction and interaction. To the Greek inventors of geometry, who did not even have a term equivalent to 'space', such talk would have seemed abstruse. Euclid's *Elements* are concerned with figures, for the most part with flat figures (on a plane). The construction of figures with certain properties was a matter of practical interest to carpenters, architects, surveyors, etc. At some point, however, members of the educated leisure class were fascinated by the subject and pursued it just for the fun of it, beyond any practical need.[19] Euclid assumes that one can execute certain elementary constructions, viz., join any pair of points by a straight line, extend a given straight beyond either endpoint, draw a circle with any point for a center and any straight segment for a radius. Thereupon he shows how to construct other figures by judiciously combining the elementary constructions and demonstrates certain features which the figures thus constructed will inevitably possess. As the book progresses the construction tasks recede, the demonstrations prevail unrivalled, and the illusion is fostered that one is reading, not about possibilities available here and now (and the constraints they are subject to), but about the properties and relations of timeless objects in another world. Demonstration follows upon demonstration in the "long chains of reasons" admired by Descartes.

Some geometrical arguments are so well knit that one may be tempted to think that the necessity displayed by them is purely verbal, that the conclusions are implicit in the very meaning of the words employed. Take, for instance, the striking fact that the three perpendicular bisectors of any triangle meet at one point. Let A, B, and C be the vertices of a triangle and let the perpendicular bisectors of AB and BC meet at Q. Since QA = QB = QC, Q evidently lies on the perpendicular bisector of the remaining side AC. But, of course, the expression 'perpendicular bisector of XY' does not designate the locus of points equidistant from points X and Y but rather the perpendicular through the midpoint of segment XY. That the latter constitutes the said locus is, however, so obvious that it may well go without saying. Just turn over the plane while holding the perpendicular bisector of XY fixed: X and Y will exchange places while every point on the perpendicular bisector stays put; hence, any such point lies at the same distance from X and Y. I have resorted here to an idea of rigid motion which can no doubt be verbalized and put forward as a definition of the subject matter of geometry. Necessary and sufficient conditions for every construction and demonstration in Euclid can also be stated in other ways.[20] But the bonds that such verbalizations disclose between the components of geometrical figures are not established by linguistic usage, let alone by linguistic agreement. Verbal necessity may link the premises to the conclusion in a geometrical proof but not the points and lines of the plane among themselves in the manner entailed by those premises.

The illustration I shall now give will, I hope, make my meaning clearer. Like the verbalizations I have alluded to, it involves the modern (Cartesian) understanding of the Euclidian plane. It also involves some—fairly simple—notions of complex analysis which, unfortunately, I cannot explain here. However, the following preliminaries should enable the reader unacquainted with those notions to see the drift of my argument. Other readers ought to proceed directly to the beginning of the next paragraph. Complex numbers were originally introduced to provide solutions for every quadratic equation. They can be readily conceived as ordered pairs of real numbers, subject to the following rules of addition and multiplication: $\langle a,b\rangle + \langle c,d\rangle = \langle a+c, b+d\rangle$; $\langle a,b\rangle \times \langle c,d\rangle = \langle ac-bd, ad+bc\rangle$. Endowed with these operations, the set $\{\langle a,b\rangle \mid a,b \in \mathbf{R}\}$ is a complete field, the field of complex numbers, usually denoted by **C**. Note that the said operations, restricted to $\{\langle a,0\rangle \mid a \in \mathbf{R}\}$, characterize a subfield of **C**, isomorphic to **R** itself. (Put $b = d = 0$ in the above definition of complex addition and multiplication.) Every complex number whose second term is equal to 0 may therefore be identified with the real number that is its first term, viz., $\langle a,0\rangle = a$; in particular, $\langle 1,0\rangle = 1$. Note further that $\langle 0,1\rangle \times \langle 0,1\rangle = \langle -1,0\rangle = -1$. Thus, $\pm\langle 0,1\rangle$ are the solutions of the quadratic equation $x^2 + 1 = 0$. The complex number $\langle 0,1\rangle$ is usually denoted by i. Evidently, every complex number can be written as a linear combination (with real coefficients) of 1 and i, thus: $\langle a,b\rangle = (\langle a,0\rangle \times \langle 1,0\rangle) + (\langle b,0\rangle \times \langle 0,1\rangle) = a + bi$. If we now assign Cartesian coordinates to the Euclidian plane, we can identify the complex number $\langle a,b\rangle$ with the point with abscissa a and ordinate b, or, better still, with the vector from the origin to that point. In this interpretation, the addition of complex numbers as defined above agrees with the standard vector addition (as displayed in the familiar rule of the "parallelogram of forces"). To interpret multiplication we go over to polar coordinates. Let ϕ be the angle formed in the positive, i.e., counterclockwise, sense between the axis of the abscissae and the vector we have identified with $\langle a,b\rangle$. Thus, $\phi = \tan^{-1}(a/b)$. Put $\rho = \sqrt{(a^2 + b^2)}$. Clearly, then, $\langle a,b\rangle = a + bi = \rho(\cos\phi + i \sin\phi)$. ϕ is the argument or amplitude of $\langle a,b\rangle$; ρ is its modulus or absolute value. If, as is usually done, we denote a complex number by a single letter, say, z, its argument is denoted by arg z and its modulus by $|z|$. A couple of trials should persuade the reader that multiplication of any complex number by another has the twofold effect of stretching (or contracting) the former by the latter's modulus and rotating it by its argument. End of preliminaries.

Consider the mappings of **C** into **C**, i.e., in the given interpretation, the mappings of the Euclidian plane into itself. (Mappings and the relevant notation were explained in Section 2.8.3.) By a region I mean a non-empty open connected subset of the plane (e.g., the interior of a polygon). A mapping $f: \mathbf{C} \to \mathbf{C}; z \mapsto f(z)$ is said to be regular on a region M if $f(z)$ changes smoothly as z ranges over M. More precisely, f is regular at $\zeta \in M$ if there is a complex number $f'(\zeta)$ which meets the following condition: for every positive real number ε there is a disk δ, centered at ζ, such that every z in this disk satisfies the inequality $|((f(\zeta) - f(z))/(\zeta - z)) - f'(\zeta)| < \varepsilon$. f is regular on M if it is regular at every $\zeta \in M$. Suppose now that Γ is a closed path in M, surrounding a simply connected[21] subregion of M, which we shall call $I(\Gamma)$ (I for interior). Then, if f is regular on M, its value at any point $z \in I(\Gamma)$ is given by

$$f(z) = \frac{1}{2\pi i}\int_{\Gamma} \frac{f(z)}{\zeta - z}\, d\zeta$$

the line integral being taken in the positive (i.e., counterclockwise) sense over the path Γ. The reader need not worry about the definition and the existence of the integral. What matters for my purpose here is that according to the theorem just quoted, the values that *any* mapping $f: \mathbf{C} \to \mathbf{C}$ takes on the boundary of a simply connected region of the plane uniquely determine its value at each and every point inside that region, provided that both the latter and its boundary are included in a region on which f is regular. (Apart from this condition, f is *completely arbitrary*.) The theorem can, of course, be proved by standard deduction from the axioms that characterize the field **C** and the definitions of the words employed. But the bond it reveals between the points of the plane should not be classified as one of "verbal" or "logical" necessity; not, at any rate, if logical necessity shapes up as inference and we do not countenance inferences from uncountably many premises. I, for one, would judge it a hypertrophy of logic to view every value that the regular mapping f takes on the uncountably infinite points enclosed by the path Γ as the conclusion of an inference which has for premises the values of f on the uncountably infinite points of Γ.

Euclid did not spell out every condition on which his constructions and demonstrations depend. It is a monument to his mathematical insight that he did make explicit one seemingly out-of-the-way requirement—the notorious Postulate 5—without which his whole project of reconstructing geometrical figures by sequences of elementary constructions would founder. As Paul Lorenzen (1984) has aptly noted, the enterprise of geometry, as the Greeks understood it, presupposes the repeatability of such sequences with homologous results. Suppose, for instance, that on the strength of Euclid's Postulates 1–3, I pick out two points A and B, draw the segment AB, draw AC equal and perpendicular to AB, bisect angle CAB, mark on its bisector β the segment AH equal to AB, draw the straight λ through H perpendicular to β and coplanar with AB, and finally verify that λ meets the extensions of AB and AC beyond B and C. (The reader is advised to sketch this exercise on a piece of paper.) The Euclidian program would plainly make no sense unless every time that the same series of constructions is repeated, starting from *any* pair of points A and B, the same verification ensues, viz., that the perpendicular λ to the bisector β at the (arbitrarily chosen) distance AB from A eventually meets the two sides of the right angle whose vertex is at A. However, in this case, repeatability is guaranteed only if the requirement put forward in Euclid's fifth postulate is satisfied.[22] The gist of that requirement can be stated as follows: If two coplanar straights λ and μ form unequal pairs of internal angles with a transversal τ, λ and μ meet on the side of τ on which they form the smaller pair of internal angles.[23] Of course, if the two pairs of angles are only slightly different, the intersection of λ and μ is apt to lie very far away. Thus, the feasibility of the Euclidian program for the construction and analysis of finite figures turns on far-flung features of the indefinitely extended plane. There are conditions equivalent to Postulate 5 which do not openly speak of matters so remote. For example: (i) Given a triangle with angles α, β, γ, it is possible to construct another, smaller triangle with the same angles (Wallis 1693). Also: (ii) It is possible to construct one rectangle

(Saccheri 1733). These apparently innocuous demands seem more germane to Euclid's overall constructive, finitistic approach to geometry than the requirement that two coplanar straights *nearly* orthogonal to a common transversal should meet. We should thank and admire Euclid—or his anonymous source—for having singled out the latter, more obviously compromising condition for explicit formulation. By thus drawing attention to the unifying structure that embraces and supports all geometrical constructions, he laid the groundwork for the modern view of geometry as concerned with space.

The repeatability of geometrical constructions is not indifferent to experimental physics. The design and interpretation of physical experiments generally presupposes some geometrical articulation of their place of occurrence. The commonest and humblest precision instruments sport simple shapes which are critically relevant to their performance, and their manufacturers assume as a matter of course that the elementary Euclidian constructions can be repeated *ad libitum* in the terms described above. Because of this, Hugo Dingler maintained that physics is irrevocably committed to Euclid's Fifth Postulate and may not toy with non-Euclidian geometries. At most, the latter can furnish useful computational tools, with no claim to reality.[24] Like so many other philosophers of science, Dingler apparently forgot that in physics every quantitative concept is blurred, i.e., is predicated with an agreed margin of error. Thus, there is no inconsistency in setting up a relativistic model of the universe with nonzero spatial Riemannian curvature on the strength of data gathered with a telescope manufactured according to Euclidian specifications, for in a small neighborhood of the observatory the Riemannian metric of the model deviates insignificantly from that of the local tangent space. (Think of an architect who plans to have a piazza tiled with marble squares, although he knows very well that a surface levelled with the assistance of terrestrial gravity is not flat and therefore, strictly speaking, does not allow such tiling.) And, of course, the overall site of the experiment at whose receiving end our telescope stands can be articulated as a fragment of a curved spacetime. Let me illustrate this with an example from relativistic cosmology. As is well known, the light from distant galaxies is redshifted, i.e., its spectrum displays familiar emission and absorption lines at frequencies lower than the ones predicted for such lines by atomic theory and measured on their analogues in terrestrial laboratories. In relativistic cosmology the redshift in question is attributed to the "expansion of the universe" and is therefore indicative of the time elapsed since the light was emitted. To ascertain the age of a given extragalactic light signal by measuring its redshift, the relativistic astronomer may not simply project the Euclidian trajectory of the incoming rays and figure out its length in light-years but must conceive the signal's history—its "worldline"—as a path in a curved spacetime. I do not see anything wrong with this. Indeed, from this standpoint the very reading of the data depends on a conjectural estimate of the spacetime curvature. But that is just one more uncertainty that one must learn to live with in our uncertain world.[25]

On the other hand, there are geometric propositions which do not admit approximation but are either exactly true or exactly false, and if true, then inescapably so. We are all acquainted with knots that become untied and others that only become tighter when one pulls the ends of the string. If a given knot belongs to one of these two kinds, so will every copy of it, no matter when or where or from what material it is made. A knot's behavior follows inexorably from the manner in which the string is threaded. Mathematicians can exactly describe such threading and also derive from first principles the kind of knot it will produce. But surely the tightening of some knots and the untying of others are not just matters of verbal necessity. Likewise, howsoever you divide the surface of a ball into patches, there will always be a manner of painting them solid red, blue, green, and yellow so that any two patches sharing a borderline display different colors; but if instead of a ball you are dealing with a ring-shaped object, you may require up to six colors to do the trick. Such properties can, of course, be inferred from a suitable definition of ball-shapedness or ring-shapedness, but again, it is no mere matter of linguistic preference how you have to define these shapes in order to license such inferences.

5.3 Mathematical physics

Given two points *A* and *B*, the existence of an intersection *C* between any two straights through *A* and *B*, the size of the angles they make at *C*, and the distance from *C* to the segment *AB* are completely determined. So much follows from the conditions spelled out in the axioms of Euclidian geometry. They impose necessary connections between the parts and features of any constructible geometrical figure whereby they both enable and constrain the human activity of constructing it. The axioms tell us what such a figure must have looked like if it was drawn, what it will look like if it will be drawn. But they do not establish any links between different times, between, say, yesterday and tomorrow. This is usually expressed by saying that geometrical relations are timeless, meaning that they hold at each particular time.

The necessity we see and feel in nature, and which we would wish to fend off and turn to our advantage, is not just timebound but timebinding: it ties the future to the present and the past, forcing change and preordaining its outcome, channelling the flux of events. At first blush, it has precious little to do with geometry. Yet the amazing contribution of modern physics to the exploitation of natural necessity by man has resulted from understanding it on the analogy of geometric necessity, through the representation of natural processes by mathematical structures. This was made possible by one of the most remarkable feats of human thought: the conception of time—past, present, and future—as a single, homogeneous linear continuum.[26]

A lucid explication of what it takes to be a linear continuum was not available until the late 19th century;[27] but the science of motion established two centuries earlier by Newton clearly presupposes that time has the same structure as the trajectory of a free particle in space. Indeed, when Galileo, in the Third Day of the *Discorsi*, let a line represent the time in which a certain space is traversed by a body in uniformly accelerated motion, and used the geometrical properties of that line to establish a functional relation between travelled distances and travel times, he must have understood that the time can be mapped bijectively onto the line, so that each segment discernible in the latter uniquely corresponds to a distinct subinterval of the former (Galileo, EN, VIII, 208–10; cf. EN, VIII, 85. See Section 3.1). A similar understanding was already implicit in the use of geometrical methods in Greek astronomy and also in the use of kinetic methods in Greek geometry itself. The ancient mathematicians did not feel bound by Euclid's short list of elementary constructions and worked on figures that cannot be drawn with ruler and compass. Some of these figures are defined by the motion of a point under certain precise conditions. For instance, the spiral studied by Archimedes in *De lineis spiralibus* is drawn by a point moving with constant speed (ἰσοταχέως) from a point, along a straight line rotating with constant speed—on a plane—about that point. The construction assigns a definite point on the plane to each instant of the motion, viz., the position reached at that instant by the moving point. The geometrical properties of the spiral depend, of course, on this definition, so their demonstration must involve a comparison between spatial magnitudes—lengths, angles, areas—and times. To establish his arguments on solid conceptual grounds, Archimedes begins his treatise by proving the following lemma: If a point moves with constant speed along any line and two lengths are taken on it, the latter have the same ratio to one another as the times in which the point traversed them. Archimedes reasons thus: Let AB be the line along which the point moves, and take on it the segments ΓΔ and ΔE. Let the time in which the point traverses ΓΔ be ZH, and HΘ the time in which it traverses ΔE. "It must be shown that the segment ΓΔ has to the segment ΔE the same ratio that the time ZH has to ΗΘ" (δεικτέον, ὅτι τὸν αὐτὸν ἔχοντι λόγον ἁ ΓΔ γραμμὰ ποτὶ τὰν ΔΕ γραμμάν, ὅν ὁ χρόνος ΖΗ ποτὶ τὸν ΗΘ—*De lin. spir.*, I). The proof proceeds by taking equimultiples of each length and the corresponding time, in accordance with the theory of proportions explained in Euclid's Book V. It is tacitly assumed that a point moving with constant speed traverses equal distances in equal times. As Dijskterhuis remarks, "it may seem that the proposition follows at once from this." However, a proof is required to show that the proposition holds also when the segments ΓΔ and ΔΕ happen to be incommensurable (Dijskterhuis 1987, p. 141).[28]

When Archimedes wrote *On Spirals*, the practice of mapping time into space was at least a century old. It was already involved in the geometrical models of planetary motions devised by Eudoxus, Plato's contemporary and friend, who also created the theory of proportions employed by Archimedes in his proof. Eudoxus associated each planet with an n-tuple of concentric spheres, such that (i) their common center is the center of the Earth; (ii) the first sphere rotates about the poles in one day; (iii) the ith sphere rotates with a constant speed of its own about an axis fixed on the $(i-1)$th sphere; (iv) the planet lies on the equator of the n-th sphere. If the number of the moving spheres and their respective axes and speeds of rotation are suitably assigned, the astronomer should be able to calculate from the planet's current position its declination and right ascension at any time. The description of a Eudoxian planetary model with given parameters can be read—on the analogy of the above definition of the Archimedean spiral—as the definition of a special kind of curve—e.g., a Venusian or a Martian trajectory—drawn by a point on any spherical surface, such as the vault of heaven or the ceiling of a planetarium. A body affixed to such a point instantiates the Eudoxian concept of a certain planet. The radius vector of the planet (i.e., the radius which carries—*vehit*—it on its tip) points in such-and-such directions at such-and-such times with the same inexorable necessity as the position vector of an Archimedean spiral reaches at prescribed times the third, the fourth, . . . the nth turn of the curve, or as any side of an equilateral triangle makes internal angles of $\pi/3$ radians with the other two. The future path of the bright spot we are now observing is fixed and timed by what it *is* at present, if indeed it *is*, say, a venus or a mars in the sense aforesaid. The success of Eudoxian astronomy never measured up to its true potential, chiefly, I suppose, due to the inaccuracy of naked eye observation and to the tremendous difficulty of estimating the appropriate parameters by manual computation (with zeroless arithmetic!). It was nevertheless sufficient to win Plato's support for the program of ascertaining "by what hypothetical uniform, ordered, and circular motions the phenomena regarding the motions of the so-called wandering stars can be preserved."[29]

Compared with Eudoxian astronomy, modern mathematical physics displays a pageant of astonishing richness. Thanks to the wonderful invention of mathematical analysis it is able to tie the present to the future and the past with "bonds of necessity" and to save the phenomena, even if their evolution is not reducible to a combination of uniform circular motions. Is the obvious difference between modern physics and ancient astronomy merely a matter of complexity—and predictive success? Or have Newton and Maxwell, Einstein and Dirac, reached for—and sometimes attained—an essentially deeper understanding of events than Eudoxus or Hipparchus? Much of the 20th century debate in the philosophy of science turns, openly or covertly, around this issue. For my own part, I am inclined to view the Eudoxian program as a paradigm of mathematical physics, *provided that*—as above—*it is interpreted realistically.*[30] There are, however, some important differences between it and the generally acknowledged Newtonian paradigm which deserve discussion. Let me mention briefly the three that come to my mind.

(1) Newton's physics is not a collection of (related) mathematical recipes for the prediction of regular occurrences in nature but a system of mathematical principles for natural philosophy. Its basic notions of time and space, mass and force, bound together in the Laws of Motion, are ultimately meant to account for *all* phenomena, not just for those observable in this or that segment of our environment.

(2) In all theories designed after the Newtonian paradigm, the states of physical systems are described and their evolution is explained in terms of purportedly universal properties of matter. Specifically, Newton's extraordinarily successful unified theory of planetary motion and free fall subsumed these two seemingly disparate families of physical processes by postulating a mutual attraction, dependent solely on mass and distance, between all pieces of matter. Second-generation Newtonians took this to mean that each piece of matter was inherently the seat of an attractive force, acting on every other piece of matter throughout the entire universe. Subsequently, physicists "discovered" further universal forces: electricity and magnetism (later unified by Ampère and, more successfully, by Maxwell), the so-called weak force manifested in radioactivity (unified with electromagnetism by Salam and Weinberg), and the strong force which keeps the positive electric charges in the atomic nucleus close together. Current research aims at unifying all acknowledged elementary forces in a single theory.

(3) In Newtonian physics the future and past states of a physical system are linked by differential equations to the forces actually working on it.

The program of universal physics was initiated ca. 1600 by Kepler and Galileo and was subsequently taken up by Descartes and his followers. It was deeply at variance with the two-tiered Aristotelian worldview favored by 16th century scholastics, but it could draw inspiration from Lucretius and other ancient writers, and ultimately from the Presocratics. Force was dismissed by the Cartesians as an obscure and indistinct idea, but it can, of course, be traced back to Greek philosophy and to prephilosophical common sense. Differential equations, on the other hand, though adumbrated by Galileo, are Newton's very own creation.[31]

Differential equations lie at the heart of mathematical physics and contain the key to the modern understanding of physical necessity. To see how they perform this role, I propose to explain the simplest form of differential equation and to briefly indicate (in notes) various ways of generalizing it. Let U be an arbitrary subset of $\mathbf{R}^{n+1}$. We consider a continuous mapping f of U into $\mathbf{R}^n$ by $\langle t, \mathbf{r}\rangle \mapsto f(t, \mathbf{r})$ $(t \in \mathbf{R}, \mathbf{r} \in \mathbf{R}^n)$. The following expression is then said to be an *ordinary differential equation of the first order*:

$$\frac{\mathrm{d}\mathbf{r}}{\mathrm{d}t} = f(t, \mathbf{r}) \tag{1}$$

We shall say that f is κ-Lipschitzian in $\mathbf{r} \in \mathbf{R}^n$ for some given positive real number κ, if, whenever $\{\langle t, \mathbf{r}_1\rangle, \langle t, \mathbf{r}_2\rangle\} \subset U \subset \mathbf{R}^{n+1}$,

$$|f(t, \mathbf{r}_1) - f(t, \mathbf{r}_2)| \leq \kappa|\mathbf{r}_1 - \mathbf{r}_2| \tag{2}$$

An *exact solution* of eqn. (1) is any mapping $\varphi: \mathbf{I} \to \mathbf{R}^n$ (where $\mathbf{I}$ is an arbitrary interval in $\mathbf{R}$) that meets the following three conditions:

(i) The first derivative φ' is defined and continuous on the interior of $\mathbf{I}$.

(ii) For every $t \in \mathbf{I}$, $\langle t, \varphi(t)\rangle \in U$.

(iii) For every $t \in \mathbf{I}$, $\varphi'(t) = f(t, \varphi(t))$.

A mapping $\varphi: \mathbf{I} \to \mathbf{R}^n$ is said to be an ε-*approximate* solution of eqn. (1)—for some real number $\varepsilon > 0$—if it meets conditions (i) and (ii) but satisfies the following requirement instead of (iii):

(iiiε) For every $t \in \mathbf{I}$, $|\varphi'(t) - f(t, \varphi(t))| \leq \varepsilon$.

If the derivative φ' becomes discontinuous on a finite subset $\{t_1, \ldots, t_n\} \subset \mathbf{I}$, φ is said to be an ε-*approximate piecewise* solution of eqn. (1) if condition (iiiε) is satisfied everywhere on $\mathbf{I}\backslash\{t_1, \ldots, t_n\}$, and for every i $(1 \leq i \leq n)$ we have that $|\varphi'_<(t_i) - f(t_i, \varphi(t_i))| \leq \varepsilon$ and $|\varphi'_>(t_i) - f(t_i, \varphi(t_i))| \leq \varepsilon$ (where $\varphi'_<$ and $\varphi'_>$ denote the derivative "from the right" and "from the left," respectively).

Suppose now that the mapping f on the right-hand side of eqn. (1) is κ-Lipschitzian in $\mathbf{r} \in \mathbf{R}^n$, for a given $\kappa > 0$. It can be proved that if φ_1 and φ_2 are two exact solutions of eqn. (1) defined on the same interval $\mathbf{I} \subset \mathbf{R}$, such that for some $t_0 \in \mathbf{I}$, $\varphi_1(t_0) = \varphi_2(t_0)$, then φ_1 and φ_2 are identical. In other words, under the stated conditions, there is on any interval about a suitable point in $\mathbf{R}$ at most one exact solution of eqn. (1) that takes at that point a given value in $\mathbf{R}^n$.[32] Suppose, further, that the domain U of the said mapping f is a closed subset of $\mathbf{R}^{n+1}$ that contains the point $\langle t_0, \mathbf{r}_0 \rangle$, and that $\mathbf{I} \subset \mathbf{R}$ is a compact interval that contains t_0. It can be shown that, if for every $\varepsilon > 0$ there is an ε-approximate piecewise solution $\varphi_\varepsilon: \mathbf{I} \to \mathbf{R}^n$ of eqn. (1) such that $\varphi_\varepsilon(t_0) = \mathbf{r}_0$, there exists an exact solution $\varphi: \mathbf{I} \to \mathbf{R}^n$ of eqn. (1) such that $\varphi(t_0) = \mathbf{r}_0$. Evidently, this exact solution is unique (by the former theorem).[33]

As I mentioned earlier, the differential equations I have been discussing are those of the simplest kind. The concept of an ordinary differential equation of the first order can be readily generalized by substituting in the above definitions the complex number field $\mathbf{C}$ for the real number field $\mathbf{R}$, or by choosing the domain U of f in eqn. (1) to be a subset of $\mathbf{R} \times \mathbf{A}$ or of $\mathbf{C} \times \mathbf{B}$—where $\mathbf{A}$ is a real and $\mathbf{B}$ a complex Banach space (see Section 2.8.4). Ordinary differential equations of the nth order involve derivatives of their solutions of order up to n.[34] Partial differential equations have solutions defined on $\mathbf{R}^m$ or on $\mathbf{C}^m$ (for some integer $m > 1$) and involve their partial derivatives.[35] Theorems on the existence and uniqueness of solutions to differential equations of these further kinds have been demonstrated under diverse restrictive assumptions. It is due to the existence and uniqueness of solutions that under suitable circumstances, differential equations provide an unequalled grasp of physical necessity.

To see this, let us go back to the simple case of eqn. (1). Suppose that $\mathbf{r}$ represents a physical quantity and that eqn. (1) expresses its time rate of change. (Remember that $\mathbf{r} \in \mathbf{R}^n$; each value of $\mathbf{r}$ is therefore a list of n real numbers and may therefore represent the distances of n particles from a given point, or the three position coordinates and three velocity components of $n/6$ particles, or the position coordinates, velocity and acceleration components, masses, and electric charges of $n/11$ particles, etc.) Suppose, moreover, that the assumptions for the existence and uniqueness of solutions are satisfied on a domain $U \subset \mathbf{R}^{n+1}$. Then, if the value of $\mathbf{r}$ is given at some time $t_0 \in \pi_1(U)$, it is thereby fixed for every time $t \in \pi_1(U)$. If the stated conditions hold, nothing—not even the Will of God—can change the course of the physical quantity represented by $\mathbf{r}$. Thus, if a physical process can be adequately conceived as the evolution in time of a quantity $\mathbf{r}$ governed by the differential equation (1) its necessity will thereby be finally grasped and understood. The same can be said, mutatis mutandis, if the adequate conception of a physical process involves a more general form of differential equation, provided, of course, that the appropriate assumptions for the existence and uniqueness of solutions are fulfilled. Note that this manner of understanding converts natural necessity into mathematical, and thus conceptual, necessity.[36] It stands to reason that some such conversion is required for an *understanding* of necessity.

The following example should further clarify the power of the mathematical physicist's approach to natural necessity, as well as its limitations. In the Hamiltonian formulation of Classical Mechanics, the evolution of an isolated system of n particles is fully determined by the current values of $6n$ functions of the time. They may be chosen to stand for the $3n$ position coordinates and the $3n$ momentum components of the particles in Euclidian space. Since we take the mechanical system to be isolated, its total energy H depends exclusively on the current position and momenta. If t denotes time and q_{hk} and p_{hk} denote respectively the hth position coordinate and the hth momentum component of the kth particle, the evolution of the mechanical system is governed by the following system of $6n$ partial differential equations of the first order:

$$\frac{\mathrm{d}q_{hk}}{\mathrm{d}t} = \frac{\partial H}{\partial p_{hk}} \qquad \frac{\mathrm{d}p_{hk}}{\mathrm{d}t} = -\frac{\partial H}{\partial q_{hk}} \qquad (h = 1, 2, 3;\ k = 1, \ldots, n) \tag{3}$$

A solution of eqns. (3) is a smooth curve in $\mathbf{R}^{6n}$ (or $6n$-dimensional Euclidian space), each point of which encodes the position coordinates $q_{hk}(t)$ and the momentum components $p_{hk}(t)$ of the n particles at a particular time t. If the conditions for the existence and uniqueness of solutions are fulfilled, there is one and only one such curve through any given point in $\mathbf{R}^{6n}$. Obviously, in that case, the position and momenta of the particles at any given time fully determine the evolution of the mechanical system before and after that time. Enraptured by this vision, Laplace stated his famous claim:

> An intelligence that knew, for a given instant, all the forces acting in nature, as well as the positions of all the things that constitute it, and who was capable of subjecting these data to analysis, would embrace in a single formula the motions of the largest bodies and those of the lightest atom. For her nothing would be uncertain, and the future, like the past, would be present to her eyes.
>
> (Laplace, OC, vol. VIII, pp. vi–vii)[37]

But even Laplace would acknowledge that human science, though "incessantly approaching" the intelligence just described, "will always remain infinitely far from it."[38] Indeed, Hamilton's equations hold in the above form (3) for an isolated system, but no physical system we will ever come across is really isolated. We do our best to seclude our experiments and our machinery from the rest of the world, at least insofar as the quantities of interest to us are concerned. But we never succeed perfectly or permanently. The Solar System lies so far away from all other significant gravitational sources that it provides a virtually insuperable paradigm of Laplacian determinism. It would, however, come apart and all predictions regarding it would be invalidated if the Sun came close to another star. Moreover, as we know, all the data of physics are more or less blurred, and of course blurred values will not determine unique solutions when substituted in differential equations such as (1) or (3). Based on this consideration, Popper (1950) has argued that contrary to the vulgar opinion, determinism is alien to classical physics. But even if one disregards the effects of blurredness, determinism encounters severe limitations in both classical and contemporary physics, as John Earman has shown in his *Primer on Determinism* (1986). I cannot summarize here this admirable book, which no budding philosopher of nature and natural science should go without. Let me just hurriedly say that universal determinism fails in a Newtonian world chiefly because there is no upper bound to the speed of signals. As a consequence of this, the spacetime region over which the values of physical quantities can be computed with certainty from an accurate and exhaustive knowledge of their values at a particular time is confined to that time alone. This limitation is overcome by Special Relativity, which offers an intellectual oasis to determinists who are ready to assume a priori that they live in true Minkowski spacetime, without missing points or other topological anomalies. However, as Einstein was quick to see, gravitational phenomena do not fit comfortably in a Minkowski spacetime.[39] The Theory of General Relativity developed by Einstein to cope with gravity provided, for the first time in history, a mathematical framework in which the deterministic evolution of the universe can be roughly yet plausible represented. But this framework admits some pretty wild spacetime topologies. Worse still, under very general assumptions, a model of General Relativity typically contains so-called singularities, which, in turn, often imply a radical breakdown of determinism.[40] Finally, with the introduction of Quantum Mechanics in the 1920s physical determinism took a surprising twist. A solution of the Schrödinger differential equation does not describe the evolution of any quantity or set of quantities one might properly be said to observe but rather that of a mathematical object from whose successive values one can calculate the current *probability* of obtaining each of the admissible values of any specific observable quantity, should one choose to measure it. But such limitations of determinism cannot, in my view, detract from the significance of mathematical physics, or its success. For its aim is, of course, to understand natural necessity where we find it, not to yield some Procrustean scheme for conceiving everything as the outcome of necessity.

5.4 Cause and law

In the following I shall often refer generically to physical systems that are conceived, to within some suitable approximation, as evolving in time according to a law expressible by a differential equation, in the manner roughly indicated in Section 5.3. Let me call them GDE-systems. I shall also speak of GDE-processes, or GDE-descriptions, etc. 'GDE' may be read as 'governed by a differential equation', provided that this expression is taken in the precise sense sketched in that section. 'DE', as in 'DE-solution', should be understood to refer to the pertinent differential equation.

There is a tendency in philosophy to regard the several states in a GDE-process as constituting a causal chain in which any two successive states are linked to one another as cause and effect. That tendency would breed no trouble if those who follow it refrained from employing the term 'cause' in its ordinary prescientific sense. Such abstinence, however, would thwart their aim, which is to present the differential equations of mathematical physics as the appropriate means for intellectually grasping causal connections in nature, and to suggest that every genuine cause-effect pair is actually embedded in some GDE-process and that the world itself is just one big connected GDE-system.[41] The tendency in question has been no doubt assisted by the aura of imprecision surrounding the ordinary notion of 'cause'. In turn, it should be held responsible for much of the obscurity and uncertainty of traditional philosophical analyses of causation. It is only recently that a few sharp-sighted philosophers have sought to break the timeworn association between cause and law (Ducasse 1924, 1969; Anscombe 1971; Cartwright 1983), but other writers still struggle to reinforce it (e.g., Tooley 1987).[42] A proper discussion of the issue could fill a whole book, but it is well that I bring this one to a close with some remarks on the difference between the commonsense idea of causation and the scientific notion of GDE-evolution, and on the complementary roles which both modes of understanding play in physics.

Let me note, first of all, that in the sentence scheme

$$\text{`}x \text{ causes } y\text{'} \tag{1}$$

one cannot, without doing violence to the English language, fill the blanks with definite descriptions of two successive momentary states of a GDE-system. As D. Gasking once observed, it would be a most unnatural and strained use of the word 'cause' to say that the movement of a falling body at 64 feet per second two seconds after it was let go *is caused* by its moving at 32 feet per second one second earlier (1955, p. 480). What goes best into the subject place marked in (1) by *x* are names (or definite descriptions) of persons or animals; though obviously one can also fill it with the name of a reputedly active thing—the Sun, the sea—or process—a hurricane, an earthquake.[43] Note that these are just the sort of natural entities that prescientific man was wont to personify. You may, no doubt, comfortably say in English that such and such an effect has been caused or can be caused by a given *machine.* Indeed one might say that is just the sort of thing machines are for. But then, perhaps this is only another way of saying that one expects machines to do—with improved efficacy—the work earlier done by men and animals.

Such prescientific but still wholly current and by no means archaic use of the verb 'to cause' is consonant with the original meaning of the Greek noun αἰτία, which the Romans rendered as *causa,*[44] whence the English 'cause'. Before being taken up by philosophy, αἰτία meant 'responsibility', mostly in a bad sense, i.e., 'blame', 'guilt', and was also used, metonymically, to designate whoever or whatever was to blame for something.[45] Inquiry into the responsibility for a given state of affairs was not confined, however, to the persons or personifiable things who authored it, but it sought to single out any abstract feature by virtue of which they were in a position to do so. Thus Democritus, in the earliest documented use of the word by a philosopher, says that "ignorance of what is better is the cause of wrongdoing" (ἁμαρτίης αἰτίη ἡ ἀμαθίη τοῦ κρέσσονος—Diels-Kranz 68.B.83). Earlier on, Herodotus had set out to record the great deeds of the Greeks and the barbarians, "as well as the cause why they warred against each other" (καὶ δι' ἣν αἰτίην ἐπολέμησαν ἀλλήλλοισι—I, 1). He also wondered "what on earth was the cause that necessitated" (τὸ αἴτιον ὅ τι κοτὲ ἦν . . . τὸ ἀναγκάζον) Thracian lions to exclusively attack camels in Xerxes' camp, although they had never before "seen or experienced the beast" (VII, 125). About the same time the Hippocratic physicians were already looking for factors of every sort—climatic, nutritional, behavioral—that could be blamed for illness and pain. It is not surprising, therefore, that when Plato's Socrates tells his young friends about his short-lived affair with natural science, the term αἰτία should have pride of place:

> ὑπερήφανος γάρ μοι ἐδόκει εἶναι, εἰδέναι τὰς αἰτίας ἑκάστου, διὰ τί γίγνεται ἕκαστον καὶ διὰ τί ἀπόλλυται καὶ διὰ τί ἔστιν.

> It seemed splendid to me to know the causes of each thing: what is that by virtue of which each thing is born, and that by virtue of which it is destroyed, and that by virtue of which it is.
>
> (*Phaedo*, 96a)

It was for dealing with just these questions that Aristotle developed his doctrine of the four types of "cause": the matter, the form, the goal, and the agent (*Phys.* II, 3). This involves such a colossal expansion of the denotation of the word αἰτία that a recent commentator has suggested that it might be rendered better as 'explanatory factor'.[46] I do not commend such ploys, which merely serve to protect young readers from becoming aware of the plasticity of human thought. But I can see, of course, that of all four Aristotelian "causes" only the agent or "source of the change" (ἀρχή τῆς κινήσεως) falls under the pre-Aristotelian meaning of αἰτία and the ordinary acceptation of 'cause'.[47] Now, Aristotle's examples of an agent are usually men: the sculptor, the builder. He does, however, emphasize that it is the builder at work as such, *the builder building* (ὁ οἰκοδόμος οἰκοδομῶν), who causes the house to rise. This might seem to adumbrate the modern philosophical view that only events can properly count as causes. But Aristotle would have been baffled by the very idea that an event—let alone an instantaneous event—might *act* to *effect* another event. It is clear to him that to bring something about, to do it or make it, is the act or work (ἐνέργεια) of a continuant.[48]

The experience of authorship, of doing a deed, is one of the primary constituents of our self-awareness. A philosopher who pretends he has no notion of it will only succeed in *literally* making a fool of himself. Nor can he hide the fact that our ordinary concept of causation is intimately bound to that notion, either by being formed as a natural extension of it, or—what seems to me more likely—by functioning from the outset as the broader category under which our own doings fall. But neither the notion of authorship nor the concept of causation of which it is a paradigm case involves the idea of law. To understand that I am doing what I am doing does not in any way imply that there is a general rule connecting my (type of) activity with its (kind of) results. Of course, in settling questions of responsibility—e.g., in a criminal trial—it is useful, nay indispensable, to consider what sort of causes usually bring about, under like circumstances, like effects. But regularity is here an indicator of causality, not a defining trait of the causal relation itself. If one clearly and distinctly sees a man pierce a hole through someone else's skull with a single blow of his fist, one understands that he made the hole, although the occurrence is unprecedented and, one hopes, will not be repeated. In fact, if succession according to a law were implicit in the very concept of causation, narrative literature—which usually employs multiple varieties of the latter without supplying even an inkling of the former—would be, for the most part, unintelligible.

But even if the extension of '*x* causes *y*' were broadened to allow physical states as causes, one big difference would still separate the relation between cause and effect, as it is ordinarily understood, from the relation between two successive states of a GDE-system: the latter relation is symmetric, while the former is not. If state σ_1 at time τ_1 and state σ_2 at time τ_2 both lie on the unique DE-solution through each, σ_1 determines σ_2 no less than σ_2 determines σ_1. Indeed, knowledge of a particular state in such a GDE-evolution can serve equally well to predict later states and to retrodict earlier ones. Causation, on the other hand, is inherently asymmetric: the stallion is not begotten by its foal; the food does not cook the fire.[49]

To my mind, however, the main discrepancy between ordinary causal thinking or etiology and the nomology of differential equations lies in that the former thrives on discreteness, while the latter assumes continuity.[50] The analysis of a causal chain is not completed until every link in it has been pinpointed in the sequence it forms with its direct and indirect neighbors. But in the differentiable manifolds in which GDE-trajectories are embedded such full pinpointing is not even conceivable. A state in a GDE-evolution does not have an immediate successor through which its power of determination is, so to speak, passed forward in the causal chain as the torch in a relay race. If two terms in a causal chain are not directly connected, one is bound to ask for the intermediate links. On the other hand, if σ_1 and σ_2 are any two states on a GDE-trajectory, they are certainly not contiguous, yet the question "What precisely does σ_1 act on, in order eventually—after uncountably many like actions—to determine σ_2?" does not make any sense. In a GDE-evolution one state determines the others not by dint of a state-to-state transmission of efficacy but formally (so to speak), through the structure in which they are all comprised.[51]

Ordinary causal thinking is man-centered and context dependent. How it singles out effects and what it pinpoints as their causes will depend in each case on human uses and views and on the purpose at hand. Pragmatic considerations also guide the choice of a specific GDE-"model" to represent a fragment of reality, but within such a "model" relations are settled once and for all, and do not vary with perspective and interest. This fosters the impression that nomological thinking comes closer to the Truth than ordinary causal thinking, and that the latter is a remnant of primitive animism which should eventually disappear. However, it does not seem likely—or even conceivable—that this could ever happen. As a matter of fact, in real life these two disparate ways of understanding work fairly well together. Thus, the best way of securing an effect E is to find a GDE-process leading to E from initial or boundary conditions C which one is able to set up. In such a case, it is appropriate to say that E is caused by C or by the act A of setting up C or by the person P who executes A. But this does not entail that the succession of nondenumerably many GDE-events joining C to E is being regarded, per impossibile, as a causal chain: in the analysis proposed, the said succession is the single link between cause C and effect E. At times, a more detailed causal analysis may be useful. Consider, for instance, a particular computer input I. Shortly after I is entered, the corresponding output O turns up on the computer screen. From a "hardware" viewpoint, entering I causes O through a continuous electrodynamic GDE-process. But from the more familiar and superficially more perspicuous "software" viewpoint, I leads to O through a finite chain of conditional commands, determined by the input itself, the computer's initial state, and the computer program, say, $I = C_0 \rightarrow C_1 \rightarrow \ldots \rightarrow C_n \rightarrow O$. Obviously, one may properly say that the execution of C_i has led to, brought about, or, indeed, caused the execution of C_{i+1} $(0 \leq i \leq n)$. By this manner of speaking one in effect singles out n computer states besides I and O, and decomposes the single GDE-process between I and O into $n + 1$ stages linking those intermediate states. Such causal analysis can certainly assist one in ascertaining what changes in the input I may be required to obtain an output different from O, but it will not contribute (except perhaps heuristically) to understanding the physical process through which, given I, the system necessarily outputs O. Causal thinking is likewise inevitable in laboratory life. Our continued quest for more detailed and accurate GDE-representations of natural phenomena depends on experiments, which must, of course, be initiated and taken notice of by persons. An experiment to test our proffered nomological understanding of some kind of phenomenon will not make sense as such if its very occurrence is in turn understood nomologically, viz., as a minor event in a major GDE-development comprising, say, the whole history of the Galaxy. Such an experiment is performed on what is purportedly a GDE-system of the sort under study. This is kept as isolated as it is humanly possible, at least insofar as the relevant physical quantities are concerned. Yet the experiment must have been contrived by men and yield results they can take stock of. The experiment's double interface appears in the nomological representation as the initial or boundary conditions and the final state of the GDE-system (cf. the passage by Margenau, quoted in note 41). But in scientific practice, it must also be understood causally, as that which the experimenter does, and as what he achieves by doing it. Thus, at the heart of physics, as in other walks of life, we resort to more than just one mode of thought. The experimental interface must be conceived from either side of it in radically different ways. A frank recognition of the irreducible difference between etiology and nomology and of their complementary roles in physics will serve us better, I dare say, than the familiar philosophical attempts at refurbishing the former to make it look like the latter.[52]

NOTES

Preface

†*Alexandri de Anima* (ed. Bruns), 89, 9–11. A. P. Fotinis (1979, p. 117) translates the same passage as follows: "This intellect will therefore be 'productive' by the very fact that it causes the being of all the intelligibles."

Chapter 1

[1] Lewis Carroll, AA, p. 158; cf. Byron, *Don Juan*, I.7.2.

[2] Ἀρχὴ ἀνυπόθετος—Plato, *Respublica*, 510b.

[3] Shapere 1982, 1985; Brown 1987; Pickering 1984a, 1984b; Galison 1987. See also Hacking 1983, 1988; Ackermann 1985; Achinstein and Hannaway 1985. The first version of the present chapter, which I read at the Pittsburgh Center for Philosophy of Science in November 1983, was written before I had seen any of these works, except Shapere 1982.

[4] If a nurse consistently obtained accurate blood pressure readings with her eyes closed, one ought to say perhaps that she practices extrasensory observation. I shall, however, ignore such a possibility, for, as far as I know, nobody speaks in this way—presumably because the thing itself has never occurred.

[5] Thus, it is wrong to say—as I did in an earlier version of this chapter—that for the result of such "an impersonal observation to become known, an observer must eventually observe the receiver with his own senses" (Torretti 1986a, p. 2).

[6] The word stems from the Greek νόος, 'intellect', 'thought', and γονεία, 'generation'. Kant (1781, p. 271; 1787, p. 327) employed it to refer to Locke's philosophy of the human understanding.

[7] Hume (THN, p. 189) asserted that our senses "convey to us nothing but a single perception, and never give us the least intimation of anything beyond," but I do not find that his claim does justice to my observational awareness. Anything that I am capable of distinguishing as a "single perception"—be it the hard, cold surface of the tiled floor under my feet or the colorful backs of the books on the shelf in front of me or the shrill sound of the car alarm from the parking lot beneath my window—points *as such* to a multitude of things beyond it. It may be argued that it does not do so of itself, but only because it is woven into a complex of experiences and expectations in the midst of which it turns up as a perception. But if I shut myself from or ignore or "forget" that complex, there would be nothing left for me to pinpoint as a "single perception."

[8] English translation by David Furley (1987, p. 117). The pair ἐτεῇ/νόμῳ ('in truth'/'by custom') clearly recalls the contrast—often discussed in 5th century Athens—between φύσις ('nature') and νόμος ('human usage').

[9] For example, Epicurus named in one breath "the shapes, colors, sizes, weights, and other

things predicated of a body as permanent attributes" in order to deny that we should regard them as nonexistent—although they are neither self-subsisting things nor incorporeal entities accruing to the body (τὰ σχήματα καὶ τὰ χρώματα καὶ τὰ μεγέθη καὶ τὰ βάρη καὶ ὅσα ἄλλα κατηγορεῖται σώματος ὥς ἀεὶ συμβεβηκότα [...] οὔθ' ὡς καθ' ἑαυτάς εἰσι φύσεις δοξαστέον [...]· οὔθ' ὅλως ὡς οὐκ εἰσιν· οὔθ' ὡς ἕτερ' ἄττα προσυπάρχοντα τούτῳ ἀσώματα—Epicurus, *Ep. Hdt.* 68; Long and Sedley 1987, **7B**).

[10] I should add that Hacker's brilliant book is highly entertaining and that if I knew that every one of my readers had spent a weekend over it I would not be writing this section. For those who have, I shall quote here a few passages followed by expressions of mild yet significant disagreement.

(1) From page 32: "If, at the oculists's, I am asked what I see written on the bottom line of the chart and I answer, 'I see only a black blur', then the 'content' of my seeing (its 'intentional object') is given by the specification 'a black blur' but this phrase in this context does not 'stand for' any entity at all." To my mind, this is a facile way of conjuring the black blur out of existence.

(2) From page 90, footnote 2: "It is immensely misleading to suggest, as J. J. C. Smart does, that thermal perception 'is less anthropocentric than our colour sense because the human being is much more like an (inefficient) thermometer than he is like an (inefficient) spectrometer' (*Philosophy and Scientific Realism* (Routledge and Kegan Paul, London, 1963), p. 85)." In my view, Smart's suggestion—if suitably amended—is not at all misleading, viz., if we compare the human being not with a thermometer in thermal equilibrium with a contiguous body but, *sit venia verbo*, with a thermorheometer. I should say that people are pretty good at estimating heat flow in and out of their skin if the temperature gradient is not noxiously large. If they sense an inward flow, they say the object they are touching is warm or hot; if an outward flow, that it is cool or cold.

(3) From page 219: "It is true that a beautiful painting does not look beautiful *to* anyone when no one is looking at it, but its appearance does not change when unobserved (only when differently illuminated). A good-looking woman does not cease to be good-looking when no one looks at her. A beautiful view does not cease to present a fine prospect when the tourists leave, but only when the sun sets. Sebastiano's Roman paintings look Michelangelesque even when the museum is closed. For an object to look thus-and-so does not require spectators—how things look when no one is looking is just as they look when someone is looking." Throughout his book Hacker fights the confusions wrought by philosophers who wilfully ignore the contexts in which expressions are normally used and apart from which they become meaningless. Is he not the victim of such a confusion here? Surely it would be true to say Sebastiano's paintings look Michelangelesque after the museum is closed, provided the lights are not turned off, but I cannot think of any occasion—except in a philosophical discussion—where it might be appropriate to say so. That looking thus-and-so remits to an unidentified potential onlooker follows at once from the fact that—even in the absence of spectators—things do not look the same from every vantage point or through every optical system.

(4) Finally, from page 128: "Normal observational conditions are no more constituents of perceptual concepts than the constant gravitational field of the earth is part of the rules of tennis. But only in something like these conditions, against this background, do the rules have a point, can the game be played as we play it. It is in the context of fairly stable sunlight, fairly constant surface structures, that we introduce our colour vocabulary and typically use it. (And were this background wildly unstable, our colour grammar would be as useless as tennis equipment in the moon.)" I fully agree with Hacker on this important point, but I do not shun the obvious implication that our "grammar" is parochial and that our modern cosmic outlook requires us to take it with a pinch of salt.

[11] My translation. The translation quoted in Hacker 1987, p. 7, omits 'moltitudini' and

glosses over 'si richiegga'.

[12] Descartes, Meditatio VI (AT, VII, 57). The Duke of Luynes rendered the stated condition in French, with Descartes' approval, as follows: "En tant qu'on les considère comme l'objet des démonstrations de Géométrie" (AT, IX, 71).

[13] Jackson 1977, p. 143. Jackson's book, though tightly argued and a joy to read, is, in my view, poorly grounded. Its main thesis is that the "immediate objects" of visual perception are mental entities. He defines: "*x* is a *mediate object of (visual) perception* (for *S* at *t*) iff *S* sees *x* at *t*, and there is a *y* such that ($x \neq y$ and) *S* sees *x* in virtue of seeing *y*. An *immediate object of perception* is one that is not mediate" (pp. 19–20). Since Jackson refers explicitly to the time *t* at which *S* sees *x*, I take it that 'sees' and 'seeing' are here forms of the tenseless present. I also understand that the negation in the definition of 'immediate object of perception' does not apply to the entire definiens of 'mediate object of perception' (for if it did, then, according to Jackson, both the Holy Ghost and the trillionth prime number would be immediate objects of perception for me now) but only to its second conjunct. Hence, *x* is an *immediate object of (visual) perception* (for *S* at *t*) iff *S* sees *x* at *t*, and there is no *y* such that ($x \neq y$ and) *S* sees *x* in virtue of seeing *y*. Now this, to my mind, is just a farfetched characterization of the empty set. Let *x* be, for instance, a red, round afterimage which I am now seeing. I see it in virtue of seeing its parts, and in virtue of seeing its contour, and also of having seen them a moment ago and continuing to see them into the next moment. Since there is no *minimum visibile*, let alone a *visibile instantaneum*, there can be no such thing as an immediate object of vision as here defined. Hardin 1988, pp. 96–109, shows that Jackson's theory of visual mental objects is incompatible with well-established results of experimental psychology.

[14] While the remarks in the last paragraph of Section 1.3 apply to many authors, I must say that I wrote them with one particular book in mind, viz., Hardin's *Color for Philosophers* (1988). This consists of a well-informed and extremely interesting exposition of current scientific knowledge of chromatic visual perception, intertwined with a philosophical argument purporting to prove that "since physical objects are not colored, and we have no good reason to believe that there are nonphysical bearers of color phenomena, and colored objects would have to be physical or nonphysical, we have no reason to believe that there are colored objects" (p. 111). Hardin musters overwhelming empirical evidence for the indeterminacy of visual shape and color under various conditions of seeing. To clinch his argument he invokes the principle that "in the absence of compelling evidence to the contrary, one should prefer a theory that regards the phenomenal realm as derivative from, or controlled by, a more fundamental domain in which determinacy prevails" (p. 108). Since our language is tailored to the indeterminacy of what Hardin calls "the phenomenal realm," his dream of determinacy goes hand in hand with a tendency to use words improperly. For instance, on p. 72, speaking of color television, Hardin says: "In a region of the screen which appears to us to be of constant color, each phosphor dot is glowing only one-thousandth of the time; at any one instant, most of the screen is dark." Now, to say that most of a screen is dark at any one instant because only a tiny fraction of it is radiating light at that very instant is about as sensible—and as good English—as saying that a glass full of water I hold in my hand is mostly empty at any one point because most of the time there is no hadron or electron precisely at that point. Of course, if the hadrons and electrons that make up the water molecules were so tightly packed as to preclude "emptiness" in the sense described, I would not be holding a glass of water but a black hole. And if the phosphor dots on the television screen emitted photons very much more frequently than they do now, the screen would promptly vanish in a blinding blaze. Thus, by Hardin's standards there would be no room (more accurately, no time) for lighted TV screens in this world.

[15] Searle 1983, p. 124, n. 9, proposes the following thought experiment to help remove the doubts of "many philosophers" who are prepared to agree with him that "causation is a

part of the experience of acting or of tactile and bodily perceptions" but who "do not concede that the same thing could hold for vision": "Suppose we had the capacity to form visual images as vivid as our present visual experiences. Now imagine the difference between forming such an image of the front of one's house as a voluntary action, and actually seeing the front of the house. In each case the purely visual content is equally vivid, so what could account for the difference? The voluntarily formed images we would experience as caused by us, the visual experience of the house we would experience as caused by something independent of us." But evidently the alleged conclusion of Searle's thought experiment follows only if we beg the question and *assume* that visual images must be experienced as caused. Otherwise, the most we can conclude from Searle's conditions is that involuntarily formed visual images would be experienced as *not* caused by us.

[16] Dewey 1938, p. 23. Perhaps the reference to the "brain" ought not to be taken literally, but should be understood to designate whatever it is beneath our skin that acts as a receiver in personal observation.

[17] In turn, the observer's properties are communicated "without matter" to the object of observation. Cf. Aristotle, *De Somnibus* 459^{b}27ff.

[18] Locke, *Essay*, editions 1 through 3, II.viii.11.

[19] Locke, *Essay*, 4th edition, II.viii.11; cf. IV.ii.11.

[20] Cartesian momentum is, indeed, a scalar, and thus conceptually different from Newtonian momentum, which is a vector, but which, in turn, cannot be simply equated with the spacelike part of relativistic four-momentum. Nevertheless, there are deep links between these concepts, which motivate the use of the same word for all three of them and which mark a contrast between a theory that uses any of them and, say, the physics of Aristotle.

[21] Let ν be the mean frequency of the eclipses of a particular Jovian moon. Let $\Delta\nu$ denote the difference between ν and the frequency actually observed at a given time. $\Delta\nu$ is positive as Jupiter recedes from the Earth and negative as it approaches our planet. If one knows the relative velocity u of Jupiter and the Earth at the time in question, one can readily compute the velocity of light c from the classical Doppler equation $c/(c+u) = (\nu + \Delta\nu)/\nu$. (Though this 19th century formula does not occur in Rømer's work, it epitomizes his reasoning.)

[22] The atmosphere of the Earth is opaque to all frequencies in an interval starting right below the requisite peak and ending well beyond it. Therefore, the peak lies outside the scope of terrestrial receivers and remained unrecorded for a good many years. On the microwave radiation background see, for example, Raychaudhuri 1979, Chapter 6. Another example—beautifully analyzed by Shapere (1982)—is the observation of the Sun's interior by means of neutrinos, in which the object and the instrument of observation are delicately suspended in a tenuous spiderweb of theories.

[23] The principle is already presupposed in the following statement from a 13th century treatise of optics by John Pecham, archbishop of Canterbury: "Nisi species rei visibilis distincte oculum sigillaret oculus partes rei distinte non apprehenderet" ("Unless the species of the visible object were to make a distinct impression on the eye, the eye could not apprehend the parts of the object distinctly"—Lindberg 1970, p. 108).

[24] For, as Frank Jackson (1977, p. 126) neatly says: "If I know that *p* would obtain whether or not *q* were the case, I cannot regard *p* as evidence for *q*." Note that, contrary to what is often said, there need not be a *causal* connection between the observed features and the receiver states that bear witness to them. A causal connection between the events observed and the receiver states that record them is neither necessary nor sufficient for the conveyance of information by the latter about the former. (i) It is not necessary. Changes in the reading of the power meter connected in parallel to your home's electric network are not caused by your electric energy consumption. The meter's motor uses a tiny (known) fraction of the

energy used in the parallel circuit. (ii) Nor is it sufficient. If a given receiver state can be caused by two or more distinct events, none of which is a necessary condition of it, a record of the state in question cannot tell which of those events has occurred. Example: A mass of 0.035672 kg is placed on a digital scale which measures mass to the nearest gram. The dial reads 0.036. But the same reading would be caused by placing on the scale a mass of 0.036211 kg.

[25] By the so-called straight rule of induction one is supposedly licensed to infer from the observation of *n* instances of a certain class *A*, *m* of which are noted to belong to the subclass *B*, that the proportion of *B*'s among *A*'s is generally *m* .

[26] *Added in proof*: This has been nicely expressed by Olivier Costa de Beauregard in his sparkling recent book on time: "Registration makes sense only to the extent that one is able to interpret it, having sufficient *a priori* knowledge. [. . .] This is to say that retrodiction is an art, and an art such that when memory or intuition that can decipher the meaning of registrations or 'fossils' is lacking, creative inspiration is needed to invent a correct theory" (1989, p. 137).

[27] Examples like the following may suggest that the above requirement is too stringent. Unbeknownst to me, John, the postman who brings the mail to my neighborhood, has an identical twin, Jack, who also works for the Postal Service. Suppose that this morning, as I went out of my building, I saw a postman across the street, whom I immediately took to be John, and indeed John he was. One could then say that I knew John as soon as I saw him, and that the information was conveyed by my eyes. Yet John's presence across the street was not a necessary condition of the state of my optical receptors when I looked at him from my door, for a state indistinguishable from it would have been effected by the presence of Jack, in the same uniform and posture, at the same spot. (Had Jack been there, my brain would no doubt have reacted in the same way to the visual stimuli, and I would have mistaken him for John.) I do not think, however, that the example proves that the requirement I stated in the text is excessive. What it shows, rather, is that the information conveyed by my eyes when I see the postman across the street suffices, *at most,* to establish that he is either John or Jack and that my correct perception of him as being John goes beyond that information and involves a happy guess (facilitated, of course, by my ignorance of Jack's existence).

[28] Dependence of awareness on grasp and of grasp on theory exacerbates the foundationist's mistrust and moves him to search for an "objective" ground of empirical knowledge outside "the subject" and her chancy freedom. In the light of our understanding of personal observation as a physical process it would seem that such a ground is ready to hand.

> What to count as observation now can be settled in terms of the stimulation of sensory receptors, let consciousness fall where it may.
>
> (Quine 1969, p. 84)

If this criterion can be made to work, the following simple definition will neatly distinguish the class of "observation sentences," which exclusively and exhaustively convey the evidence on which science rests:

> An observation sentence is one on which all speakers of the language give the same verdict when given the same concurrent stimulation.
>
> (Quine 1969, pp. 86–87)

But there is no way of finally establishing that the class thus defined is not empty. Moreover, not only are we as yet unable to say "in general what it means for two subjects to get the same stimulation" (Quine 1969, p. 160), but we cannot normally ascertain in a particular instance that such is at least approximately the case except by appealing to the awareness of the subjects, shared in conversation.

Chapter 2

[1] See also Miller 1987, p. 106: "Explanation is a central goal of science, perhaps the central one." Miller, however, dismisses the inferential view of explanation shared by Nagel, Hempel, and Popper and defines explanation as "an adequate description of underlying causes helping to bring about the phenomena to be explained" (1987, p. 60).

[2] Deductive-nomological explanation was contrasted by Hempel and others with so-called inductive-statistical explanation. While in the former the explanandum is inferred by standard logical deduction from an explanans that includes one or more universal laws, in the latter the explanandum is derived from statistical or probabilistic laws by means of "inductive logic." I cannot go into this matter here, but I shall make a few dogmatic statements so that the reader can see where I stand. (i) Plans for an "inductive logic" are closely linked to the logical interpretation of probability invented by Keynes (1921) and painstakingly elaborated by Carnap (1950, 1962). I find no merit in this interpretation. (ii) Let *L* be the statement that the probability that an event of type *A* occurs under conditions of type *C* is 0.55. I do not think that *L* will (statistically or otherwise) explain the fact that here and now an event of type *A* has actually taken place under conditions of type *C*. (iii) Suppose, on the other hand, that in a large collection of instances of *C*, events of type *A* have occurred in about 55% of all cases. This is, in itself, a (complex) particular event. Call it *B*. I believe that *L* can be invoked to explain *B*. The explanation provided differs from the usual deductive-nomological explanation involving universal laws in that *L* does not bestow logical certainty on *B* but only a great probability, amounting to practical certainty if the collection in question is large enough. (See Sections 4.2 and 4.3.) However, this follows from *L* and the stated conditions by standard logical deduction. So whatever the explanans does for the explanandum in this kind of explanation it does by means of standard logic. Railton's (1978) proposal for a deductive-nomological model of probabilistic explanation involves a different approach to this matter which I find quite attractive and very much worth exploring.

[3] One requirement that was much discussed is that *L*, *C*, and *F* must be true. Since this cannot be readily certified—at any rate in the case of *L*—one must normally settle for inferences that *would be* explanatory *if* the explanans were true. In such a potentially explanatory inference, the premise *L* must be a "lawlike" statement but might well not be a true law. Lawlikeness should not be difficult to ascertain, if everyone would but agree on a precise definition of this property.

[4] See Scheffler 1963, pp. 58–59. Scheffler considers how a deductive explanation is affected by the "availability of totally disparate ways of singling out the same concrete entities" (l.c., p. 66). However, his sole concern here is that he may be forced to acknowledge abstract, "intensional" entities (events-under-a-description) as the true referents of explananda. To avert this peril he equates the objects of explanation with mere utterances or inscriptions.

[5] There were several opinions as to how the moving agent—e.g., the gunpowder—continued to drive the projectile after the latter had ceased to be in direct contact with it. For strict Aristotelians, the projectile was then moved by the neighboring air, either (i) because the air pushed aside by the advancing projectile rushed to fill the gap in the wake of it and thereby in turn pushed it from behind, or (ii) because the air, not being itself heavy

by nature, derived from the original agent a power to move the projectile *contra naturam*. The 6th century Aristotle commentator John Philoponus thought the aforesaid views preposterous and claimed that "an incorporeal motive force is imparted by the projector to the projectile." This expendable power, known to the Latin scholastics as *impetus*, was further characterized by Philoponus as ἐνέργεια τις κινητική, "a kinetic activity of some sort." See Clagett 1959, pp. 505ff.

[6] Galileo, EN, VIII, pp. 272ff. On Galileo's natural philosophy and its historical background, see Clavelin 1968.

[7] On the above assumptions, the centripetal acceleration at time t is equal to $\mathbf{r}(t)\omega_t$, where ω_t stands for the Moon's angular velocity about the Earth at t. In his actual reasoning Newton simplified the matter even further. He employed the *average* Earth-Moon distance r. If T is the Moon's period, its average centripetal acceleration is $4\pi^2 rT^{-2}$. Substituting this known quantity in the l.h.s. and $\mathsf{g}(\rho/r)^2$ in the r.h.s. of eqn. (1), one can figure out the acceleration g with which the Moon would fall—under the stated hypotheses—if it were brought down to the surface of the Earth. Newton found the value of g given by this calculation to be in good agreement with that obtained from the observation of pendular motion (Newton, *Principia*, Bk. III, Prop. IV).

[8] The trajectory calculated from eqn. (1) would satisfy the first two Keplerian laws; but Kepler himself had already found it necessary to introduce a correction in his lunar theory, which he thought could be accounted for by a force from the Sun. See Dreyer 1953, p. 404; Newton, *Principia*, Bk. III, Prop. XXII.

[9] "The velocity which a given force can generate in a given matter in a given time is directly as the force and the time, and inversely as the matter" (Newton, *Principia*, Bk. II, Sec. VI, proof of Prop. XXIV). Plainly, this amounts to $\Delta\mathbf{v} = \mathbf{F}\Delta t/m$, whence $\mathbf{F} = m\Delta\mathbf{v}/\Delta t$. In the proof of Prop. I of Bk. I, Newton had already gone to the limit $\Delta t \to 0$.

[10] Write $F \propto Mmr^{-2}$ for the magnitude F of the force exerted by a particle of mass M on a particle of mass m when r is the distance between them. F must be proportional to m or else, by the Second Law of Motion, the acceleration of a freely falling body would vary inversely with its mass. F must be proportional to M for, by the Third Law of Motion, F is also the magnitude of the force which the second particle exerts on the first. Finally, F is proportional to r^{-2} if Kepler's Third Law of Planetary Motion holds good. (Note, by the way, that according to the very Law of Universal Gravitation for which it supposedly provides inductive grounds Kepler's Third Law can only hold approximately.)

[11] Einstein's prediction agrees with the observed motion if the Sun is very nearly spherical. But if Dicke and Goldenberg's controversial measurement of the Sun's oblateness is correct, the quadrupole moment generated by the Sun's equatorial bulge would cause Mercury's perihelion to advance by some 3.4″ per century, and Einstein's prediction would be perceptibly off the mark. See Will 1981, pp. 176ff.

[12] According to Einstein's theory of gravitation, the periastron of an orbiting particle in a spherically symmetric field attributable to a solar mass advances by 8.6″ per century if the distance from the particle to the field's center is like that from Venus to the Sun, and by 3.8″ per century if the said distance is like that from the Sun to the Earth. These numbers would seem to agree tolerably well with the observed trajectories of Venus and the Earth, respectively (see Weinberg 1972, p. 198); but the effect is small and, had not Einstein's theory directed our attention to it, we would not dare to count it as a violation of Newton's law.

[13] Revised laws of gravity, in which Newtonian attraction is made to depend on $r^{-2\pm\varepsilon}$ for some positive $\varepsilon << 0$, were proposed in the late 19th century to account for Mercury's perihelion shift. The second course was followed by Einstein.

[14] "Dasjenige [. . .] welches macht, daß das Mannigfaltige der Erscheinung in gewissen

Verhältnissen geordnet werden kann" (Kant 1787, p. 20). Cf. Refl. 4673: "Die *Ordnung der Dinge, die neben einander* seyn, ist nicht der Raum, sondern der *Raum ist das, was eine solche Ordnung oder besser coordination* nach bestimmten Bedingungen möglich macht" (Kant, *Ak.*, XVII, p. 639).

[15] "The manifold of representations can be given in an intuition which is merely sensible, that is, nothing but receptivity; and the form of this intuition can lie a priori in our faculty of representation, without being anything more than the way in which the subject is affected. But the *combination* (*coniunctio*) of a manifold in general can never come to us through the senses and cannot, therefore, be already contained in the pure form of sensible intuition. For it is an act of the spontaneity of the power of representation; and since the latter, to distinguish it from sensibility, must be called understanding, all combination [. . .] is an operation of the understanding [*eine Verstandeshandlung*]." (Kant 1787, pp. 129f.; Kemp Smith translation, slightly modified.)

[16] "Appearances, so far as they are thought as objects according to the unity of the categories, are called *phænomena*" (Kant 1781, pp. 248f.; Kemp Smith translation).

[17] In nonrelativistic Quantum Mechanics, the temporal evolution of the dynamic state of an isolated physical system is governed by the Schrödinger equation, which is a first-order partial differential equation. Given the theorems on existence and uniqueness of solutions to such equations, it is unquestionable that the theory conceives the said evolution deterministically. But the dynamical state of a quantum system *S* is so defined that it yields, at any given time *t*, for each observable physical quantity *Q*, a probability-weighted spectrum of alternative values that might be recorded if *Q* is measured on *S* at *t*. Thus, quantum-mechanical determinism is a far cry from the kind of determinism that Kant had in mind, which, of course, requires that the same laboratory preparations always lead to the same measurements.

[18] An object is said to be reducible (*zurückführbar*) to one or more objects if all statements about the former can be reformulated as statements about the latter (Carnap 1961, p. 1).

[19] Carnap 1961, p. 84. The said alternatives are mentioned only as examples. The respective fundamental relations are described as follows: (*a*) the spatial and temporal relations between the elementary charged particles; (*b*) the topological relations between spacetime points and the "one-to-many coordinations between real numbers and spacetime points, corresponding to the several components of the *potential functions*, i.e., the electromagnetic 4-vector field and the tensor field of gravitation"; (*c*) "*coincidence* and the *proper time function*."

[20] Of course, the person from whose life such cross sections are taken—e.g., Carnap himself or the reader—is not one of the ground elements but one of the many objects purportedly reducible to them.

[21] "An object (or, respectively, a kind of object) is said to be *epistemically prior* to another [. . .] if the latter is known through the mediation of the former, whence our cognition of the latter presupposes cognition of the former" (Carnap 1961, p. 74). The four main domains of objects stand in the following relations of epistemic priority: first come the objects pertaining to my own mind; next come physical objects; then, objects pertaining to other minds; finally, cultural [*geistige*] objects (Carnap 1961, p. 79).

[22] For a rigorous elucidation of the meaning and use of the material and the formal modes of speech, see Carnap 1937, §§64–65. The following example may suggest what is intended: "It is a pseudo-thesis of idealism and older positivism, that a physical object (e.g., the moon) is a construction out of sense-data. Realism on the other hand asserts, that a physical object is not constructed but only cognized by the knowing subject. We—the Vienna Circle—neither affirm nor deny any of these theses, but regard them as pseudo-theses, i.e., as void of cognitive meaning. They arise from the use of the material mode, which speaks about 'the

object'; it thereby leads to such pseudo-questions as the 'nature of this object', and especially as to whether it is a mere construction or not. The formulation in the formal idiom is as follows: 'A physical object-name (e.g., the word 'moon') is reducible to sense-data predicates (or perception predicates)'" (Carnap 1936, pp. 428–29).

[23] Carnap offers the following illustrations: "The predicate 'red' is observable for a person N possessing a normal color sense. For a suitable argument, namely a space-time-point *c* sufficiently near N, say a spot on the table near N, N is able under suitable circumstances—namely, if there is sufficient light at *c*—to come to a decision about the full sentence 'the spot *c* is red' after a few observations—namely by looking at the table. On the other hand, the predicate 'red' is not observable by a color-blind person. And the predicate 'an electric field of such and such an amount' is not observable to anybody, because, although we know how to test a full sentence of this predicate, we cannot do it directly, i.e., by a few observations; we have to apply certain instruments and hence to make a great many preliminary observations in order to find out whether the things before us are instruments of the kind required" (Carnap 1936, p. 455). The attentive reader will not fail to notice that a spot on a table which can be observed a few times cannot be a spacetime *point* as Carnap says but should be described as a worldline; better still, as a three-dimensional world-*tube.*

[24] Carnap does not explicitly say that L_O is an interpreted first-order language, but he does stipulate that L_O is a nominalistic, extensional language (1956, p. 41).

[25] *T* stands for a theory in Carnap 1956, p. 45, line 8 from below. Two lines earlier, in line 10 from below, *T* stands for the postulates of a theory. *T* stands for the *conjunction* of such postulates on p. 43, line 2. See also the passage quoted in footnote 14. By the way, I do not understand why Carnap requires a scientific theory to be *finitely*, not just *recursively*, axiomatizable.

[26] "The specification [. . .] of the postulates *T* is essential for the problem of meaningfulness. The definition of meaningfulness must be relative to a theory *T*, because the same term may be meaningful with respect to one theory but meaningless with respect to another" (Carnap 1956, p. 48).

[27] Anyway, as the following remarks will suggest, Newton-Smith's third criterion for "more observational" terms does not apply to contemporary scientific discourse:

> It is the experiential language of the physical sciences [. . .] that is difficult to understand, much more so for the outsider than the theoretical language. There is, I believe, no comparison between a philosopher's cognitive difficulty in reading theoretical articles in quantum mechanics and his difficulty in reading current experimental articles in any developed branch of physics. The experimental literature is simply impossible to penetrate without a major learning effort.
>
> (Suppes 1984, p. 122)

[28] Jerry Fodor has also come out in defense of "theory neutral observation" against "the holism story" that "what you observe is going to depend comprehensively upon what theories you hold because *what your observation sentences mean depends comprehensively on what theories you hold*" (Fodor 1984, p. 27). Fodor grants that what one does in fact observe depends on one's background knowledge. But he proceeds to show that not all the background knowledge available to the observer can influence personal observation. It is as if the cognitive function at work in human perception only had access to a certain part of the information at the perceiver's disposal. Fodor illustrates this point with the Müller-Lyer illusion. To verify just how stubborn this illusion can be the reader ought to examine Fig. 1.a on p. 338 of the *Oxford Companion to the Mind* (Gregory 1987). Should this book not be ready at hand, just take a

piece of graph paper and mark at four grid intersections the four vertices of a square, about two inches high. Label the vertices, counterclockwise, with the first four letters of the alphabet and draw the two sides joining A with B, and C with D, respectively. Now draw equal segments AM, AN, BP, BQ, CS, CT, DU, and DV, each 1/2 inch long, in such a way that ∠MAB = ∠NAB = ∠PBA = ∠QBA = 30° and ∠SCD = ∠TCD = ∠UDC = ∠VDC = 150°. Then, although AB = CD by construction, and the reader can check this at any time by looking at the grid printed on the paper, he will inevitably see AB as being shorter than CD, and his knowledge will have no power to change this perception. There is a difference, however, between "the fixation of appearances—what I'm calling observation—" (Fodor 1984, p. 40) and "the perceptual fixation of belief." While the former may well be impervious to knowledge not comprised in the "module" that governs it, the latter normally depends on every piece of information the observer can recall. To show that information is immune to the effects of theory change, Fodor reminds us that centuries after Kepler and Newton, we still obdurately see the Sun rise and set. This habit may not be curable by education, "because it may be that the inaccessibility of astronomical background to the processes of visual perceptual integration is a consequence of innate and unalterable architectural features of our mental structure" (Fodor 1984, p. 40). Fodor's point is well taken; but 'observation' in his sense is at most the starting point of empirical knowledge, not its court of last appeal, let alone its only source of meaningful ideas. For vigorous criticism of Fodor 1984, see Churchland 1988 (followed by Fodor 1988, a spirited defense).

[29] Indeed, what is "normal" illumination? For viewing a flower, no doubt, it is daylight. But for viewing a star? My slide projector has a switch that can be lowered to halve the light intensity. At what position of the switch can the color I see on the upper right-hand corner of a particular slide properly be said to be such-and-such? No wonder we have no words for such nuances.

[30] The reader ought to try a hand at criticizing the correspondence rules proposed by Carnap seven pages earlier in the same book (Neurath, Carnap, and Morris 1971, vol. I, p. 200). Another example, connecting the kinetic energy of the molecules of a gas to its temperature—"measured by a thermometer and, therefore, an observable in the wider sense"—is given in Carnap 1966, p. 233. In *The Structure of Science,* Ernest Nagel discusses a rule of correspondence he ascribes to Bohr's theory of the atom of 1913 and by virtue of which "the *theoretical* notion of an electron jump is linked to the *experimental* notion of a spectral line." The connection, however, is not effected here by a freely adopted stipulation but is prescribed, as Nagel duly notes, by the accepted theory of optical phenomena: "On the basis of the electromagnetic theory of light, a line in the spectrum of an element is associated with an electromagnetic wave whose length can be calculated, *in accordance with the assumptions of the theory,* from experimental data on the position of the spectral line. On the other hand, the Bohr theory associates the wave length of a light ray emitted by an atom with the jump of an electron from one of its permissible orbits to another such orbit" (Nagel 1961, p. 63; my italics).

[31] Reichenbach 1924, pp. 39, 67. According to Definition 20, the "external forces" mentioned in Definition 19 do not include "metrical forces," i.e., forces that act on all materials in the same way and against which there is no shielding. On such forces, which Reichenbach subsequently dubbed "universal," see Ellis 1963/64; Torretti 1983, pp. 236–38.

[32] Cf. also the "coordinative definition" of 'probability' given in Reichenbach 1949, p. 69.

[33] Einstein 1905b, p. 892, characterizes his "stationary system" as a "coordinate system [...] in which the equations of Newtonian mechanics hold"; a condition blatantly at odds with the subsequent development of the paper.

[34] Laue 1955, p. 3. This is the 6th edition of Laue's book. Lange's definitions of an inertial frame and an inertial time scale do not occur in the original version (Laue 1911), which was

the first textbook of Special Relativity. They are given in English in Robertson and Noonan 1968, p. 13. Thomson and Lange's work on the foundations of mechanics is the main subject of a masterly Ph.D. dissertation by Robert diSalle (1988).

[35] A Poincaré transformation is the product of a (homogeneous, i.e., origin-preserving) Lorentz transformation and a translation. The Poincaré group is sometimes called the full or inhomogeneous Lorentz group.

[36] Of course, for fixed v and large values of T the distance Δ_i between the points of occurrence of E and E_i is very large too. For fixed T and v, Δ_i is minimal in the direction of the relative motion of F and F'; if the straight line joining the locations of E and E_i in F is parallel to that direction, $\Delta_i = (T/v)\sqrt{(c^2 - v^2)}$ and therefore converges to 0 as $(v/c)^2$ increases to 1.

[37] Shapere 1966 offered this as a reductio ad absurdum of Feyerabend's thesis concerning the incommensurability of physical theories. "How could two such theories be relevant to one another? How is criticism of a theory possible in terms of facts unearthed by another if meaning depends on, and varies with, theoretical context, and especially if there is *nothing* in common to two theories" (Shapere 1984, p. 73).

[38] The time coordinate function Einstein is here looking for must not only provide numerical labels—dates—by which to identify events. The numerical relations between the labels assigned to different events are supposed to convey physically significant information. It was only much later, when he was working on his theory of gravity, that Einstein reluctantly gave up the idea that time and space coordinates must be physically meaningful. Reichenbach (1928) claimed that the relation of simultaneity between distant events determined by Einstein time is a purely conventional one, for it is not linked to causal relations. He bases his contention on the fact that the class of events Einstein-simultaneous in an inertial frame F with a given event E is only a proper subset of the class of events with which E cannot be connected by signals. However, a definition of Einstein simultaneity in terms of connectibility-by-signals had been given by Robb (1914). Indeed, it is the *only* nontrivial equivalence between events which can be defined in such terms in the Minkowski spacetime geometry of Special Relativity (Malament 1977a). For a gallant defense of Reichenbach's standpoint, with some windmill tilting at "the mathematical fallacy" in the arguments of his critics (including myself), see Havas 1987.

[39] Viz., since Ole Rømer (1644–1710) noted small but significant variations in the period of the Jovian moons when timed by this method.

[40] It has another grave disadvantage which Einstein does not mention: a time coordinate function defined by this procedure is not an inertial time scale in the sense of Neumann and Lange; therefore, a frame of reference endowed with it is not one in which the Principle of Inertia, let alone "the equations of Newtonian mechanics," will hold good.

[41] On the epistemic status of "established" physical theories with known validity limits, see Rohrlich and Hardin 1983.

[42] The Rayleigh-Jeans law might with greater propriety be called the Einstein mock-law, for, as T. S. Kuhn recalls, it was derived in Einstein 1905a, a paper "submitted for publication in March 1905, a month before the beginning of the correspondence in *Nature* through which Rayleigh and Jeans produced the law since known by their names. [. . .] Einstein pauses over it only long enough to note its impossible consequence: infinite energy in the radiation" (Kuhn 1978, p. 180).

[43] Putnam 1988, p. 130, says that he first presented his account of meaning and reference in lectures at Harvard in 1967–68, and in lectures at Seattle and the University of Minnesota the following summer. The dating has some historical interest, for a theory of meaning akin to Putnam's was independently developed by Saul Kripke more or less at the same time.

[44] Indeed, it has become so pervasive that some writers who do not explicitly endorse it apparently take it for granted. For example, Richard W. Miller (1987, p. 403) says that "what

Aristotle referred to in using *hudor* (the word everyone translates as 'water') does, strictly and literally, exist. His belief that there is hudor in the world was strictly and literally true." Now, I would concede that what Aristotle's housekeeper called ὕδωρ does exist and pours out of the tap in my lavatory. But Aristotle meant by ὕδωρ one of the four simple bodies that make up the sublunary world, viz., the one out of which everything that melts is made (*Metaphysica,* Δ, 4, 1015^a11; Δ, 24, 1028^a29), and I do not think that Miller would wish to assert that such a body "strictly and literally" exists. (Note, by the way, that what pours out of the water tap is not strictly and literally what we mean by H_2O.)

[45] He invokes them in order to refute the notion that the meaning of a word is a unique mental representation associated with the word, a point on which I fully agree with him. And, of course, I also agree with Putnam's final word on meaning and reference in this book: "Reference is not just a matter of 'causal connections'; it is a matter of *interpretation* [. . .]. And interpretation is an essentially holistic matter [. . .]. Knowing what the words in a language mean (and without knowing what they mean, one cannot say what they *refer to*) is a matter of grasping the way they are *used.* But use is holistic; for knowing how words are used involves knowing how to fix beliefs containing those words, and belief fixation is holistic" (Putnam 1988, pp. 118–19).

[46] Although the words 'denote' and 'connote' have a long history in the English language, their contemporary philosophical acceptations can be accurately traced to J. S. Mill's *System of Logic* (1843). "The name is said to signify the subjects *directly,* the attributes *indirectly;* it *denotes* the subjects, and implies, or involves, or indicates, or as we shall say henceforth, *connotes* the attributes. [. . .] Whenever the names given to objects convey any information, that is, whenever they have properly any meaning, the meaning resides not in what they *denote,* but in what they *connote*" (Mill, SL, pp. 32, 34). C. I. Lewis 1946, p. 39, defines the "denotation or extension" of a term as "the class of all actual things to which the term applies," while its "connotation or intension" is to be identified "with the conjunction of all other terms each of which must be applicable to anything to which the given term would be correctly applicable." The definitions I gave above differ from Lewis' in two respects: (i) I equated the connotation of a term with certain objective conditions, not with the terms that convey them; for it may well occur that the conditions in question—viz., the necessary conditions for the term being applicable—are not fully analyzed and that we lack the words to express them. (ii) Because I am chary of set-theoretical paradoxes and uncertain about Julius König's distinction between paradox-prone sets and paradox-immune classes (König 1905 in Heijenoort 1967, pp. 148–49), I refrained from collecting the denotata of a term into a set or class.

Putnam (PP, vol. II, p. 216) correlates the pair intension-extension with Frege's *Sinn* ('sense') and *Bedeutung* (usually rendered as 'reference'). This correlation suggested to me the title of the present section, but I shall henceforth avoid it, because the Fregean *Bedeutung* of a general term is not its extension or denotation but the Fregean concept expressed by it, i.e., the mapping from objects to truth-values that takes the value 'the True' on the term's extension and the value 'the False' outside it (Frege, NS, pp. 128–29).

I shall usually say that a general term *refers* to the objects it *denotes.* It does not seem to me that K. Donnellan's clever distinction between 'denoting' and 'referring' applies to general terms. The distinction was meant to apply to definite descriptions, as in the following example: If the Democratic mayor of New York toasted the health of "our next President" in October 1988, he presumably *referred* to Mr. Michael Dukakis but in fact he *denoted* Mr. George Bush. Cf. Donnellan 1966, §6.

[47] As a further exercise in conceptual relativity the reader may reflect on the following: While Kuratowski and Wiener applied their ingenuity to reducing the concept of order to that of set membership, in our own time mathematicians have had to devise methods that

make set-theoretical predicates and relations manageable by digital computers, which access data items sequentially. See Ball 1982, p. 65.

[48] Plainly, some such organ is implied by the metaphor of 'grasping', just as the mind's eye was implied—at any rate in Latin and Greek—by the metaphors of 'intuiting' and 'theorizing'. As the reader will have noticed, in the idiolect of this book it is objects—i.e., denotata—that are grasped *by* or *with* concepts. Thus intensions reside in the grasping hand of the soul as its fingers, or rather as its patterns of action.

[49] Of course, he would now. "The reference of a word like 'gold' is fixed by *criteria* known to experts" (Putnam 1988, p. 36; my italics).

[50] In the original version of Putnam's story, *XYZ* "tastes like water" and "quenches thirst like water" (Putnam, PP, vol. II, p. 223). Later, Putnam changed the Twin Earthian liquid to a mixture of water (80%) and grain alcohol (20%) and assumed that Twin Earthians were so constituted "that they do not get intoxicated or even taste the difference between such a mixture and H_2O" (Putnam 1981, p. 23). Since, on the other hand, humans would at once be able to tell one liquid from the other, the situation is not symmetric. In my opinion, this revision destroys any philosophical interest that the story may have had. In the new version Twin Earthians are merely insensitive to the presence of alcohol in their water, just as we remain indifferent to the presence of chlorine in ours. But an English-speaking community of germs would not recast their semantics merely because we persist in calling 'water' what to them is a dangerously toxic mixture of H_2O and Cl, inter alia. Anyway, Putnam has subsequently gone back to his original story, and he now describes *XYZ* as a substance "you couldn't tell [from water] by the appearance or taste or after-effects, or by washing clothes in it" (Putnam 1988, p. 31).

[51] The following, rather more realistic variant of Putnam's Twin Earth story occurred to me while reading Kripke's *Naming and Necessity* (1972, 1980). Imagine two remote planets called Saganus and Twin Saganus. Both are much like the Earth and contain silver in reasonable quantities, but all the silver in Saganus is ^{107}Ag, while that in Twin Saganus is ^{109}Ag. (In our environment pure silver is a mixture of approximately 51.83% of the former and 48.17% of the latter isotope.) Suppose that each planet is inhabited by English-speaking peoples who base their respective international monetary systems on the (local) silver standard. Suppose further that Saganian astronauts are about to reach Twin Saganus and to open trade between the planets. Will Saganians acknowledge ^{109}Ag as 'silver'? Will Twin Saganians allow the homonymous word in their dialect to apply to ^{107}Ag? More to the point, will either civilization accept payments in the alien isotope? Without filling in many more details of the story there is no plausible way of telling exactly how the denotation of these otherwise seemingly rigid "natural kind" designators might on such a critical occasion shrink or grow or twist or bend under the forces of interest and prejudice (including scientific preconceptions) or by sheer historical accident.

[52] The inclusion of 'natural kind' among the semantic markers of 'water' suggests that the above analysis of meaning applies also to words which are not natural kind terms. However, Putnam does not elaborate this suggestion.

[53] The causal connection may be indirect and mediated by other users, but obviously there must be some users for whom the connection is or has been direct. The idea that one must be causally related to the things one talks about in order to successfully refer to them must appear incredibly pedestrian to anyone familiar with the history of modern physics and chemistry. Mendeleev, for example, introduced the terms 'ekaluminum' and 'ekasilicon' for two hypothetical elements that, in his periodic table, would occupy the slots between zinc and arsenic, next to aluminum and silicon. He gave fairly precise descriptions of their main properties. The subsequent discovery of such elements—now known, respectively, as gallium and germanicum—was received as a remarkable confirmation of Mendeleev's ideas

concerning the relations between the properties of elements and their atomic weights. To the extent that Mendeleev did in effect conjecture the existence and anticipate the discovery of Ga and Ge he must have succeeded in denoting these elements by those terms of his which so accurately connoted their characteristic properties.

[54] Putnam contends that, because natural kind terms are indexical, their extensions cannot be determined by their respective intensions. It is indeed characteristic of an indexical word—such as 'you', 'there', 'tomorrow'—that when it is uttered by somebody, it denotes whatever meets a certain condition relative to that utterance. This condition, which we may plausibly call the intension of the indexical word, evidently does not fix its denotation, for it is the same on all occasions on which the word is used, while the denotation changes. As J. R. Searle aptly puts it: "What is special about indexical expressions is that the lexical meaning of the expression by itself does not determine which object it can be used to refer to, rather the lexical meaning gives a rule for determining reference relative to each utterance of the expression" (1983, p. 222). But the indexicality of natural kind terms, if such there be, does not work in quite this way. Let Australians call 'water' whatever is the same chemical compound—"give and take some impurities"—as the stuff in the South Pacific, while Scots call 'water' whatever is the same chemical compound as the stuff in the North Atlantic. Then the term 'water' used in Scotland and the homonymous term used in Australia are coextensive if and only if the conditions for being the same compound as the stuff in the South Pacific are identical with the conditions for being the same compound as the stuff in the North Atlantic. (This holds, of course, provided that the category signified by the semantic marker 'chemical compound' is understood in the same way throughout the preceding sentences. Otherwise, the homonyms in question may or may not be coextensive, but pursuant to Putnam's analysis, they do not share the same meaning, for they carry different semantic markers.)

[55] Aristotle made no allowance for so-called intensive quantities. For him, big and small and bigger and smaller (τὸ μέγα καὶ τὸ μικρὸν καὶ μεῖζον καὶ ἔλαττον) are admissible attributes of quanta (*Metaphysica*, Δ, 13, 1020^{a}23), but a quantity does not admit of more or less, like a quality does (τὸ πόσον οὐκ ἐπιδέχεται τὸ μᾶλλον καὶ τὸ ἧττον—*Categories* 6, 6^{a}26; cf. 8, 10^{b}27). This remark of Aristotle contradicts Putnam's characterization of quantities as being "capable of more or less," thus showing that categories which are normally termed homonymously—at least in translation—have been very differently understood in the course of history. Little is gained, therefore, for the stability of reference by divorcing it from intensions if it remains bound to semantic markers and the shifting categories signified by these.

[56] The distinction is made in practically the same terms by Richard Feynman in his justly celebrated *Lectures on Physics* (Feynman, Leighton, and Sands 1963, vol. I, p. 11-5).

[57] For a proof, see Krantz et al. 1971, p. 81. The independence of the five axioms is proved on pp. 77ff.

[58] Of course, if we repeatedly encounter as yet unobserved instances of a physical magnitude and our means of observing it manifestly admits improvement, standard English usage requires us to say that the actual instances of that magnitude are being partially and imperfectly recorded. However, this does not imply that a perfectly accurate and complete record of those instances would inevitably disclose their division into disjoint classes of instances of equal size. For it could well be that the common assumption that extensive physical magnitudes can be weakly ordered by size is only an illusion—or, for those who are not fooled by it, a useful convention—made possible by the very imperfection and imperfect perfectibility of our observations.

[59] The Krantz-Luce-Suppes-Tversky axioms for an extensive structure with an essential maximum are designed to accommodate the so-called Einstein Rule for the Addition of

Velocities. Let A, B, and C be three bodies moving uniformly with respect to one another along parallel lines in the space of an inertial frame. In relativistic kinematics the velocity $v(A,C)$ of A relative to C is computed from the velocity $v(A,B)$ of A relative to B and the velocity $v(B,C)$ of B relative to C according to the following rule:

$$v(A,C) = v(A,B) * v(B,C) = \frac{v(A,B) + v(B,C)}{1 + v(A,B)v(B,C)c^{-2}}$$

where c stands for the speed of light in vacuo. One verifies easily that '*'—as here defined—stands for a binary associative and commutative operation on **R** which maps the square $[-c,c] \times [-c,c]$ onto its side. For any real numbers x and y such that $|x| \leq |y| \leq c$, we have that $|x * y| \leq c$. In this sense, c can be described as an "essential maximum" for the extensive structure $\langle[-c,c],*\rangle$. However, before jumping to the conclusion that relativistic velocities reside in a Krantz-Luce-Suppes-Tversky extensive structure with an essential maximum, one ought to bear in mind that the peculiar rule of addition reproduced above is merely the result of applying to the very special case under consideration the relativistic rule for transforming the velocity vector of a particle from one inertial reference frame (endowed with Einstein time) to another.

[60] Metaphysical realists may relish the following idea: if the ultimate indivisible carriers of a given magnitude *in rerum natura* sport only certain real values $a_1, \ldots, a_n$, then the abstract structure realized by that magnitude is not the semigroup $\langle \mathbf{R}^+,+\rangle$ but the substructure of it generated by $\langle a_1, \ldots, a_n\rangle$. I doubt, however, that a working scientist would see any point in it. He is happy to represent any magnitude in the ideal structure $\langle \mathbf{R}^+,+\rangle$ and to leave open the question as to what values of it are actually instantiated. But then, of course, metaphysical realism, the offspring of medieval theology, is quite foreign to the spirit and the practices of modern science.

[61] This fact was used for a different philosophical purpose by Philipp Frank, who was Einstein's successor in Prague and a vigorous advocate of logical empiricism (Frank 1946, in Neurath et al. 1971, p. 455). Krajewski 1977, p. 57, argues against Frank and Feyerabend that by treating mass as "a function of two variables (body and its velocity)" the relativist does not necessarily abandon the classical concept; for biologists surely do not depart from it when they treat the mass of young animals as a function of two variables, viz., the creature's body and the time elapsed since its conception or since its birth. I am unable to see the point of Krajewski's argument. I take it that classical mass is independent of time only for bodies through whose boundaries there is no net flow of mass; but that for bodies that eat and drink and sweat and breathe classical mass is, of course, time dependent. For such bodies, however, relativistic mass is what Krajewski would call a function of three variables.

[62] The term 'mass' has been used to express still other concepts in the literature of Relativity. In his first paper on the subject, Einstein defined the electromagnetic force on a charged particle as a space vector which has the same components in all inertially moving Cartesian systems with parallel axes and is therefore equal in magnitude and direction to the electrostatic force exerted by the field on the particle in the latter's momentary inertial rest frame. Then, says Einstein, "if we maintain the equation mass × acceleration = force" and measure the acceleration in an inertial frame with respect to which the particle moves with velocity **v**, we shall find that the mass depends both on the particle's speed $v = |\mathbf{v}|$ and on the angle which **v** makes with the force **F**. In particular, if **v** and **F** are collinear, we obtain the *longitudinal mass*:

$$m_\lambda = \mu(1 - v^2c^{-2})^{-3}$$

whereas, if **v** and **F** are perpendicular we obtain the *transversal mass*:

$$m_{\tau} = \mu(1 - v^2c^{-2})^{-1}$$

In the above formulae, μ denotes the particle's mass in its momentary inertial rest frame, and c stands for the vacuum speed of light. Einstein remarks that these results are valid for every ponderable particle, for they do not depend on the amount of charge, and an arbitrary particle can always be regarded as a charged particle with "an *arbitrarily small* charge." He also notes that "with a different definition of force and acceleration we would obtain other values for the masses" (Einstein 1905b, p. 919).

In his excellent monograph *Special Relativity*, W. G. Dixon introduces the concept of *inert mass*, defined as the proper mass m_0 of a particle in a conventionally chosen reference state, somewhere in the middle between the states of maximum and minimum proper mass in the range of states under study. Relative to this reference state we can define the relativistic internal energy U of a state of proper mass m by

$$U = (m - m_0)c^2$$

In any collision that preserves particle identity, the law of conservation of energy can then be expressed as the conservation of internal energy U plus kinetic energy T, thus removing from consideration the summand $\Sigma m_0 c^2$ (summation intended over the colliding particles). This is useful because the contribution of this summand, which is constant, "is normally far larger than either of the variable contributions ΣT or ΣU" (Dixon 1978, p. 116). As far as I can judge, Dixon's 'inert mass' has a far stronger claim to succeeding the classical term 'mass' than 'longitudinal', 'transversal', 'relativistic', or even 'proper mass'. The concept of inert mass is, indeed, all but useless in elementary particle physics, "as the interchange that occurs between kinetic and rest energies in particle collisions and decays is too large" (Dixon 1978, p. 116). But nobody in his right mind would claim that Newtonian terms had referents at this level.

[63] In his recent book *The Emperor's New Mind* (1989), p. 220, Roger Penrose makes this point with inimitable clarity and conciseness: "One might try to take the view that [the rest mass] would be a good measure of 'quantity of matter'. However, it is not additive: if a system splits in two, then the original rest-mass is not the sum of the resulting two rest-masses."

[64] See Maier 1949, Chapter 2, "Das Problem der quantitas materiae," from which I draw most of the following information. In Proposition XLIV of his *Theoremata de corpore Christi* (1276), Giles of Rome discusses an interesting physicotheological problem. In the Eucharist, the substances of bread and wine have become the body and blood of Christ, but the *accidentia* of bread and wine remain. If one of these *accidentia* changes—e.g., if the white color of the wafer becomes greenish or if the taste of wine turns into the taste of vinegar—the holy body and blood of Christ cannot be affected. The change in some *accidentia* must therefore inhere in some other, more fundamental *accidens* of the vanished substances bread and wine, and ultimately in the quantity of their matter, which persists even as their color, taste, smell, etc., vary. In the tradition of Aristotelian scholasticism to which Giles belonged, *quantitas materiæ* was conceived as the only *accidens* of a body which contains in itself its own ground for division ("habet in se propriam rationem divisionis"—Aquinas in Boetii de Trin., qu.IV a.2, ad tert.), and was normally equated with bulk. But the floury and winy bulk of the Eucharist can decrease by condensation or increase by rarefaction. What is the permanent in which such changes inhere? Surely not the body and blood of Our Lord. Giles solves the difficulty by distinguishing a twofold quantity and two sorts of dimension in the matter of bread and wine and of all generable and corruptible things, namely, a quantity by virtue of

which the matter is "so much" ("tanta et tanta"), and another by virtue of which it takes up so much room ("occupat tantum et tantum locum"), the latter being grounded upon the former as upon a subject ("in prima quantitate . . . tanquam in subjecto fundatur alia quantitas"). Giles returned to this issue on the twofold quantity of matter in his Commentary to the Physics of Aristotle, where he writes:

> In materia duplex est genus quantitatis, unum per quod habet materia quot sit tanta et tanta, ut quod sit multa vel pauca, aliud per quod habet materia quod occupet tantum et tantum locum, ut magnum vel parvum. Nec est idem materiam esse tantam et tantam et eam occupare tantum locum, nam si ex aqua fiat aer, tanta materia quanta est in uno pugillo aquae erit in decem pugillis aeris. Remanebit ergo ibi tantum de materia, quia nihil ibi deperditur, sed non remanebit ibi occupatio tanti loci.
>
> (*Phys.* IV, text. 84; quoted by Maier 1949, p. 30)

[65] A useful explanation and discussion will be found in Lakoff 1987, Chapter 15. I do however exhort the reader to study Putnam 1980, now reprinted in vol. III of his *Philosophical Papers* and in the second edition of *Philosophy of Mathematics,* edited by Benacerraf and Putnam (1983).

Putnam's reasoning has been severely criticized by various authors. Those I have read—Merrill 1980, Pearce and Rantala 1982a and 1982b, Lewis 1984—have not impressed me. Merrill notes that the Löwenheim-Skolem Theorem applies to interpretations in freely restructurable domains, while realists normally view the world as a richly and rigidly structured collection of objects. Merrill is right, of course, but Putnam's point is that, in order to capture the purportedly preestablished structure of the world, the realist would need a direct grasp of essences, which he does not appear to have. Lewis follows Merrill, while at the same time anticipating that Putnam would retort as I have just suggested. In Lewis' judgment, "the realism that recognises a nontrivial enterprise of discovering the truth about the world needs the traditional realism that recognises objective sameness and difference, joints in the world, discriminatory classifications not of our own making [. . .] an objective inegalitarianism of classifications, in which grue things (or worse) are not all of a kind in the same way that bosons, or spheres, or bits of gold, or books are all of a kind" (Lewis 1984, pp. 228–29). Personally, I do not see how such inegalitarianism of classifications can make much of a difference in the context of Lewis' own variety of superrealism. For surely his philosophy must make allowance for "possible worlds"—call them *classificationally aberrant*—in which the "counterparts" of the individual objects intended in my discourse are grouped into "privileged classes" that differ in important ways from those to which the objects themselves actually belong. Thus, in the light of current physics, we could cite the following examples of classificationally aberrant worlds: (i) Aristotelian worlds, in which the counterparts of the Sun, the Moon and Venus belong to the same privileged class, which does not contain any gravitating body; (ii) worlds patterned after classical physics ca. 1910, in which the counterparts of our protons belong to the class of ultimate constituents of matter, a proper subclass of the class of things governed by Newton's Laws of Motion; (iii) pre-Weinberg/Salam worlds, in which the counterparts of electromagnetic and weak interactions between elementary particles belong, respectively, to two totally unrelated classes of events. In Lewis' parlance, a "possible world" *W* is *doxastically accessible* to me if and only if I believe nothing, either explicitly or implicitly, to rule out the hypothesis that *W* is the world where I live (Lewis 1986, p. 27). Since some of my beliefs are doubtless wrong, the actual world—i.e., the "possible world" that contains me—is certainly not doxastically accessible to me. Therefore it may well be that the worlds now doxastically accessible to me are in effect classificationally aberrant.

The scope of Putnam's model-theoretic argument is limited by the fact that, as I stressed above, the Löwenheim-Skolem Theorem holds only for first-order languages (in which bound variables range over individuals, but not over classes and relations). Such languages are notoriously inadequate for conveying the mathematics of mathematical physics (see, for example, Shapiro 1985). Hacking (1983, p. 105) raised this point against Putnam's argument. A reply is already implicit in Putnam's (1980) lengthy discussion of alternative (incompatible) set theories: a realist would have to choose one among them as he settles for a definite second-order language, and barring intellectual intuition of the eternal nature of sets, his choice would be purely conventional (cf. Section 2.8.7).

[66] This can be readily seen if we substitute 'scheme' for 'network' in Kuhn's characterization of a scientific revolution as "a displacement of the conceptual network through which scientists view the world" (Kuhn 1962, p. 101; quoted in Section 2.3) and reread the saying that "after a revolution scientists are responding to a different world" (Kuhn 1962, p. 110; quoted in Section 2.3) in the light of Putnam's dictum "*We* cut up the world into objects when we introduce one or another scheme of description."

[67] In his presidential address "On the Very Idea of a Conceptual Scheme," Donald Davidson (1974) dismissed the dualism of scheme and content as "the third dogma" of empiricism (an unveiled allusion to Quine 1951). I note in passing that, much as I admire Davidson's brilliant paper, I cannot agree with his matter-of-course identification of conceptual schemes with languages, nor with his assumption that, in order to "express the same scheme," two languages must be intertranslatable (Davidson 1984, p. 185). For languages sport structural features which do not always reflect their speakers' modes of thought. Thus, when I *say* that

It rains more often in London than in Madrid, (E)

I do not *think* that there is in both cities something—denoted by the pronoun 'it'—that does the raining, as the syntax of (E) might suggest. To me (E) is just an odd way of putting what I can state without any such suggestion in my own language:

Llueve más a menudo en Londres que en Madrid. (S)

In fact I have never felt that I have to change my habitual manner of thinking when I pass from Spanish to English, although I am keenly aware that the two languages are not generally intertranslatable. Full accurate translation between living languages is unattainable even if their educated speakers have roughly the same intellectual outlook, because languages are not molded only by the patterns of understanding they articulate, but also and mainly by the richly nuanced social relations they sustain, and these differ deeply from one linguistically identifiable group of peoples to another. Intertranslatability cannot therefore be tested on bloodless classroom examples like (E) and (S). Indeed to share Davidson's optimism about translation I should first have to see, say, a Spanish production of *A Streetcar Named Desire* or an English production of *La casa de Bernarda Alba*, which would convey to me the same dramatic situations as a good staging of these plays in their original language. It is, of course, practically impossible to faithfully repeat in the same language a stage production with a different cast. But if two languages are involved, one cannot even imagine what a faithful repetition would consist in.

[68] I am happy to note that the usual assumption that "conceptual systems are monolithic, that is, that they provide a single, consistent world view" is being challenged also by George Lakoff. See Lakoff 1987 (the quotation is from p. 317).

[69] The doctrine goes back to Aristotle. See *Metaphysica*, Z, 1; 1028^{a}14ff.: "That which 'is'

primarily is the 'what', which indicates the substance of the thing. [. . .] And all other things are said to be because they are, some of them, quantities of that which *is* in this primary sense, others qualities of it, others affections of it, and others some other determination of it" (Ross translation). The rest of Book Z of Aristotle's *Metaphysics* is devoted to a painstaking but far from conclusive elucidation of 'substance' (οὐσία).

[70] "Comme je conçois que d'autres Estres peuvent aussi avoir le droit de dire *moy*, ou qu'on pourroit le dire pour eux, c'est par là que je conçois ce que j'appelle *la substance* en general" (Leibniz, GP, VI, 502). "Imo rem accurate considerando dicendum est nihil in rebus esse nisi substantias simplices et in his perceptionem atque appetitum; materiam autem et motum non tam substantias aut res quam percipientium phænomena esse, quorum realitas sita est in percipientium secum ipsis (pro diversis temporibus) et cum cæteris percipientibus harmonia" (Leibniz, GP, II, 270).

[71] Had he acknowledged a greater degree of incoherence to each, he might have succeeded better in keeping them together.

[72] Note that this is true also of physical cosmology, which accounts for certain special phenomena by means of hypotheses concerning the large-scale structure and the early stages of spacetime, but in no way intends—like the old philosophical cosmologies—to come to grips with the fullness of human experience (see Torretti 1984).

[73] On ordered pairs, see Section 2.6.1.

[74] Readers for whom the foregoing statements are not obvious may wish to read the following proof (which, by the way, should also assist them in improving their grasp of the notions involved).

DEFINITION: If σ is a scheme for the echelon construction of a set, the *complexity* of σ is the number of occurrences of the symbols $\mathcal{P}$ and $\times$ in σ.

THEOREM: Let A be an echelon set over the set of sets $\mathbb{S} = \{S_1, \ldots, S_m\}$. If $\mathbb{S}' = \{S'_1, \ldots, S'_m\}$ is a set of sets equinumerous with $\mathbb{S}$, there is one and only one echelon set over $\mathbb{S}'$ which is homologous to A.

PROOF: Let σ be the scheme for the echelon construction of A. We proceed by induction over the complexity of σ. If σ has complexity 0, then $\sigma = i$ for some integer i such that $1 \leq i \leq m$. Hence, $A = S_i$ and the one and only echelon set over $\mathbb{S}'$ that is homologous to it is S'_i. Suppose now that the theorem is satisfied by every scheme for echelon construction τ of complexity equal to or less than n ($n \geq 0$), and that σ has complexity $n + 1$. Then, either (i) $\sigma = \mathcal{P}\tau_0$, where τ_0 is a scheme of complexity n for the echelon construction of an echelon set over $\mathbb{S}$; or (ii) $\sigma = \times\tau_1\tau_2$, where τ_1 and τ_2 are schemes of complexity equal to or less than n for the echelon construction of echelon sets over $\mathbb{S}$. Let B_κ and B'_κ denote the unique echelon sets over $\mathbb{S}$ and $\mathbb{S}'$, respectively, with scheme τ_κ ($\kappa \in \{0,1,2\}$). Then, either (i) $A = \mathcal{P}B_0$ and $\mathcal{P}B'_0$ is the one and only echelon set over $\mathbb{S}'$ that is homologous to A; or (ii) $A = \times B_1B_2$, and $\times B'_1B'_2$ is the one and only echelon set over $\mathbb{S}'$ that is homologous to A.

[75] The reader ought to verify that the examples of species of structure given in the text are in effect characterized by transportable conditions. Contrast them with the following set of conditions for the (1,2)-list of structural components $\langle a,b \rangle$:

C*. (i) $x \in a$ if and only if x is a male born in Corsica between 1700 and 1800;
(ii) b = Napoleon Bonaparte.

Clearly, the set of conditions C* is not transportable. For if a bijection f maps a onto an equinumerous set of Antarctic penguins a', neither is every $x \in a'$ an 18th-century Corsican male nor is the penguin $f(b)$ = Napoleon.

[76] Let H_f be the graph of f. Then, following Bourbaki's identification, $f = \langle H_f, G, G\rangle$. Since $G \in \mathcal{P}(G)$ and $H_f \subset G^2$, $f \in (\mathcal{P}(G^2) \times (\mathcal{P}(G))^2)$, which is plainly an echelon set over G. Readers ought to satisfy themselves that g is likewise an element of an echelon set over G.

[77] In order to state the Axiom of Choice, let us consider an arbitrary set of non-empty sets that we shall denote by A. The *union* of A, $\bigcup A$, is the set of all elements of the sets in A. By a *choice function* for A we shall mean a mapping $f{:}A \rightarrow \bigcup A$ such that for every $X \in A$, $f(X) \in X$. (The choice function f sends each set in A to an element of that very set, which f may be said to *choose* as a representative.) The Axiom of Choice says that every set of non-empty sets has a choice function.

[78] A weak order $\langle S, \leq\rangle$ well-orders the set S if every non-empty subset of S has a first element in $\langle S, \leq\rangle$, i.e., if for every non-empty $A \subset S$ there is an $a \in A$ such that $a \leq x$ for every $x \in A$. Zermelo's well-ordering theorem says that for every non-empty set S there exists a weak order that well-orders S. Zermelo's well-ordering theorem evidently entails the Axiom of Choice. (In the notation of note 77, if the set $\bigcup A$ can be well-ordered, the subset of $\bigcup A$ consisting of the elements of a given set $X \in A$ has a first element x_1; put $x_1 = f(X)$.) Since the well-ordering theorem can in turn be inferred from ZFC, it may be taken, in the context of ZF, as a substitute for the Axiom of Choice.

[79] Mac Lane 1986, pp. 386–406, gives a short and very readable introduction. The standard textbook is Mac Lane 1971. For a concise yet highly instructive survey of the basic concepts of category theory, see Bell 1988, pp. 1–48.

Chapter 3

[1] The treatise is entitled "On Local Motion" (*De motu locali*). The epithet 'local' is motivated by the Aristotelian and medieval use of 'motion' (κίνησις, *motus*) to refer to any change of an attribute of a substance, be it its place, its qualities, or its size. The modern use of 'motion' was probably promoted by the belief that all forms of change could ultimately be reduced to changes of place.

[2] My references to Galileo's writings are to Antonio Favaro's Edizione Nazionale (EN). The page numbers of this edition are given in the margins of Stillman Drake's English translation of the *Discorsi* (Galileo 1974), which I usually follow.

[3] He was anticipated in this by some 14th century authors, notably Nicole Oresme. The following passage from his *Tractatus de configurationibus qualitatum et motuum* lucidly argues for the geometric representation of physical magnitudes: "Quamvis tempus et linea sint incomparabiles in quantitate, tamen nulla proportio reperitur inter tempus et tempus que non inveniatur in lineis et econtra [. . .]. Et similiter est de intensione velocitatis, videlicet quo omnis proportio que reperitur inter intensionem et intensionem velocitatis reperitur etiam inter lineam et lineam [. . .]. Ideoque in notitiam difformitatum velocitatum possumus devenire per ymaginationem linearum ac etiam figurarum" (Claggett 1968, pp. 288, 290; cf. Claggett 1959, Chapters 4–6).

[4] All inferences from the definition of uniformly accelerated motion are made via Proposition I:

> The time in which a certain space is traversed by a moveable in uniformly accelerated motion from rest is equal to the time in which the same space would

> be traversed by the same moveable carried in uniform motion whose degree of speed is one-half the maximum and final degree of speed of the said uniformly accelerated motion.
>
> (Galileo, EN, VIII, 208)

The proof of this proposition is regarded by many as invalid (e.g., by Clavelin 1983, pp. 38ff.). But I do not see that one can find fault with it if one's attention is directed to the physical concepts involved rather than to the geometrical concepts by which Galileo chooses to represent them. We must bear in mind that Galileo conceived speed as a quantifiable physical property of a body, by virtue of which the latter moves as it does, and not merely as a quotient or as the limit of a sequence of quotients between two such properties. A body in motion traverses a certain distance in a certain time because, while it moves, it has at each instant a certain speed. Let a uniformly accelerated body B depart from rest at time 0 and attain the final speed $2V$ at time $2T$. Proposition I asserts that B will have covered the same distance $2VT$ that would have been traversed by a body moving with constant speed V during the whole time interval $(0,2T)$. By hypothesis, B's speed increases linearly with time. Hence, B must attain speed V at time T. For each time t less than T there is a "degree of speed" v less than V such that B has speed $V-v$ at time $T-t$ and speed $V+v$ at time $T+t$. Since the excess of the latter speed over V is exactly compensated by the defect of the former, B must, as a consequence of having just these speeds at the said times, move precisely as it would if at both times it had the speed V. Since this holds for every pair of times equidistant from time T in the time interval $(0,2T)$, and B actually has speed V at time T, it is clear that the distance travelled by B thanks to the increasing speeds it successively attains during the interval $(0,2T)$ will be equal to the distance it would travel if it had the constant speed V throughout that interval. Q.E.D. The interested reader should collate this reconstruction with the original text in Galileo, EN, VIII, 208–9.

[5] If we are not put off by anachronism, we may characterize Galileo's concept à la Bourbaki as follows: A *uniformly accelerated motion* is a structure $\langle I,s \rangle$, where I is an interval open in $\mathbf{R}$ with greatest lower bound equal to 0, and s is a smooth mapping of I into $\mathbf{R}$ by $t \mapsto s(t)$, such that $d^2s/dt^2 = \text{const}$. A real motion instantiates this concept if, when its duration is equated with I and the distance traversed in time t is measured by $|s(t) - s(0)|$, the said second-order differential equation is satisfied for all $t \in I$.

[6] This fact may have encouraged Clavelin to assert that the definition of uniformly accelerated motion "does not play any role in the subsequent development of Galileo's argument" (Clavelin 1983, p. 36). To my mind, its role is secure enough if, as it happens, Propositions I and II and the two important corollaries that follow them are inferred from the definition.

[7] This conclusion is not explicitly drawn by Galileo but it readily follows from his Proposition V (which in turn is derived from the postulated constraint I mentioned in the text):

> The ratio of times of descent over planes differing in incline and length, and of unequal heights, is compounded from the ratio of lengths of those planes and from the inverse ratio of the square roots of their heights.
>
> (Galileo, EN, VIII, 220)

Let a_i, h_i, and α_i, respectively, denote the length, height, and inclination of plane P_i; let t_i stand for the time in which a freely falling body reaches the bottom of P_i after departing from rest at the top $(i = 1, 2)$. Proposition V says that, for any two inclined planes, P_1 and P_2, $(t_1/t_2)^2 = (a_1/a_2)^2(h_2/h_1) = (a_1 \sin \alpha_2)/(a_2 \sin \alpha_1)$. If P_1 is an initial segment of P_2, $\alpha_1 = \alpha_2$, so that $(t_1/t_2)^2 = (a_1/a_2)$. This is precisely the import of Proposition II, which follows from the definition of uniformly accelerated motion:

> If a moveable descends from rest in uniformly accelerated motion, the spaces run through in any times whatever are to each other as the duplicate ratio of their times; that is, are as the squares of those times.
>
> (Galileo, EN, VIII, 209)

Galileo was of course well acquainted with the simple trigonometry used in the above argument.

[8] It is worth noting that Proposition VI, like all those that Galileo proves from the postulated constraint, is inferred from it via Proposition III:

> If the same moveable is carried from rest on an inclined plane, and also along a vertical of the same height, the times of the movements will be to one another as the lengths of the plane and the vertical.
>
> (Galileo, EN, VIII, 215)

The proof of Proposition III can be reformulated as follows: Each point p on an inclined plane P has a height h_p which it does not share with any other point on P. By the postulated constraint, the speed $v(p)$ which a body freely falling along P has as it passes p is equal to the speed $v(h_p)$ it would attain on reaching height h_p as it fell vertically from the top of P. Since this equality matches one-to-one every point on P with every height on the vertical from the top of P, Galileo concludes that the body descends along either trajectory with equal speeds. Hence, the times of motion are to one another as the spaces traversed (cf. Proposition I of Part I; EN, VIII, 192). This argument reminds one of the proof of Proposition I (of Part II) which I reformulated in note 4. Both proofs seek to make the theory of uniform motion serviceable in the study of non-uniform motion. However, while the proof of Proposition I achieved this goal by showing how to calculate an equivalent uniform speed for any given uniformly accelerated motion, the proof of Proposition III does nothing of the sort. Moreover, it rests on a universal quantification over the *heights*, not the *times*, at which the body attains equal speeds on either trajectory. It is therefore doubtful that my vindication of the former proof can also back the latter. Since all the testable consequences invoked by Galileo in support of the aforesaid constraint follow in fact from Proposition III, they confirm this proposition no less and indeed, in view of the stated doubt, much better than they do the constraint. Why, then, didn't Galileo directly postulate Proposition III as a constraint on free falls? I can think of only one answer to this question: Proposition III is not made plausible by the dynamical considerations that Galileo adduced for the constraint actually postulated by him (see the text above); and, despite his spokesman's arrogant promise of "la verità assoluta," he would not place all his eggs in the one basket of hypothetico-deductivism.

[9] Galileo says that the ascending plane constitutes an obstacle on striking which a falling body would lose some of its impetus. However, he adds, "if the obstacle that prejudices this experiment were removed, it seems to me that the mind understands that the impetus, which in fact takes its strength from the amount of the drop, would be able to carry the moveable back up to the same height" (EN, VIII, 207–8).

[10] Condition (ii) is not stated by Galileo, but without it one may not claim—as he does in his proof—that a body falling from a height h on a plane that makes an angle α with the vertical travels (from the top) a distance $h \cos \alpha$ in the same time it would require when falling vertically to reach the ground.

[11] Thus, for example, in the core of Galileo's theory of free fall there is no place for a concept of force. Yet Newton believed that Galileo had derived his Law of Fall from the Newtonian Second Law of Motion (Newton, *Principia,* p. 64: scholium to the Laws of Motion). I. B. Cohen (1967) regards this as evidence that Newton had not read the *Discorsi.* I shall not dispute Cohen's scholarly findings. However, if Newton had read the 1655 edition of Galileo's book, he might well have seen a special case of his own concept of an impressed force in the—doubtless extratheoretical—notion of an "impetus of descent" which, on the one hand, is measured by the static force that is exactly sufficient to balance it and, on the other hand, is proportional to the acceleration.

[12] Newton, *Principia,* p. 298; Bk. I, Prp. 79, Scholium (my emphasis). Newton adds that when these two stages of inquiry will have been carried through "we shall at last be entitled to dispute more reliably about the species, causes and grounds of the forces" ("tum demum de virium speciebus, causis et rationibus physicis tutius disputare licebit"—*Principia,* p. 298). This third stage of inquiry, characteristic of the older philosophy of nature, has remained outside the purview of mathematical physics ever since Newton declared that he had not named the cause of gravity (*causam gravitatis*) because he had not yet been able to derive from phenomena the reason or ground of its recorded properties, and he would not frame hypotheses ("rationem vero harum gravitatis proprietatum ex phænomenis nondum potui deducere, et hypotheses non fingo"—*Principia,* p. 764). Let me observe, by the way, that the occurrence of *rationem* in this passage from the General Scholium of the *Principia* indicates, in my view, that the *rationes physicæ* mentioned in the passage from p. 298 quoted in this note stand for reasons or grounds, in contradistinction to the *rationes illæ, sive proportiones mathematicæ* named earlier on the same p. 298, which are, of course, ratios or quotients.

[13] In Andrew Motte's translation (revised by Cajori), the descriptions read as follows:

> PHENOMENON I. That the circumjovial planets, by radii drawn to Jupiter's centre, describe areas proportional to the times of description; and that their periodic times, the fixed stars being at rest, are as the 3/2th power of their distances from its centre.
>
> [Phenomenon II is the same as I, with 'circumsaturnal' and 'Saturn' substituted for 'circumjovial' and 'Jupiter', respectively.]
>
> PHENOMENON III. That the five primary planets, Mercury, Venus, Mars, Jupiter, and Saturn, with their several orbits, encompass the sun.
>
> PHENOMENON IV. That the fixed stars being at rest, the periodic times of the five primary planets, and (whether of the sun about the earth, or) of the earth about the sun, are as the 3/2th power of their mean distances from the sun.
>
> PHENOMENON V. That the primary planets, by radii drawn to the earth, describe areas in no wise proportional to the times; but the areas which they describe by radii drawn to the sun are proportional to the times of description.
>
> PHENOMENON VI. That the moon, by a radius drawn to the earth's centre, describes an area proportional to the time of description.
>
> (Newton, *Principia,* pp. 556–63)

Evidently, no observation can yield the exact equalities asserted in these descriptions. But in laying them down, Newton tells us, he neglects the minutiae of imperceptible errors ("errorum insensibiles minutias in hisce phænomenis negligo"—*Principia,* p. 563, line 18; the sceptical reader will not fail to note that the observed deviations from the stated equalities are anything but *insensibiles*).

[14]The reader who skipped Section 2.8 on mathematical structures is advised to take a look at it now.

[15]As a matter of fact, the identification of a physical theory with its set of models, instead of its set of statements, had been championed much earlier by the Dutch logician E. W. Beth (1948/49). His program was taken up and divulged by Bas van Fraassen (1970, 1972) about the same time as Sneed's book appeared.

[16] The second "half-volume" of Stegmüller's *Theorie und Erfahrung* (1973) has been translated into English as *The Structure and Dynamics of Theories* (1976). More recently, still a third "partial volume" has been added to *Theorie und Erfahrung*, devoted to "the development of the new structuralism since 1973" (Stegmüller 1986). For a bibliography of Sneedian structuralism, see Diederich, Ibarra, and Mormann 1989.

[17]The contention that using category theory would involve greater "technical complexity" (Balzer, Moulines, and Sneed 1987, p. xxii) makes sense only for someone, like myself, who has learned mathematics in the standard set-theoretical setting and is a novice in category theory. In Twin Earth, where Twin Cantor came after Twin Mac Lane, the Bourbakian approach is justifiably accused of needless complexity.

[18]Balzer, Moulines, and Sneed 1987, p. xxiii. The theory-element of Balzer, Moulines, and Sneed is the direct successor of Sneed's original TMP, or *theory of mathematical physics* (Sneed 1971). I have summarily explained the latter notion in Torretti 1986b, pp. 184ff.

[19]Less innocuous examples of set-theoretical nonsense will be found in Balzer, Moulines, and Sneed's treatment of "the diachronic structure of theories" in Chapter 5 of their book. Here we meet an "injective mapping" $\mathbf{g}\colon \mathbf{H} \times \mathbf{C} \to \mathcal{P}(\mathbf{S})$, where **H** stands for the "set" of historical periods, **C** for the "class" of all scientific communities, and **S** for the "class" of scientists. Such excesses are not uncommon in paramathematical literature. They remind me of Aristotle's remark: πεπαιδευμένου . . . ἐστιν ἐπὶ τοσοῦτον τἀκριβὲς ἐπιζητεῖν καθ' ἕκαστον γένος, ἐφ' ὅσον ἡ τοῦ πράγματος φύσις ἐπιδέχεται (*Eth. Nich.* I, 3, $1094^{a}23$–25).

[20]Think, for example, of the usual distinction between an overall geochronometrodynamic approach to gravitation, which seeks to explain gravity by some geometric property of the spacetime manifold, dependent on the distribution of matter, and Einstein's General Theory of Relativity, which conceives the spacetime manifold precisely as a semi-Riemannian manifold whose metric tensor is coupled to the distribution of matter by the Einstein field equations.

[21] For example, if $\langle S,f\rangle$ is a topological space as defined by conditions $\mathrm{T_K}$ in Section 2.8.5, the closure f, which is a mapping from $\mathcal{P}(S)$ to $\mathcal{P}(S)$, will induce homonymous mappings from $\mathcal{P}^n(S)$ to $\mathcal{P}^n(S)$ for every $n > 1$; f will also determine its own range $C \in \mathcal{P}^2(S)$—recall that $x \in C$ if and only if x is a closed set of $\langle S,f\rangle$—and hence the power sets $\mathcal{P}^n(C) \in \mathcal{P}^{n+2}(S)$; the existence of all these definite entities is entailed by, and is therefore a necessary condition of, our specification of $\langle S,f\rangle$ by conditions $\mathrm{T_K}$.

[22]After writing the above I was pleased to find that a similar conclusion had been reached within the Sneedian school by Thomas Bartelborth in the excellent doctoral dissertation on the logical reconstruction of classical electrodynamics he wrote under Moulines. He notes that when the four Maxwell equations are written in the familiar vector form, three of them are "laws" in the Sneedian structuralist sense, but the equation $\mathrm{div}\,\mathbf{B} = 0$ is a mere "characterization"—a lack of symmetry for which there is no physical justification. On the other hand, all four equations can be encoded as a single "characterization" in the tensor formulation $\partial_{[\mu}\mathrm{F}_{\nu\lambda]} = 0$ (Bartelborth 1988, p. 21). I should add, however, that the distinction between "characterizations" and "laws" by Balzer, Moulines, and Sneed is not syntactic as Bartelborth says (disparagingly), but properly semantic, for it rests on the number of distinct structural components *named*—i.e., referred to—in either sort of statement.

[23] Note that this seemingly restrictive approach, followed by Sneed (1971), does not involve any real loss of generality, for any monadic or polyadic predicate P is uniquely associated with a real-valued function χ_P on the set of objects to which P can be meaningfully ascribed. χ_P—the *characteristic function* of P—takes the value 1 at every object of which P is true and the value 0 at every object of which P is not true.

[24] For example, Sneed has this to say about the Newtonian force function in Classical Particle Mechanics:

> All means of measuring forces, known to me, appear to rest, in a quite straightforward way, on the assumption that Newton's second law is true of some physical system, and indeed also on the assumption that some particular force law holds.
> (Sneed 1971, p. 117)

[25] "Classical categories" are—or are matched one-to-one by—concepts defined by necessary and sufficient conditions (Lakoff 1987, p. 286). Most common nouns and noun phrases employed in describing the world do not denote the extension of such classical categories. Their connotation, whatever it may be, can plausibly be termed a nonclassical category. General features of nonclassical categories are summarized in part in Lakoff 1987, p. 56, and discussed passim (see Lakoff's subject index, s.v. 'category').

[26] As Gähde aptly notes: "The condition that the same inertial mass should always be assigned to a given object, in whatever application of classical mechanics it turns up in, is by no means an empirical statement but rather a basic postulate [*eine Grundforderung*] that is immediately tied to the physical concept of mass" (1983, p. 75).

[27] Condition (ii) was not mentioned by Sneed when he first introduced constraints (Sneed 1971, p. 170, D28), but this omission was corrected by Balzer and Sneed (1977, p. 196, D0 and n. 4). However, Balzer, Moulines, and Sneed (1987, p. 47) treat condition (ii) as the characteristic property of a special kind of constraint—termed *transitive* constraints—but do not include it in their definition of constraints in general. The reason they give for this is that "although transitivity will hold in most examples, there are cases of constraints not being transitive." They name constraint $\mathbf{C}_4$(**SETH**) of Simple Equilibrium Thermodynamics (as reconstructed in Chapter III of the same book), but this is patently transitive. However, Professor Moulines has kindly let me know that the index 4 is a misprint (unfortunately not noted in the list issued with the book) and that the object they actually had in mind is $\mathbf{C}_6$(**SETH**). Now, $\mathbf{C}_6$(**SETH**), which is **SETH**'s version of energy conservation, does indeed fail to meet condition (ii), so it is not a transitive constraint; but it also fails to meet condition (i), so that in fact it is not a constraint at all. For, by the definition of $\mathbf{C}_6$(**SETH**), if x is a potential model of the theory-element **SETH** with nonconstant internal energy and x belongs to a family $F \in \mathbf{C}_6$(**SETH**), there must be another potential model y of **SETH** such that the "concatenation" $x \circ y$ has constant internal energy, and F must also contain both y and $x \circ y$ (Balzer, Moulines, and Sneed 1987, p. 144). It follows at once that, if x is as described, the singleton $\{x\} \notin \mathbf{C}_6$(**SETH**), in violation of Balzer, Moulines, and Sneed's condition *DII2* (a/3) for constraints (which is identical with condition (i) in the text above).

[28] Balzer, Moulines, and Sneed 1986; 1987, Chapter 8. Of course, the word 'science' should be taken here with a pinch of salt, for the structuralist treatment only extends to scientific disciplines patterned after the paradigm of mathematical physics. The idea that a jurist or a linguist could grasp their subject matters as instances of Bourbakian species of structure seems to me utterly preposterous.

[29] The addition of the global link **GL**(**T**) to the theory-core **K**(**T**) leads to a revised statement of the empirical claim of **T**. Instead of (3), we now have

$$\mathcal{P}(\mathbf{I}(\mathbf{T})) \subset \mu(\mathbf{GC}(\mathbf{T}) \cap \mathcal{P}(\mathbf{M}(\mathbf{T})) \cap \mathcal{P}(\mathbf{GL}(\mathbf{T}))) \qquad (3^*)$$

For brevity's sake, one may define with our authors the *content* of **T**, $\mathbf{Cn}(\mathbf{T}) = \mu(\mathbf{GC}(\mathbf{T}) \cap \mathcal{P}(\mathbf{M}(\mathbf{T})) \cap \mathcal{P}(\mathbf{GL}(\mathbf{T})))$, and state the empirical claim of **T** simply as $\mathbf{I}(\mathbf{T}) \in \mathbf{Cn}(\mathbf{T})$ (Balzer, Moulines, and Sneed 1987, pp. 90, 91). Note, by the way, that this shortened statement is equivalent to (3*) if, but only if, all constraints are transitive, for otherwise you could have a subset of **I**(**T**) which does not belong to $\mu(\mathbf{GC}(\mathbf{T}))$ although **I**(**T**) does.

[30] Sneed and his associates consistently use '**T**-non-theoretical term' for a term that is not **T**-theoretical. But to me that is like calling a radio signal that is not frequency-modulated an FM-non-radio signal.

[31] From the vantage point thus attained we can also make good sense of some seemingly paradoxical statements often met in philosophical literature. E.g. the statement that Newton's Laws of Motion are not laws, but definitions—indeed, they do serve to characterize the framework under which any particular physical situation must be grasped in order merely to conceive it as a candidate for Newtonian treatment. Or the statement that any grand theory of physics has an irrefutable "hard core"—viz., the framework element, whose potential models are at the same time its models.

[32] On the content and the publication history of this greatly overrated paper, see Truesdell 1984, pp. 516ff.

[33] Balzer, Moulines, and Sneed 1987, pp. 182–83, eloquently defend the exclusion of the Third Law from **CPM**. They argue that (i) in certain applications of **CPM** the Third Law reaction of some objects on their surroundings is absolutely ignored, while (ii) in other applications "there are forces that cannot even 'in principle' be considered to be counterbalanced by another equal and opposite force." The examples adduced in support of their contentions deserve comment. In connection with (ii) they mention the Lorentz force on charged particles in motion in an electromagnetic field. Now, one may indeed figure out the deflection of a beam of such particles by treating each particle as an intended model of **CPM** subject to a postulated force which—one assumes—happens to act right there where the particle is momentarily placed. But if you wish to understand what is going on and how it could be modified, say, by changing the electromagnetic field sources, you will do better to set **CPM** aside and view the situation as an intended model of a suitable theory-element of electrodynamics, in which the particles interact—i.e., exchange energy and momentum—with the field. In connection with (i), Balzer, Moulines, and Sneed recall that "when we study the motion of a projectile near the earth's surface we always consider the force of attraction exerted by the earth on the projectile, but nobody cares about the supposed 'reaction' due to the 'attraction' exerted by the projectile on the earth." Orthodox Newtonians may tell us that this can be done because the latter is negligible, but as a matter of fact "the talk about the counterbalancing force is a piece of metaphysical ornament with no bearance whatsoever upon the real-life applications and calculations of the theory." On this point I only wish to note that by expunging Newton's Third Law as a superfluous metaphysical ornament from the dynamics of projectiles one tacitly reinstates the ancient physical gap between projectiles and moons.

[34] Stegmüller 1986, p. 260, D8-16 (3). Cf. Sneed 1971, p. 114, D7 (3); Stegmüller 1973a, p. 108, D1 (3). The definition of **CPM** in Balzer, Moulines, and Sneed 1987 (pp. 29f.) does not specify what T is, but prescribes that it can be mapped bijectively onto **R**.

[35] The **CPM** position function in McKinsey, Sugar, and Suppes 1953 assigns to each particle of a mechanical system at any given time an ordered triple of real numbers satisfying the standard requirements for Cartesian coordinates. But coordinates, of course, are physically meaningless unless they are tied to a physical frame (Bunge 1967, p. 104). In

Classical Mechanics, as I noted above, a non-inertial frame can be used only if it is in turn referred to the inertial ones. Indeed Streintz pointed out over a century ago that the coordinates which occur in the differential equations of motion must be linked to a nonrotating, force-free "fundamental body" (Streintz 1883, as reported by diSalle 1988, p. 89).

[36] Balzer, Moulines, and Sneed 1987, pp. 267ff., show how classical rigid body mechanics can be "reduced" to **CPM**.

[37] Pursuant to the definition of $\mathbf{M}_p(\mathbf{CPM})$ by Balzer, Moulines, and Sneed 1987, pp. 29–30, a partial potential model of **CPM** should simply be defined as a 5–tuple of non-empty sets $\langle P, T, S, N, R \rangle$, such that P is finite, N is countably infinite, and T, S, and R are uncountable.

[38] See my summary of Lange's views in Torretti 1983, pp. 17ff. See also diSalle 1988, Chapters IV and V.

[39] At first blush it may seem possible to confer non-**CPM**-theoretical status to time and position simply by conceiving **CPM** as a specialization of a still more basic theory-element **CPK** (Classical Particle Kinematics), dependent on Newton's First Law but not on the Second. Lange's inertial frame and Neumann's inertial clock would no doubt be models of **CPM**. However, in order to make sense of measurements of time and distance referred to them, it would not be necessary to *assume* this, but only that they are models of **CPK**. The distinction between **CPK** and **CPM** would not carry us very far towards a resolution of Stegmüller's dreadful circle but would anyway vindicate Sneed's claim that the only **CPM**-theoretical concepts are force and mass. The matter is rather more involved, however, for the only criterion available in Newtonian mechanics for telling free particles from particles acted on by a constant force involves the Third Law. Thus, Newton's wisdom is visible not only in the statement of the First Law as a separate axiom (and not just as a special case of the Second Law) but also in his conceiving all *three* laws as the mutually complementary principles of a single theory.

[40] Gähde's work has received careful and respectful attention from the more orthodox members of the Sneedian school. See Balzer 1985, 1986. Cf. also Stegmüller 1986, pp. 155–89; Balzer, Moulines, and Sneed 1987, pp. 73ff.

[41] Gähde (1983, pp. 129–30) very aptly observes that nonmechanical procedures for mass measurement yield values of **CPM** mass only insofar as they are known or assumed to agree with mechanical procedures. (Recall the similar remark I made above concerning non-mechanical clocks.)

[42] The meaning of invariance under Galilean transformations in Newtonian mechanics is explained in most modern textbooks. I have dealt with the subject in Torretti 1983, pp. 15–31.

[43] Note that each $k \in \lambda$ must be assigned exclusively to one and only one object of the family, or it would not do its job as a label. If the same object carries two (or more) different labels, it counts as two (or more) different members of the family. These conditions are nicely captured if we define a family $\{\alpha_k\}_{k \in \lambda}$ as a mapping f of λ into some set from which every object α_k is drawn, or as the graph $\{\langle f(k), k \rangle \mid k \in \lambda\}$ of such a mapping.

[44] Note that the mapping μ: $\mathbf{M}_p(\mathbf{T}) \to \mathbf{M}_{pp}(\mathbf{T})$ introduced in (1) of Section 3.3 can be equated with the σ-complementary projection ρ_σ on $\mathbf{M}_p(\mathbf{T})$ if we put $\sigma = \{m+1, \ldots, n\} \subset \{0, \ldots, n\}$. Indeed, it is not necessary to place the **T**-theoretical functions at the end of the list of distinguished components of $\mathbf{M}_p(\mathbf{T})$ as we did there. We could take the very names of the structural components of $\mathbf{M}_p(\mathbf{T})$ as the index set (i.e., label the terms of each potential model of **T** by their respective names, as one would normally do) and put $\sigma = \{x \mid x \text{ is the name of a } \mathbf{T}\text{-theoretical term}\}$.

[45] Suppose that α_r is one such family member; in other words, suppose that $r \in \sigma$, and that for some $x \in \rho_\sigma(\mathbf{F})$ and some $y \in \varphi_\sigma(x)$, α_r belongs to y. One may then say, following Gähde, that α_r is *supplementary* (*ergänzend*) with respect to $\varphi_\sigma(x)$. Since Gähde deals specifically with indexed families of functions, the expression actually defined by him is 'ergänzende Funktion bezüglich $e_\sigma(x)$' (or, in English and using my notation, 'supplementary function with respect to $\varphi_\sigma(x)$'). In Gähde's idiolect, the fiber maps φ_σ ($\sigma \subset \lambda$) are "zulässige Ergänzungsfunktionen" ("admissible supplementing functions") for the original family **F**, which is why he designates them by the letter 'e' (instead of φ).

[46] $\langle \mathbf{M}(\mathbf{T}), \mathbf{M}_p(\mathbf{T}), \mathbf{GC}(\mathbf{T})\rangle$ is what Stegmüller 1986, p. 159, D6-1, calls a "generalized theory-core" ("verallgemeineter Kern für eine Theorie"). We could also throw into it the global link **GL(T)**, but the concept of an intertheoretic link apparently had not yet been introduced when Gähde wrote his book, for it is not mentioned in it.

[47] Perhaps the following remarks are apposite:

(α) I let σ range over $\{1, \ldots, n\}$ and not over the full set of indices $\{0, \ldots, n\}$ because we are not interested in complementary projections that cut off the collection $\mathcal{D}$ of base sets (which was assigned the index 0).

(β) Condition (i) does the job of the following somewhat esoterically worded Gähdian condition:

(i′) There is a $z \in \rho_\sigma(\mathbf{M}_p(\mathbf{T}))$ such that α_r is supplementary with respect to $\varphi_\sigma(z)$.

I defined the key term in note 45; in the light of that definition, it is clear that if $r \in \sigma$, α_r must be supplementary with respect to $\varphi_\sigma(z)$.

(γ) Gähde does not refer to the alternative, explicitly excluded by my condition (ii), that $\varphi_\sigma(x) \cap \mathbf{M}(\mathbf{T})$ could be empty. Note, by the way, that this alternative is barred if the proposal on theory-nets I made at the end of Section 3.3 is adopted *and* **T** is the framework element of a theory-net, for in that case, by definition, $\varphi_\sigma(x) \subset \mathbf{M}_p(\mathbf{T}) = \mathbf{M}(\mathbf{T})$.

[48] For example, Gähde shows that mass alone does not qualify as **CPM**-theoretical by his criterion but does so qualify as a member of the set {mass, force}.

[49] I have taken some liberties with Balzer, Moulines, and Sneed's typography. I write $\mathbf{K}_1/\mathbf{K}_2$ instead of K', K—substituting boldface for italics and a virgule for their unperspicuous comma. I also substitute the Greek letters α and β for the original i and j.

[50] On November 9, 1987, Professor Moulines kindly pointed out to me that "the current version of structuralist metatheory does not deny the possibility of 'self-interpreting' theories—on the contrary, it contemplates it explicitly." Indeed, the said possibility is implied by the loop of interpreting links displayed in Fig. VIII-1 of Balzer, Moulines, and Sneed 1987, p. 407, for linkage is transitive according to the definition *DVIII-1* (5) on p. 389 of the same book. I heartily welcome this development.

[51] Balzer, Moulines, and Sneed 1986, p. 297. I am in deep sympathy with the antifoundationist attitude conveyed by these words. Cf. also the highly suggestive reflections on "Foundationalism versus Coherentism" that form the concluding section of Balzer, Moulines, and Sneed 1987, pp. 411–23.

[52] For a brief account of the redshift in the spectra of extragalactic light and its cosmological interpretation, with references to the literature, see Torretti 1983, Section 6.3, in particular eqn. (6.3.7). Professor Moulines would rather say that relativistic cosmology *explains* the redshift and that what it tells us about it is in turn *interpreted* by the atomic theory (private communication, printed in Moulines and Torretti 1989). I will not quarrel about words. What I find so striking in this example is that the anomalous position of the lines

recorded in the spectra of very old radiation should falsify the atomic theory were it not that we are able to read them in the light of relativistic cosmology. Thus, in this case, the "interpreting" atomic theory cannot be properly brought to bear on the phenomenon in question without the guidance of the "interpreted" cosmological theory.

[53] "Eine mathematische Theorie, von uns immer kurz mit *MT* bezeichnet, ist definiert als eine Ansammlung von Zeichen nach gewissen Spielregeln" (Ludwig 1978, p. 17). This characterization is objectionable on at least three counts: (i) not every collection of symbols governed by rules can pass for a mathematical theory; (ii) none of the major mathematical theories of physics has ever been fully and truly formalized; (iii) in the light of Gödel's findings of 1931, the formalization of mathematics faces a dilemma—either elementary arithmetic (and every theory containing it) cannot be fully formalized, or the "rules of the game" must be so devised that not even a computer endowed with infinite memory can determine in every case whether those rules have or have not been followed.

[54] In a paper published in English, Ludwig (1984a) says 'pictorial' instead of 'iconic'. But I think that the latter is a better rendering of the word he uses in German.

[55] Current methods for the observation of elementary particles rely on their electromagnetic interaction with a macroscopic receiver. Obviously, only the presence of charged particles can be recorded by such means.

[56] Recall, however, that Balzer, Moulines, and Sneed would now countenance self-interpreting physical theories. See note 50.

[57] Normally, one simply says 'induction' to refer to what Ludwig calls here 'incomplete induction'. The unusual epithet marks the contrast between this bastard form of argument and Aristotle's "complete" induction, or *epagoge,* in which a general conclusion flows from the consideration of *every* relevant particular (ἡ γὰρ ἐπαγωγὴ διὰ πάντων—*An. Pr.* II, 23, 68^{b}29; cf. 69^{a}17).

[58] Ludwig 1978, p. 200. In another relevant passage, Ludwig says: "Der Grundbereich einer Theorie wird erst von der Theorie *und* ihrer Anwendung her allmählich in seiner Abgrenzung sichtbar" (1974, vol. II, p. 15). This can very roughly be translated as follows: "The fundamental domain of a theory only gradually becomes visible within its boundaries, from the perspective offered by the theory *and* its application." The passage might be taken to mean that the theory and the practice of applying it gradually teach us *what* real-texts belong to it, among the many which lie there beforehand neatly distinguished and arranged. But in actual fact physical theories preside over the experimental production of new, unprecedented real-texts within their purview and also direct our attention to hitherto neglected features of those already familiar. For instance, the steady-state cosmological theory of Bondi and Gold gave a curious reading of the dark night sky, presenting it prima facie as a blazing hemisphere from which vast tracts had been mysteriously erased. This reading follows from the theory's Perfect Cosmological Principle, according to which the universe in the large must look more or less the same at all times and places. The mysterious disappearance of most of the blazing light that ought to be there was explained by postulating the expansion of the universe (independently supported by Hubble's Law). See Bondi 1961, Chapter III; for a more plausible reading of the phenomenon, not based on the said principle, see Harrison 1965, 1977.

[59] About 1685 Newton used the inverse square optical law to measure the distance to Sirius by comparing its apparent brightness with that of the Sun. He figured out that Sirius is 1,000,000 times farther than the Sun, which is less than twice the currently accepted value.

[60] Ludwig 1970; cf. Ludwig 1978, §6. Moulines 1976 used uniformities for explicating approximation in physics within the Sneedian framework. See also Moulines 1981; 1982,

§2.7; Moulines and Jané 1981; Stegmüller 1986, pp. 227–68; Balzer, Moulines, and Sneed 1987, pp. 323–85.

[61] The concept of a uniformity on a set was introduced by André Weil (1937) as a generalization of the concept of a metric (i.e., of a distance function). The structure $\langle S, U \rangle$, where U is a uniformity on the set S, is called a *uniform space*. U induces a canonic topology on S, which can be characterized as follows: for each $x \in S$ and each entourage V of $\Delta(S)$, let the *ball* $B(x, V)$ be the set $\{z \mid z \in S \;\&\; |x - z| < V\}$; then, a subset A of S is open in the topology induced by U on S if and only if for every $x \in A$ there is a $V \in U$ such that $B(x, V) \subset A$.

[62] In the discussion of U3 in the main text, the expression 'conceivable approximation' should be understood, not in an absolute sense, but in the context of accepted physical ideas. Thus, e.g., the conviction that electric charge exists only in integral multiples of a finite quantity sets an upper bound to the conceivable improvement of approximations of any iconic relation involving test charges (i.e., charges so small that their presence does not affect the electromagnetic field).

[63] Schwarzschild 1916; Droste 1916. The original treatment of Mercury's orbit as an application of General Relativity by Einstein (1915) used an *approximate* solution of the Einstein field equations in vacuo.

[64] Traditionally, the correction of Mercury's orbit for the gravitational influence of the other planets is calculated by the perturbation methods of classical celestial mechanics, on the doubtless valid assumption that Newton's theory of gravity provides a sufficiently good approximation to General Relativity for the purpose at hand. But one could also apply the methods of the so-called Parametrized Post-Newtonian (PPN) approximation, which have been specially developed for use with chronogeometric theories of gravity (see Weinberg 1972, Chapter 9; Will 1981, Chapter 4).

[65] This use of 'model' in current philosophical writing must be carefully distinguished from the standard mathematical usage of the same word. In the mathematical sense, *models* are models of structures, and a model of a structure of a given species is any set endowed with structural features satisfying the requirements of that species. Thus, the set **R** of the real numbers, as structured by addition and multiplication, is a model of the type of structure known as a *field*. This is the sense in which I have used the word 'model' throughout my discussion of Sneed's work. The other, common philosophical sense of 'model' is closer to ordinary speech. As a warning signal, I shall surround the word, when used in this sense, in double quotation marks (colloquially referred to as shudder quotes). A "model" in this sense is a representation of an individual or generic object by an object of a different sort. Just as one can speak of a cardboard model of the Parthenon, one speaks nowadays of a computer "model" of the American economy. A model in the mathematical sense can be as concrete as one desires without thereby clashing with the abstract structure modelled by it. The latter's features are further specified by the model's peculiar properties. Thus, the fact that all Cauchy sequences of real numbers converge in **R** does not follow from its being a model of a field but is quite compatible with it. On the other hand, a "model" in the other sense usually has properties of its own that one will do well to ignore in order to see it as a "model"—the cardboard Parthenon will collapse under heavy rain; a computer "model" of a market economy does not produce any tangible goods or evils (except perhaps indirectly, through the use that can be made of it to influence the real economy), etc.

[66] I certainly do not think that Ludwig's "mapping principles" can be equated with the "correspondence rules" of Carnap and Reichenbach (see Section 2.4.3), as proposed by Kamlah 1981, p. 74. The realm of experience must be already articulated in order that the correspondence rules can bestow an observational meaning on theoretical discourse. On the other hand, Ludwig's mapping principles are instrumental in the very spelling and

reading of real-texts.

[67] Dudley Shapere (1964, 1966) ably made this point against Kuhn and Feyerabend. See Shapere 1984, pp. 45–46, 73.

[68] See also Pickering 1984b, pp. 180–95. After I wrote the above summary of Pickering's story, the discovery of weak neutral currents was admirably retold in great detail by Peter Galison (1987, pp. 135–241).

[69] We obtain a model M of the real projective plane by equating its "points" with the straight lines through a given point P in Euclidian space. Three such "points" are said to be "collinear" if they lie on the same plane through P. As a collection of "collinear points" every plane through P is automatically equated with a projective "line." We obtain another model M' of the real projective plane by equating its "points" with the planes through P. Three such "points" are "collinear" if they meet on the same straight line through P. As a gathering of "collinear points" every Euclidian straight line through P becomes in M' a projective "line." M and M' are realizations of the same abstract structure (the real projective plane) in the same domain, namely, the straight lines and planes through P, coupled by the symmetric relation of incidence between straights and planes. But M and M' differ in the manner in which they map that structure on this domain.

[70] Let M be an n-dimensional differentiable manifold. A *congruence* in M is a collection of curves in M such that each point of M lies in one and only one of them. A *foliation* of M is a collection of hypersurfaces (i.e., $(n-1)$-dimensional submanifolds) of M such that each point of M lies on one and only one of them.

[71] The following definition of the fiber derivative will be sufficient for our purpose. Let E be a vector bundle over a manifold M, and let E_p denote the fiber of E over $p \in M$. E_p is then a vector space. If ζ is any vector in E_p, p is said to be the projection $\pi\zeta$ of ζ into M. Clearly, then, for any $v \in E$, $v \in E_{\pi v}$. Consider a smooth real-valued function f on E. Let f_p stand for the restriction of f to E_p, and $Df_p(v)$ for the derivative of f_p at $v \in E_p$. Let $\mathcal{L}(E_p,\mathbf{R})$ denote the linear space of all real-valued functions on E_p. The *fiber derivative* of f is the mapping of E into $\bigcup_{p\in M}\mathcal{L}(E_p,\mathbf{R})$ by $v \mapsto Df_{\pi v}(v)$. The reader will recall that, if T_pM and $T_p{}^*M$ designate, respectively, the tangent and the cotangent space at $p \in M$, then $T_p{}^*M = \mathcal{L}(T_pM,\mathbf{R})$ and T_pM can be identified with $\mathcal{L}(T_p{}^*M,\mathbf{R})$. Since the base set of the tangent bundle TM is none other than $\bigcup_{p\in M}T_pM$ and the base set of T^*M is likewise $\bigcup_{p\in M}T_p{}^*M$, it is clear that the fiber derivative of the Lagrangian $L: TM \to \mathbf{R}$ is a mapping of TM into T^*M.

[72] The *causal past* $J^-(x)$ of a spacetime point x is defined by the following condition: $y \in J^-(x)$ if and only if y is the origin of a future-directed null or timelike spacetime curve that ends in x. The causal past $J^-[\gamma]$ of a timelike curve γ is the union of the causal pasts of the points on γ.

[73] The *chronological past* $I^-(x)$ of a spacetime point x is defined by the following condition: $y \in I^-(x)$ if and only if y is the origin of a future-directed timelike spacetime curve that ends in x. The chronological past $I^-[\gamma]$ of a timelike curve γ is the union of the chronological pasts of the points on γ. In the manifold topology of a relavitistic spacetime the chronological past of a point or a curve is always an open set, whereas their causal past can be neither open nor closed.

[74] I follow Malament 1977b, p. 63. The definition in Glymour 1977, p. 52, essentially amounts to the same but seems to me less perspicuous.

[75] Let A and B be two topological spaces. A is a *covering space* for B if there exists a *covering map* $f: A \to B$, that is, a *continuous mapping of A onto B* that meets the following condition: every point $p \in B$ has an open neighborhood U whose inverse image $f^{-1}(U)$ is a union of disjoint open sets of A, everyone of which is mapped homeomorphically onto U by f. (A *homeomorphism* is an isomorphism of topological spaces.)

[76] Namely, orientability, spatial orientability, inextendibility, noncompactness, causality and strong causality, the existence of a global time function, and the existence of a Cauchy surface. A spacetime is causal if it contains no closed future-directed causal (i.e., null or timelike) curves; it is strongly causal if every neighborhood of each spacetime point p includes a neighborhood of p into which no causal curve enters more than once. A set S of spacetime points is a Cauchy surface if no point of S lies in the chronological past of another point of S and for every spacetime point z every inextendible timelike curve through z meets S. In the absence of a Cauchy surface, causal influences can always unpredictably "rush in from infinity." Malament shows that the condition of noncompactness need not be shared by two spacetimes when they are o.i., in Glymour's strong sense. Since the latter relation is symmetric, compactness, too, need not be shared by o.i. or, consequently, by w.o.i. spacetimes. We may therefore add compactness to the above list. On the other hand, I have deleted the first property listed by Malament, namely, temporal orientability, because the very definition of weak observational indistinguishability presupposes that *both* terms in the relation contain future-directed curves, and this cannot be the case unless *both* are temporally orientable.

[77] In an earlier version of this chapter, I suggested that the existence of o.i. spacetimes in General Relativy confirms that so-called scientific realism is irrelevant to scientific discourse. A referee took strong exception to this. My point will perhaps become clearer if we recall Hilary Putnam's characterization of metaphysical realism in the same issue of *Synthese* which carried Glymour's "Confessions of a Metaphysical Realist" (1982):

> What makes the metaphysical realist a *metaphysical* realist is his belief that there is somewhere, One true theory. [. . .] In company with a correspondence theory of truth, this belief in One true theory requires a *ready-made* world—the world itself has to have a "built-in" structure since otherwise theories with different structures might correctly "copy" the world (from different perspectives) and truth would lose its Absolute (non-perspectival) character.
>
> (Putnam 1982a, p. 147)

What I find so fascinating about Glymour's discovery is that it shows how—if we abide by the realist idiom—theories with different structures may be said to correctly "copy" the world from the *same* perspective, and indeed from *every single* perspective. But, of course, it is misleading to speak in this way, for, as Ronald Giere bluntly puts it, *"metaphysical realism plays no role in modern science"* (1988, p. 98; for Giere 'metaphysical realism' is, in Putnam's words, the doctrine that "there is one true and complete description of 'the way the world is'").

[78] Empiricist philosophers subsequently made a great fuss about Mercury's perihelion advance, an "anomaly" of Newton's theory of gravity which Einstein mentioned as early as 1907 as an outstanding problem and which his General Theory of Relativity (GR) allegedly solved. In this connection it is worth noting (i) that in his 1913 address to the Society of German Natural Scientists Einstein declared that Newton's theory could not be faulted for its account of known gravitational phenomena, but only for its incompatibility with Relativity; (ii) that Einstein's work on Mercury's perihelion advance was not done on behalf of GR, but of a quaint earlier theory of his, which shares with GR the vacuum field equations $R_{\mu\nu} = 0$; (iii) that the evidence for these equations furnished by the said phenomenon depends on questionable assumptions about the shape of the Sun.

[79] Paraphrased from Kemeny and Oppenheim 1956, as reprinted in Brody 1970, p. 313. The authors say that a theory is well systematized if it sports a good mix of simplicity and strength. However, they do not tell us when such a mix is better or worse than another one.

[80] Schaffner 1967, p. 144. I have changed Schaffner's indexing of the T's to adjust it to mine.

[81] Spector 1978 (a much expanded version of Spector 1975) contrasts the usual analysis of intertheoretic reduction with his own "concept replacement." He describes a physical theory as a set of sentences closed under deducibility in a formal language. According to the usual analysis, such a theory T_1 is reduced to another theory T_2 if T_1 can be deduced from $T_2 \cap B$, where B is a set of sentences (bridge laws) which state conditions for the predicates in the vocabulary V_1 of T_1 in terms of predicates in the vocabulary V_2 of T_2. But in Spector's analysis T_1 is reduced to T_2 "if and only if for each *term* of V_1 one can construct a function of terms of V_2 such that, upon *replacement* of the former by the latter in the *laws* of T_1, one obtains transformed statements which can be shown (in one way or another) to be laws of T_2" (Spector 1978, p. 33; my lettering). Let L_1 be a law of T_1 which has thus been *re*duced without being *de*duced. To the question, "Has the law L_1 been *explained* on the basis of theory T_2?" Spector replies:

> I think the answer is clearly Yes. The phenomena previously described by L_1 have been accounted for by showing how they can be correctly redescribed on the basis of the conceptual apparatus of [. . .] T_2. If this isn't a brand of explanation, then I do not know what is. But the point at stake is more fundamental than that of the deduction of L_1. It hardly matters whether one wishes to say that L_1 has or has not been 'deduced' from the 'laws' of theory T_2. For whether it is correct or not, the crux of the issue of *explanation* does not lie with that. In a *reductive explanation* of the kind under consideration, it is, rather, our certified ability to *redescribe on the basis of concepts of a deeper theory* which is the heart of our gain in understanding. And here, it does not matter where the certification of the correctness of the redescription comes from.
>
> (Spector 1978, pp. 65–66)

I fully agree.

[82] As I noted in Section 3.6, the Sneedian structuralists have adapted Ludwig's treatment of approximation to their conceptual setup. Balzer, Moulines, and Sneed (1987) distinguish between *empirical* and *full* intertheoretic equivalence. Two theories are empirically equivalent if, regardless of any differences in their conceptual structures, they both "yield the 'same' explanations and solve the 'same' problems for the 'same' systems, i.e. if they have the 'same' empirical contents" (1987, p. 284). The difficulty with this idea of equivalence is that, as our authors incisively remark, "we have no more access to a set of conceptually unstructured systems than we have to Kant's *Ding an sich*. And the notion of seeing the 'same system' in two different conceptualizations (i.e. as two systems) is poorly understood." (p. 285). Two theories are fully equivalent "if their full theoretical structures are in some sense 'isomorphic', and if both theories are about the 'same' phenomena" (p. 285). The authors do some formal work on a concept of full intertheoretic equivalence and produce an admittedly trivial example of it. They observe that it is not easy to find simple examples of equivalent theories that are not just two versions of the same theory, and go on to suggest "that the concept of equivalence in reality does not have the weight philosophers and philosophers of science sometimes attach to it" (p. 303).

Chapter 4

[1] Cf. Garber and Zabell 1979. The *Oxford English Dictionary* gives, s.v. 'probable', the following obsolete senses: "2. Such as to approve or commend itself to the mind; worthy of acceptance or belief; rarely in bad sense, plausible, specious, colourable. (Now merged in modern sense 3.) b. Of a person: Worthy of approval, reliable." The "modern sense 3" is this: "Having the appearance of truth; that may in view of present evidence be reasonably expected to happen, or to prove true; likely." Curiously, one of the examples illustrating sense 3, drawn from Thomson (Lord Kelvin) and Tait's *Treatise on Natural Philosophy*, involves the expression 'probable error', by which these authors do not mean an error that is likely to happen, but a quantity such that the actual error of an observation is "as likely to exceed as to fall short of it in magnitude." Apparently, the "modern sense 3," which, as we saw, can be traced back to ancient Greek and Latin usage, was merging in the lexicographer's mind with a purely modern sense, rooted in the mathematical theory of probability. (Compare the same dictionary, s.v. 'probability', 3.) The link between senses 2 and 3 is brought out very clearly in the following passage from Cicero: "Quis enim potest ea quae probabilia videantur ei non probare?" ("Who can refrain from approving that which to him appears probable?"—*De finibus* , 5.76; quoted by Garber and Zabell 1979, p. 45.)

[2] ὀυδέποτε φαντασία μονοειδὴς ὑφιστάται ἀλλ' ἁλύσεως τρόπον ἄλλη ἐξ ἄλλης ἤρτεται—Sextus, *Adv. Math.*, 1.176. Philosophical programs for reducing experience to an aggregate of self-contained mental or "logical" atoms disregard this plain fact.

[3] Galileo, EN, VIII, pp. 591–94. There is an English translation in David 1962, pp. 192–95. Historians have found forerunners of the mathematical analysis of chances in the Renaissance and the Middle Ages. I was particularly impressed by the Latin poem *De Vetula*, which is believed to have been composed in France in the first half of the 13th century. The poem clearly and correctly describes, as in Galileo's paper, the different kinds of throw that can be made with three dice. Although the purpose of this exercise is not explained, it is hard to imagine that it was not that of calculating odds. The relevant lines are reprinted in Kendall 1956 and can also be found in Pearson and Kendall 1970, pp. 33–34.

[4] 'Chance' and 'probability' are curiously intertwined in the definitions that de Moivre places at the beginning of his book:

> 1. The probability of an Event is greater or less, according to the number of Chances by which it may happen, compared with the whole number of Chances by which it may either happen or fail.
>
> 2. Wherefore, if we constitute a Fraction whereof the Numerator be the number of Chances whereby an Event may happen and the Denominator the number of all Chances whereby it may either happen or fail, that Fraction will be a proper designation of the Probability of happening. Thus if an Event has 3 Chances to happen, and 2 to fail, the Fraction 3/5 will fitly represent the Probability of its happening, and may be taken to be the measure of it.
>
> (De Moivre 1756, pp. 1–2)

Plainly, 'chances' stands here for equally easy—or, as one would say today, *equiprobable*—alternatives.

[5] Leibniz, SS, VI.ii, p. 492. We ought to be grateful to Ian Hacking (1975, p. 128) for finding and reproducing this beautiful saying.

[6] As noted by Stigler 1986, pp. 66–67, our current understanding of Bernoulli's Theorem

does not quite agree with his original statement. To be true to the latter we must allow ε to represent, in the context of eqn. (6), not a fixed, arbitrarily small positive real number, but the variable fraction $1/n$. However, Stigler "hasten[s] to add that the limitation in Bernoulli's [. . .] statement was not intrinsic in his proof; the proof was rigorous even for a later formulation."

[7] It is worth noting that Bernoulli's Theorem does not say anything about the probability that the relative frequency of *r*-valued outcomes in a particular sequence of increasing length will converge to $\mathbf{P}(A_r)$ as more and more trials are made. The said frequency could come within some small distance ε of $\mathbf{P}(A_r)$ at some stage of the game and diverge thereafter. The probability that this may happen is given by the strong law of large numbers, proved in various forms in the 20th century (see, for instance, Révész 1968). The strong law presupposes the wider concept of a probability space defined towards the end of this section and is stated in note 11.

[8] To be quite precise, consider the infinite sequence of finite probability spaces $\langle T,\mathbf{P}\rangle$, $\langle T^2,\mathbf{P}_2\rangle, \ldots, \langle T^n,\mathbf{P}_n\rangle, \ldots$. Eqn. (6) defines the typical term of an infinite sequence in the union $\bigcup_n \mathcal{P}(T^n)$ of the power sets of the sample spaces $T, T^2, \ldots, T^n, \ldots$ (In other words, eqn. (6) gives the value, at an arbitrary positive integer n, of a mapping of the set of all positive integers into $\bigcup_n \mathcal{P}(T^n)$.) Bernoulli's Theorem says that the infinite sequence of values of the functions $\mathbf{P}, \mathbf{P}_2, \ldots, \mathbf{P}_n, \ldots$ at the matching terms of the sequence defined by (6) converges to 1 (for every choice of the parameters r and ε).

[9] A proof of this statement will test the reader's understanding of the concepts involved. Let $s_1, \ldots, s_n$ be any particular sequence of n throws. Take as sample space the set T^n of all such sequences. Assume that the comparative facility of events in this sample space is measured by a probability function $\mathbf{P}^*$. Let S_k denote the event consisting of throw s_k at the kth place, preceded and followed by any throws whatsoever at the remaining places. In other words: $S_k = T^{k-1} \times \{s_k\} \times T^{n-k}$. Obviously $\mathbf{P}^*(S_k) = \mathbf{P}(\{s_k\})$. Hence, $\mathbf{P}(\{s_1\}) \ldots \mathbf{P}(\{s_n\}) = \mathbf{P}^*(S_1) \ldots \mathbf{P}^*(S_n)$. Let **A** stand for the left-hand side of this equation, and **B** for the right-hand side. We thus have that $\mathbf{A} = \mathbf{B}$. We also have that $P^*(S_1 \cap \ldots \cap S_n) = \mathbf{P}^*(\{s_1\} \times \ldots \times \{s_n\})$, an equation we shall abbreviate as $\mathbf{C} = \mathbf{D}$. By eqn. (4) in Section 4.2, $\mathbf{A} = \mathbf{D}$ if and only if $\mathbf{P}^* = \mathbf{P}_n$. By eqn. (3) in Section 4.2 and the remark made right after it, $\mathbf{C} = \mathbf{B}$ if and only if the events S_k $(1 \leq k \leq n)$ are all mutually independent, or, equivalently (since $\mathbf{P}^*(S_k) \neq 0$ for all k), if and only if the probability of obtaining a particular throw at a particular place in the sequence is not affected by the nature of the throws obtained at the remaining places. Since $\mathbf{A} = \mathbf{B}$ and $\mathbf{C} = \mathbf{D}$, $\mathbf{A} = \mathbf{D}$ if and only if $\mathbf{C} = \mathbf{B}$. Thus, $\mathbf{P}^* = \mathbf{P}_n$ if and only if the events S_k $(1 \leq k \leq n)$ are all mutually independent, which was the statement we set out to prove.

[10] κρατερὴ γὰρ Ἀνάγκη πείρατος ἐν δεσμοῖσιν ἔχει—Parmenides, in Diels-Kranz, 28 B 8.30f.

[11] By repeatedly performing the above construction on n copies of an arbitrary probability space $\langle T,\mathcal{F}(T),\mathbf{p}\rangle$ we obtain the nth product space $\langle T^n,\mathcal{F}_n(T),\mathbf{p}_n\rangle$. In order to formulate the strong law of large numbers mentioned in note 7 we must construct from $\langle T,\mathcal{F}(T),\mathbf{p}\rangle$ an infinite product space that I shall designate by $\langle T,\mathbf{F},\rho\rangle$. Here T^∞ is, of course, the set of all infinite sequences in T. If α belongs to $\mathcal{F}_n(T)$ (so that α is a set of n-tuples of elements of T), the cylinder set [α] generated by α in T^∞ is the collection of all infinite sequences in T whose initial n-tuple belongs to α. **F** is the smallest Borel field that contains all the cylinder sets generated in T^∞ by sets in $\mathcal{F}_n(T)$, for every positive integer n. The probability function ρ is defined on cylinder sets by the condition $\rho([\alpha]) = \mathbf{p}_n(\alpha)$; this condition, coupled with the KPS axioms, is sufficient to define ρ on all **F**. (I shall occasionally refer to ρ as the canonic measure on T^∞ relative to the probability function **p**.) Consider now any set A_r in $\mathcal{F}(T)$. Let us say that a cylinder set in T^∞ is A_r-based if it is generated by the Cartesian product of m copies of A_r and $n-m$ copies of its complement A'_r. Let T_r be the subset of T^∞ constituted by all those

infinite sequences in T in which the relative frequency of terms belonging to A_r converges to the limit $\mathbf{p}(A_r)$. T_r is a union of (infinite) intersections of A_r–based cylinder sets and hence belongs to **F**. Therefore, so does its complement T'_r. The strong law of large numbers asserts that $\rho(T_r) = 1$ and $\rho(T'_r) = 0$. Of course, this does not imply that $T_r = T^\infty$ and that T'_r is the empty set.

[12] The Axiom of Choice was stated in note 77 to Chapter 2. As mentioned in Chapter 2, note 78, the Axiom of Choice holds for all sets if and only if every set can be well-ordered, i.e., linearly ordered in such a way that every one of its subsets has a first element. This is a pretty strong requirement and some very distinguished mathematicians have scorned it. The Axiom of Choice is a powerful tool, but it produces a few unpleasant results, the most significant of which is perhaps the one noted in the main text above and further explained in the following note. R. M. Solovay (1970) showed that this result does not follow from a weakened version of the axiom that, he claims, suffices to prove all the pleasant consequences of the standard version, which most mathematicians are loath to forgo. Unfortunately, Solovay's proof requires the hitherto unproven consistency of Zermelo-Fraenkel set theory with the following assumption: there exists a weakly inaccessible uncountable cardinal, i.e., a cardinal number $\kappa > \aleph_0$ such that κ is neither the successor of another cardinal nor the sum of fewer than κ cardinals smaller than κ itself (see Wagon 1985, pp. 208–9).

[13] For greater precision, let us define the support of a probability function **p** as the smallest set U such that $\mathbf{p}(U) = 1$. Let $\langle S, \mathcal{F}(S), \mathbf{p}\rangle$ be a probability space. If the countably additive probability function **p** has finite or countable support, it can be readily extended to a countably additive probability function $\mathbf{p}'$ defined on the full power set $\mathcal{P}(S)$. However, it follows from the Axiom of Choice that **p** cannot be thus extended if it is one of the standard probability functions with uncountable support, for then there will always be some subsets of S at which the extended function $\mathbf{p}'$ cannot be defined lest it should forgo countable additivity. Indeed, at the time of writing, no instance was known of a probability function with uncountable support which could be extended in the said manner, although no general proof had yet been given that such a function does not exist. I am grateful to Jorge López for instructing me on this matter.

[14] "The separate consideration of first the *space* (*without the measure*, or any other kind of structure), and then all the possible *measures*, not only, and most importantly, meets the need of the subjective conception by providing [probability functions] $\mathbf{p}_i$ which are possibly different for each individual i ('tot capita, tot sententiae'), but also satisfies other more 'neutral' requirements (probabilities conditional on different hypotheses, or different states of information, or 'mixtures', and so on" (de Finetti 1974, vol. 2, p. 258). Thus, "the subject of the calculus of probabilities is no longer a single function **P**(E) of events E, that is to say, their probability considered as something objectively determined, but the set of all functions **P**(E) corresponding to admissible opinions" (de Finetti 1937, in Kyburg and Smokler 1980, p. 64).

[15] ἐδιζησάμεν ἐμεωυτόν—Heraclitus, in Diels-Kranz, 22.B.101.

[16] More precisely, within 976,000 years. The half-life of ^{234}U is set at 2.44×10^5 years.

[17] This procedure is not available when there are infinitely many distinct alternatives. However, the assumption that they are equally easy was also decisive in the earlier attempts to measure chances in such cases. Thus, for example, it was thought obvious that if it is equally easy for a dimensionless particle to fall on any point of a unit square, the chance that it will do so on a given part of the square must be measured by the area of that part. See note 19.

[18] See the passage from de Moivre (1756) quoted in note 4. In the famous "Philosophical Essay on Probabilities" prefixed to his *Théorie Analytique des Probabilités*, Laplace says that "the

theory of chances consists in reducing all events of the same kind to a certain number of cases that are equally possible [. . .] and ascertaining the number of cases favorable to the event whose probability is being sought." There follows the much derided 'classical definition' of probability as "a fraction whose numerator is the number of favorable cases, and whose denominator is the number of all possible cases" (Laplace, OC, vol. 7, pp. viii–ix). It is worth noting that, although Laplace himself called this "la définition même de la probabilité" (OC, vol. 7, p. xi), he did not mean it as such, but only as the *measure* of the probability when the prospective outcomes can be partitioned into equiprobable cases. "If all cases are not equally possible, one shall determine their respective possibilities, and then the probability of the event will be the sum of the probabilities of each favorable case" (Laplace, OC, vol. 7, p. 181; an almost identical statement occurs in the "Philosophical Essay" right after the remark on "the very definition of probability").

[19] Bayes' postulate of uniformity can be represented by a probability space $\langle [0,1], \mathcal{F}, \mathbf{p} \rangle$, where $\mathcal{F}$ is the smallest Borel field that contains every interval in [0,1], and **p** assigns equal values to intervals of equal length. Then, for any $x \in \mathcal{F}$, $\mathbf{p}(x)$ is the probability that $q \in x$.

[20] Other, less simple assignments of probability to individual drawings can also yield an equality between the ratio of white balls to all balls and the probability that the next ball drawn will be white. But this probability cannot depend on that ratio alone. The proportion of white balls in the urn is relevant to the probability of drawing white only if conjoined with a distribution of probabilities among the various possible alternative drawings.

[21] In practice, a weaker requirement will do: the sampling procedure should have no bias for or against individuals with characteristics relevant to the hypothesis that is being tested. The above definition of random choice rests on our understanding of chance as facility of occurrence. If we give it up and seek to explain chance in terms of randomness, we must find another definition for this term or else acknowledge with Harald Cramér (1946, p. 138) that it cannot be defined. Note that de Finetti regards the adjective 'random' as synonymous with 'unknown' (de Finetti 1974, vol. 1, p. 28).

[22] One of Méré's problems is similar to that discussed by Galileo and was solved along the same lines. The following statement of the difficulty, copied with two slight amendments from Todhunter 1865, p. 11, is a translation from Pascal's letter of 29 July 1654: "If we undertake to throw a six with one die the odds are in favor of doing it in four throws, being as 671 to 625; if we undertake to throw two sixes with two dice the odds are not in favor of doing it in twenty-four throws. Nevertheless 24 is to 36, which is the number of faces in two dice, as 4 is to 6, which is the number of faces in one die." Pascal commented: "Such was his great scandal, which made him say haughtily that propositions lacked constancy and that arithmetic was inconsistent [*se démentoit*]." I take it that Méré's difficulty did not arise, as the one dealt with by Galileo, from a perceived conflict between arithmetical calculations and a "long observation" of actual dice games but rather from the incompatibility of two ways of calculating chances, both of which Méré believed to be warranted by arithmetic. Méré's other problem is subtler and caused Fermat and Pascal to tread new ground. (Similar problems had been raised and tackled in the 16th century by Luca Paccioli and G. F. Peverone, but they failed to solve them. See Kendall 1956, in Pearson and Kendall 1970, p. 27.) Suppose *A* and *B* agree to play for a prize *C* under the following rules: Up to three games will be played; the prize goes to the player that wins two games; draws are not counted. How ought *C* to be divided between *A* and *B* if they have to stop playing after *A* wins the first game? If they went on playing one of the following three mutually exclusive cases must occur: (i) *A* wins the second game; (ii) *A* loses the second game but wins the third; (iii) *A* loses the next two games. *A* carries the prize in cases (i) and (ii); *B* only in case (iii). Therefore, if both players have the same chance of winning each game, then, after the first was won by *A* the odds are 3 to 1 that *A* will carry the prize. Hence, if they must stop playing at that time, $0.75C$

ought go to A and $0.25C$ to B.

[23] This ratio is set at 1/2, "forasmuch as such Women, one with another, have scarce more than one Child in two years" (Graunt 1676, p. 81).

[24] "The next inquiry will be, In how long time the City of *London* shall, by the ordinary proportion of Breeding and dying, double its breeding People? I answer, In about seven years, and (*Plagues* considered) eight. Wherefore, since there be 24000 pair of Breeders, that is 1/8 of the whole, it follows, that in eight times eight years the whole People of the City shall double, without the access of Forreiners: the which contradicts not our Account of its growing from two to five in 56 years with such accesses. According to this proportion, one couple, *viz., Adam* and *Eve,* doubling themselves every 64 of the 5610 years, which is the *Age* of the World according to the *Scriptures,* shall produce far more people than are now in it. Wherefore the World is not above 100 thousand years older, as some vainly imagine, nor above what the *Scriptures* make it" (Graunt 1676, pp. 85–86).

[25] Graunt (1676, p. 84) explains his method as follows: "Whereas we have found, that of 100 quick Conceptions about 36 of them die before they be six years old, and that perhaps but one surviveth 76; we having seven *Decads* between six and 76, we sought six mean

Of an hundred there die within the first six years	36
The next ten years, or Decad	24
The second Decad	15
The third Decad	9
The fourth	6
The next	4
The next	2
The next	1"

proportional numbers between 64, the remainder, living at six years, and the one, which survives 76, and find that the numbers following are practically near enough to the truth; for men do not die in exact proportions, nor in Fractions, from whence arises this Table following. *Viz.*

[26] "Il n'y a point de hasard à proprement parler mais il y a son équivalent: l'ignorance où nous sommes des vraies causes des événements" (quoted from d'Alembert by M. G. Kendall in Pearson and Kendall 1970, p. 31, without any further indication of the source).

[27] Cf. the following remark by the Dutch physicist A. Pannekoek, a contemporary of Poincaré: "We speak of chance in nature, when small variations in the initial data occasion considerable variations in the final elements, because we cannot observe those small variations" (quoted by Brush 1976, p. 654, n. 53, from the Proceedings of the Netherlands Academy, Sect. Sci. (1903), vol. 6, p. 48).

[28] Objectivity subject to radical rethinking is not precisely what Poincaré had in mind in his explication of chance when he referred to a hypothetical state of affairs in which "the natural laws have no further secret for us" (SM, p. 68). It is, however, the only kind of physical objectivity known to us, the best that can be claimed both for the deterministic systems and for the chance setups.

[29] A clear and concise explanation of the concepts of recursiveness and Turing computability is given by Martin Davis, in Edwards' *Encyclopedia of Philosophy,* s.v. 'Recursive function theory'. The interactive computer program *Turing's World* by Barwise and Etchemendy should be a great help in understanding those concepts. For a more technical introduction the reader may turn to Enderton 1977 and Bell and Machover 1977, Chapters 6–8. Davis'

excellent textbook *Computability and Unsolvability* (1958) is now available in an inexpensive Dover reprint.

[30] Dissatisfaction with Church's proposal of 1940 issues from an example published by J. Ville in 1939. Ville's example is a sequence σ of 1's and 0's that combines the following two features: (i) the relative frequency of either label converges to the same limit—the von Mises probability of that label—in σ and in every subsequence obtained from σ by place selection in Church's sense; (ii) the pattern of convergence of the relative frequency of either label to the respective von Mises probability violates the law of the iterated logarithm that governs the almost sure convergence of the relative frequency of successes in an infinite sequence of Bernoulli trials to the chance of success in a single trial. (In any finite initial segment of σ the relative frequency of zeroes is equal to or greater than 1/2.) The existence of such a sequence constitutes a fatal objection to the von Mises-Church concept of the collective only if it is assumed that collectives must share all the properties proved for infinite sequences of identically distributed independent random variables in the established mathematical theory of probability (viz., the theory determined by the Kolmogorov axioms).

[31] The monster-barring condition VMC 2 is stated in von Mises 1964, p. 12, but not in the more popular book, von Mises 1957.

[32] On the other hand, **p*** need not satisfy the requirement of complete additivity (KPS 2). Indeed, according to standard set theory, **p*** generally *cannot* satisfy this requirement unless the label set *L* is finite (in which case the requirement is trivial, for there are no countable collections of mutually exclusive subsets of *L* over which to sum **p***). Recall, however, that as I pointed out in notes 12 and 13, this limitation follows from the Axiom of Choice and vanishes with it. (The counterexamples proposed in van Fraassen 1977, pp. 133–34, are not von Mises collectives, or even semicollectives, for the first two do not comply with VMC 2 and the third one fails to meet VMC 1.)

[33] See, e.g., Kamke 1932, p. 147. Indeed, when such objections were first voiced, no restrictions had yet been imposed on place selections. Now, if *every* order- preserving numerical sequence ζ is admissible for this role, it is not hard to show that VMC 1 and VMC 3 cannot be jointly satisfied except in the trivial case in which **p** takes only the values 0 or 1. (See Schnorr 1970b, p. 15, Lemma 2.5.) Reichenbach evaded the difficulty by dropping the axiom of randomness from his version of frequentism. His "probability sequences," which satisfy the axiom of convergence, are therefore semicollectives in our sense. Thus, Reichenbach's 'probability' is coextensive with von Mises' 'chance'. However, Reichenbach's most interesting results and applications concern what he calls "normal sequences," which satisfy VMC 1 and VMC 3 with place selection redefined as follows: An order-preserving sequence ζ is said to be a *place selection* for the sequence σ in the label space *L* if, for some definite *k*-tuple $\langle\lambda_1, \ldots, \lambda_k\rangle$ $(\lambda_i \in L)$ and every positive integer *n*, $X_\zeta(n) = 1$ if and only if $\langle\sigma_{n-k}, \ldots, \sigma_{n-1}\rangle = \langle\lambda_1, \ldots, \lambda_k\rangle$ (that is to say, ζ is a place selection if $\sigma \circ \zeta$ selects any term of σ that follows immediately upon a certain label-"word"). Popper complained in 1935 that Reichenbach had not proved the existence of normal sequences (Popper 1959a, p. 171, n. 4.); but Wald's consistency proof of 1937 implies it (see Schnorr 1970b, p. 23, Theorem 3.2).

[34] To see this, consider a semicollective σ, with label set {1,0}, in which the relative frequencies $f_1, f_2, \ldots$ of either label converge to positive probabilities. If there is a rule for computing σ_n for every positive integer *n*, that rule provides at the same time the characteristic function of the range of a place selection we may denote by ρ. Clearly, $\sigma \circ \rho$ is a semicollective in which the relative frequencies $f_1(0), f_2(0), \ldots$ are identically zero; thus σ violates VMC 3 and is not a collective. The limitation to two labels does not involve any loss of generality. If the label set from which the terms of σ are drawn had countably many labels, the above reasoning would still apply to the sequence σ* obtained by equating one of the labels in σ with 0 and mixing all the rest into 1. If σ* is not a collective, neither is σ.

[35] Let p_n and c_n denote the *n*th prime number and the *n*th composite number, respectively. Consider the sequence σ defined by $\sigma_k = \mathbf{H}$ if $k = 1$ or if k is prime; $\sigma_k = \mathbf{T}$ if k is composite. σ is a semicollective in which the relative frequency of **H**'s converges to the limit 0. Apply now the place selection ζ given by $\zeta(1) = 1$, $\zeta(2n) = c_n$, $\zeta(2n+1) = p_n$ (for every positive integer n). $\sigma \circ \zeta$ is the semicollective **HTHTHT**. . . in which the relative frequency of **H**'s converges to the limit 0.5.

[36] Note, by the way, that on the hypothesis that the probability of heads on each toss is 1/2, the finite sequence given above is not a whit less probable than, say,

HHTHHTTHTHTTTHHHTTTHTTHHTTHTTH.

The odds are 1,073,741,823 to 1 against either one. But in stark contrast with the former sequence, there is no alternative hypothesis known to me in which the latter is more likely to occur. I might turn suspicious, however, if the second list of **H**'s and **T**'s agrees with the binary expression of the experimenter's social security number.

[37] I suppose this is what R. C. Jeffrey had in mind when he described frequentism as "a doomed attempt to define probability in such a way as to turn the laws of large numbers into tautologies" (Jeffrey 1977, p. 220). Note, however, that as mathematical theorems, the laws of large numbers unquestionably are "tautologies" (in Jeffrey's very broad sense). But frequentism does substitute trivial *logical* truths for the purely *practical* certainties warranted by the laws of large numbers. (See the following note.)

[38] Wolfgang Stegmüller, whom I have largely followed in my presentation of this objection, believes that it is fatal (*tödlich*) to frequentism. He aptly remarks that what is merely a *practical certainty* according to the strong law of large numbers becomes in the theory of collectives a matter of *logical necessity* (1973b, p. 37; cf. de Finetti 1974, vol. 2, pp. 38–39). I am not aware that von Mises ever considered the objection in the form presented here. He did reply, however, to a variant of it, involving not the strong but the weak law of large numbers, i.e., Bernoulli's Theorem (von Mises 1957, pp. 110ff.). Now, the apodosis of Bernoulli's Theorem does not speak of the probability of a relative frequency but of the limiting frequency of a sequence of probabilities, viz., the probabilities $\mathbf{p}_n$ ($n \in \mathbf{Z}^+$) that the relative frequency of a certain outcome in an *n*-tuple of trials falls within a given, arbitrarily small neighborhood of the probability **p** of that outcome on each trial (see Section 4.2). The frequentist restatement of the theorem does not, therefore, involve a metabasis from practical to logical certainty. Von Mises acknowledges that Bernoulli's Theorem is a consequence of his axioms VMC 1 and 3, as indeed it ought to be if the theory of collectives is in fact a theory of probability. But he goes on to argue that it is not, as his critics claimed, a trivial corollary of VMC 1 alone, for there are semicollectives for which the theorem does not hold. He proposes the following example: Let σ be the sequence of square roots of integers, calculated to seven decimal places. The relative frequency of items in σ whose fourth digit is less than 5 demonstrably converges to the limit 1/2. σ can therefore be regarded as a semicollective with a label space consisting of this attribute—call it "success"—and its negation—or "failure." However, as the sequence progresses it displays increasingly long runs of consecutive successes (followed by long runs of failures). Von Mises takes this to mean that the probability $\mathbf{p}_n$ of success in an *n*tuple of square roots does not converge with increasing n to a given, arbitrarily small interval centered on 1/2.

[39] De Finetti 1974, vol. 1, p. x; in vol. 2, p. 42, he refers sarcastically to "the enchanted garden wherein, among the fairy-rings and the shrubs of magic wands, beneath the trees laden with monads and noumena, blossom forth the flowers of *Probabilitas realis.*" Cf. de Finetti 1931, p. 302: "La probabilità oggettiva non esiste mai." In 1989, *Erkenntnis* devoted its September issue (vol. 31, nos. 2–3) to Bruno de Finetti's philosophy of probability. It contains an English translation of a long philosophical essay of his and articles by Richard

Jeffrey, Isaac Levi, Brian Skyrms, and other distinguished scholars. Unfortunately it reached me too late to be considered here.

[40] "In the conception we follow and sustain here only *subjective* probabilities exist—i.e., the *degree of belief* in the occurrence of an event attributed by a given person at a given instant and with a given set of information" (de Finetti 1974, vol. 1, pp. 3–4). Cf. de Finetti 1931, p. 301: "E'appunto *quel grado affatto soggetivo di affidamento* che nel linguaggio corrente si designa col nome di probabilità, e la mia opinione è proprio questa: che il concetto espresso dal linguaggio ordinario abbia, una volta tanto, un valore assolutamente superiore a quello dei matematici, che da secoli si affaticano inutilmente per vedervi un significato che non esiste" (my italics).

[41] In a sense, however, there is a partial ordering of *possibilia,* inasmuch as an event A can be judged 'no less possible' than another event B if A is a consequence of B (de Finetti 1974, vol. 1, p. 71n.).

[42] Proof: By definition, $\inf X = 0$. For simplicity's sake, put $\sup X = 1$. Let ε be a positive real number and choose $\mathbf{P}(X) = -\varepsilon$. Then, your loss equals $k\varepsilon^2$ if E fails and $k(1 + \varepsilon)^2$ if E obtains. But had you chosen $\mathbf{P}(X) = 0$, your loss would be 0 if E fails and k if E obtains. Thus, you will be better off, whatever happened, if you make $\mathbf{P}(X)$ non-negative. Choose $\mathbf{P}(X) = 1 + \varepsilon$. Then, your loss equals $k\varepsilon^2$ if E obtains and $k(1 + \varepsilon)^2$ if E fails. But had you chosen $\mathbf{P}(X) = 1$, your loss would be 0 if E obtains and k if E fails. Thus, you will be better off if you make $\mathbf{P}(X)$ not greater than 1.

[43] "In wishing to consider as a *definition* of prevision some relation connecting it with probability, one is led into an extremely unnatural position. In other words, one makes it appear as though the elementary notion of prevision presupposes a knowledge of something much more complicated and delicate; that is, the probability *distribution* itself. Because it is unnatural, the situation is also dangerous, in the sense that it leads one to think that the definition to be made in this way, *ex novo,* allows a certain element of arbitrariness. In other words, that it requires, or permits, a choice of conventions, which are inspired by considerations of convenience" (de Finetti 1974, vol. 2, pp. 262–63).

[44] I note in passing that Keynes (1921) and Renyi (1970) take conditional probability as a primitive. De Finetti (1974, vol. 1, p. 134) admits that "every prevision, and, in particular, every evaluation of probability, is conditional; not only on the mentality or psychology of the individual involved, at the time in question, but also, and especially, on the state of information in which he finds himself at that moment." However, according to him, "something which in itself is so obvious, and yet so complicated and vague to put into words, is clearer if left to be understood implicitly rather than if one thinks of it condensed into a symbol."

[45] On the other hand, there is no question that from de Finetti's standpoint you can form a prevision regarding the status of a complicated truth-function, when you do not know whether it is tautological, contingent, or contradictory, just as you can make a bet on the trillionth digit in the decimal expansion of e (although there is an algorithm for computing it). It has been noted that the interpretation of probability as prevision encounters some grave difficulties due to our lack of logical omniscience; but I cannot go into this matter here. For a stimulating discussion, see Garber 1983.

[46] Indeed, in connection with a different problem, de Finetti proclaims that the prevision $\mathbf{P}(X)$ "will henceforth have the meaning we have assigned to it; it will make no sense to set up new conventions in order to re-define it for this or that special case" (de Finetti 1974, vol. 1, pp. 246–47). On the other hand, he speaks without qualms about the probability—i.e., the prevision—of "the sure event" (de Finetti 1972, p. 76).

[47] In a footnote to the passage just quoted, de Finetti warns us that the arbitrary set χ, on which the probability function $\mathbf{P}$ is supposed to be defined at some moment and in some

case, need not be "reducible to a ring (or to an σ-ring) of events." Now, χ is a ring (or field or Boolean algebra) of events if and only if (i) χ contains the sure event, (ii), for every event E belonging to it, χ contains also the negation or complement of E (i.e., the event that obtains if and only if E fails), and (iii) for every pair of events E_1 and E_2 belonging to it, χ contains the join of E_1 and E_2, i.e., the event that obtains unless E_1 and E_2 fail together. As we shall soon see, whenever **P** is defined at two mutually incompatible events, de Finetti's conditions of coherence determine the value of **P** at their union and at their respective complements. Thus, the domain of **P** inevitably extends to a ring of events, provided only that it can be partitioned into mutually exclusive alternatives. So this was the assumption that de Finetti's caveat was presumably meant to preclude. As to his additional warning that the domain of **P** need not be a σ-ring, I ought perhaps to mention the following. A σ-ring is what in Section 4.2 I called a Borel field, i.e., a Boolean algebra closed under *countable* union. In other words, if χ is a σ-ring and χ contains a sequence $E_1, E_2, \ldots$ of events, χ also contains the join of all events in the sequence, i.e., the event that obtains if and only if at least one event in the sequence obtains. As I said in note 13, a real-valued function satisfying the Kolmogorov conditions KPS 1 and KPS 2 cannot be defined, without some restrictive requirements, on the power set of an infinite set (unless we give up the Axiom of Choice). This raises a difficulty when one considers an infinite spectrum of alternatives while believing with de Finetti that every well-defined event can be assigned a probability. But in order to deal with this difficulty one need not deny that a probability function is definable on the power set of the set of alternatives (which is, of course, a σ-ring). It is sufficient to weaken the requirement of complete additivity imposed on probability functions by KPS 2 to the requirement of simple additivity (FPS 2). De Finetti himself opted and repeatedly argued for this solution; see, in particular, de Finetti 1972, pp. 93ff.

[48] As a matter of fact, he went so far as to say that an evaluation of probabilities that violates the constraints imposed by the conditions of coherence contains "an intrinsic contradiction" (de Finetti 1937, in Kyburg and Smokler 1980, p. 83; cf. de Finetti 1974, vol. 1, p. 73). This claim is preposterous. Under de Finetti's definition of probability as prevision the assignment of probability to an event cannot *contradict*—i.e., be the *denial of*—the assignment of probability to another event.

[49] Consider the amounts your opponent ought to receive from you under the stated betting conditions after the three possible outcomes of the game: (i) $E_1 = 1, E_2 = 0$; (ii) $E_1 = 0, E_2 = 1$; and (iii) $E_1 = E_2 = 0$. (Recall that E_1 and E_2 are incompatible, so that $E_1 = E_2 = 1$ is not a possible outcome.) A short calculation shows that in all three cases you will owe your opponent an amount equal to $-cr$, which is of course positive if c and r have opposite signs.

[50] See de Finetti 1974, vol. 1, pp. 90, 186–91, where a statement that implies (2) is derived from the second condition of coherence by a beautiful geometrical method.

[51] The question of complete additivity is discussed at length in de Finetti 1972, pp. 84ff. As Bas van Fraassen (1979, p. 346) aptly remarks, the issue is otiose for the working statistician, who applies probability to industrial or medical problems, but not for the physicist, who represents experimental situations by means of structures defined on infinite sets.

[52] 'Likelihood' is to be understood here in the precise sense in which it was used by R. A. Fisher: the *likelihood* of the statistical hypothesis H in the light of data E is equal to the probability of E conditional on H. Elsewhere, I do use 'likely'—but generally avoid using 'likelihood'—in its loose ordinary sense.

[53] The concept of exchangeability can be readily extended to random quantities. "We shall say that $X_1, X_2, \ldots$ are exchangeable random quantities if they play a symmetrical role in relation to all problems of probability, or in other words, if the probability that $X_{i_1}, X_{i_2}, \ldots, X_{i_n}$ satisfy a given condition is always the same, however the distinct indices $i_1, i_2, \ldots, i_n$ are

chosen" (de Finetti 1937, in Kyburg and Smokler 1980, p. 82). De Finetti also studied the somewhat less restrictive condition of 'partial exchangeability', which holds if the events under consideration can be partitioned into several classes so that exchangeability holds within each class. If the events are partially exchangeable, the probability that out of $n = n_1 + \ldots + n_k$ events (n_i of the ith class), $h = h_1 + \ldots + h_k$ events (h_i of the ith class) will obtain, is the same, no matter how the $n_1, \ldots, n_k$ events are chose in their respective classes, and no matter which of them are among the $h_1, \ldots, h_k$ successes (de Finetti 1974, vol. 2, p. 212). For a sample of recent work on exchangeability, see Koch and Spizzichino 1982.

[54] The Representation Theorem was communicated to the International Congress of Mathematicians held in Bologna in September 1928 (de Finetti 1930). A proof and an extensive discussion of its philosophical significance will be found in de Finetti 1937 (in Kyburg and Smokler 1980, pp. 78–107; see also de Finetti 1972, pp. 211ff.; 1974, vol. 2, pp. 211ff.). Feller 1971, pp. 228–29, and Loève 1978, vol. 2, p. 31, give concise and illuminating proofs using resources previously developed in these books. R. C. Jeffrey's "Probability Measures and Integrals" (in Carnap and Jeffrey 1971) is a rigorous presentation *ad usum philosophorum* of the mathematical ideas involved in de Finetti's Representation Theorem, which is stated without proof in the article's concluding section.

[55] A probability function **P** can be defined on the power set $\mathcal{P}(S)$ if, following de Finetti, we do not require it to be completely additive or if, following Solovay 1970, we put up with a weakened Axiom of Choice (see note 13). Readers faithful to both statistical and set-theoretical orthodoxy may regard **P** as a function defined on the smallest Borel set that contains all the subsets $\{x \in S \mid E_i = 1\}$ $(i = 1,2, \ldots)$.

[56] A probability function **P** is a mixture of the probability functions $\mathbf{P}_1, \ldots, \mathbf{P}_n$ (defined on the same domain as **P**) if **P** is a linear combination of the $\mathbf{P}_i$ weighted by a suitable set of coefficients (more accurately, if there is a list $\langle a_1, \ldots, a_n \rangle$ of non-negative real numbers such that $\Sigma a_i = 1$ and $\mathbf{P} = \Sigma a_i \mathbf{P}_i$). This idea of a mixture is extended in a natural way to the case in which **P** depends on a continuum of probability functions $\mathbf{P}_x$, indexed by a parameter x ranging over some interval $[a,b]$. Let $f\colon \mathbf{R} \to [0,1]$ be a nowhere negative integrable function, identically 0 outside $[a,b]$. **P** is a mixture of the $\mathbf{P}_x$ if

$$\mathbf{P} = \int_a^b \mathbf{P}_x f(x) \mathrm{d}x$$

The "components" of the mixture are weighted by the "density" f. The function F defined by

$$\mathsf{F}(x) = \int_{-\infty}^{x} f(x) \mathrm{d}x$$

is clearly a normed distribution, concentrated on $[a,b]$, in the sense explained in note 57. By a further, not unnatural extension of the original idea, we say that **P** is a mixture of the $\mathbf{P}_x$ if there is a distribution F concentrated on $[a,b]$ such that

$$\mathbf{P} = \int_a^b \mathbf{P}_x \mathrm{d}\mathsf{F}(x)$$

[57] By a normed distribution on the real line, concentrated on a closed interval $[a,b]$, I mean a nondecreasing right-continuous function $\mathsf{F}: \mathbf{R} \to [0,1]$, such that $\mathsf{F}(a) = 0$ and $\mathsf{F}(b) = 1$. About the (Lebesgue-Stieltjes) integral in eqns. (7) and (8) see, e.g., Cramér 1946, Chapter 7. Following Cramér, I write it so that

$$\mathsf{F}(x) = \int_{-\infty}^{x} d\mathsf{F}(x)$$

even if the derivative $d\mathsf{F}(x)/dx$ does not exist. Note that if F is a normed distribution on the real line, concentrated on $[a,b]$, and if $\mathcal{B}(a,b)$ denotes the Borel sets of $[a,b]$ (i.e., if $\mathcal{B}(a,b)$ is the smallest Borel field on $[a,b]$ that contains all the subintervals in $[a,b]$), a function M: $\mathcal{B}(a,b) \to [0,1]$ is uniquely defined by putting, for any two real numbers u and $v \geq u$ in $[a,b]$,

$$\mathsf{M}([u,v]) = \int_{u}^{v} d\mathsf{F} = \mathsf{F}(v) - \mathsf{F}(u)$$

By the definition of F, $\mathsf{M}([a,b]) = 1$. By the definition of the integral, M is completely additive. Therefore $\langle [a,b], \mathcal{B}(a,b), \mathsf{M} \rangle$ is a probability space (in Kolmogorov's sense).

[58] M can be extended in many ways to a finitely additive probability function on every subset of $[0,1]$. Such functions agree with one another on all intervals (Savage 1954, p. 53).

[59] Compare the following passage in Braithwaite 1957, pp. 9–10:

> De Finetti in fact has shown that the type of dependence between events required for the use of an eductive formula [i.e., a rule for predictive inference, in Carnap's sense] is explicatable in terms of sets of independent and equiprobable events. He has effected an *analysis* of such dependent events in terms of independent events in the same sense of 'analysis' as that in which Fourier analysed all periodic functions in terms of simple harmonic functions. It follows from Fourier's Theorem that any musical note (i.e., any periodic sound) can be produced by a suitable combination of tuning forks each yielding a 'pure' note. Similarly de Finetti's theorem proves that any phenomenon of dependence permitting the application of an eductive formula can be produced by suitably combining phenomena of independence: the principle of the dependence arises, in a sense, out of the principle of the combination involved (in technical terms, out of the 'measure' used in the differential element of the Stieltjes integral concerned). De Finetti, while protesting against treating the throwing-an-irregular-coin situation as being analogous to the drawing-from-many-bags situation, has provided the best possible argument for treating them on a parity, as we frequentists do.

The last sentence in the quotation alludes to an argument by de Finetti that I examined in Section 4.3. Braithwaite evidently regards "the type of dependence between events required for the use of an eductive formula"—viz., dependence-cum-exchangeability—as an objective relation evinced by the phenomena of dependence he mentions.

[60] In the notation we have been using, this corollary may be stated as follows:

$$\lim_{n \to \infty} \mathrm{P}(E_{n+1} = 1 \mid n^{-1} \sum_{i=1}^{n} E_i = k) = k$$

[61] Let $\mathbf{P}_\tau$ denote our probability function at time τ. Let E be an event unknown at times $\tau = 1$ and $\tau = 2$, and let A stand for all the information relevant to E that becomes known to us between $\tau = 1$ and $\tau = 2$. According to the statement made above, $\mathbf{P}_2(E) = \mathbf{P}_1(E|A)$. De Finetti never proved it (perhaps he thought it was "intuitively obvious"?); but an argument by David Lewis, reported in Teller 1973, pp. 222ff., shows that, if we are ready to bet in the manner prescribed by de Finetti's first method for eliciting previsions but do not follow the policy of equating $\mathbf{P}_2(E)$ with $\mathbf{P}_1(E|A)$ whenever E and A are as above, our opponent can place us in a position in which we lose at any event. Thus, the statement in question follows from the first condition of coherence. The argument is, of course, inapplicable if, as it generally happens in real life, the information acquired between $\tau = 1$ and $\tau = 2$ includes facts for which $\mathbf{P}_1$ was not defined or leads to a revision of the very scheme in terms of which the domain of $\mathbf{P}_1$ was conceived.

[62] Popper 1957. Popper could not attend the meeting, and his paper was read by P. K. Feyerabend. Popper's frequentism, as presented in *Logik der Forschung* and further elaborated in some of the appendices added to its English translation (Popper 1959a, pp. 318–58), differed significantly from the doctrine of von Mises. It is said to have prompted Abraham Wald's work on the consistency of the concept of a collective (Popper 1983, p. 361). In his essay "Three Views of Human Knowledge," published in 1956, Popper made a passing, noncommittal reference to probability as propensity (Popper 1963, p. 119). By then he had already written the first draft of the detailed discussion of the matter contained in his *Postscript* to *Logik der Forschung*, which was set in type in 1956–57, but remained unpublished until 1983 (Popper 1983, pp. 281–401; the text printed on pp. 348–61 formed the bulk of Popper 1959b).

[63] Right after the passage quoted by Braithwaite, Peirce had added that

> the 'would-be' of the die is presumably as much simpler and more definite than a man's habit as the die's homogeneous composition and cubical shape is simpler than the nature of the man's nervous system and soul; and just as it would be necessary, in order to define a man's habit, to describe how it would lead him to behave and upon what sort of occasion—albeit this statement would by no means imply that the habit *consists* in that action—so to define the die's 'would-be' it is necessary to say how it would lead the die to behave on an occasion that would bring out the full consequence of the 'would-be'; and this statement will not of itself imply that the 'would-be' of the die *consists* in such behavior. Now in order that the full effect of the die's 'would-be' may find expression, it is necessary that the die should undergo an endless series of throws from the dice box, the result of no throw having the slightest influence upon the result of any other throw.
>
> (Peirce, CP, 2.664f.)

This passage should be sufficient to offset D. M. Mackay's objection to the priority claim put forward by Braithwaite on Peirce's behalf, viz., that "habit is a backward–looking word, propensity is a forward–looking word," and that Popper's proposal was to conceive the probability distribution not as an epitome of the experimental arrangement's past behavior but rather as "a structural property here and now" (Körner 1957, p. 82). Some genuine differences between Popper's view and Peirce's are noted by W. W. Bartley in Popper 1983, p. 282, note 2.

[64] In Popper 1983, p. 353, a new clause has been added to this passage, so that it ends as follows: "Clearly, we shall have to say, with respect to each of these few throws with this fair die, that the probability of a six is 1/6 rather than 1/4, in spite of the fact that these throws

are, according to our assumptions, *members of a sequence* of throws with the statistical frequency 1/4, and in spite of the fact that two or three throws cannot possibly influence the frequency 1/4 of the long sequence."

[65] The same passage is quoted by van Fraassen (1980, p. 187).

[66] Van Fraassen defines a member of a good family as an ordered pair $\langle\sigma, G\rangle$, where σ is a sequence in the set K of possible outcomes—i.e., a mapping of the natural numbers into K—and G is a partition of K—i.e., an exhaustive classification of K into mutually exclusive subsets. By combining a given sequence σ with different such partitions or a given partition G with different such sequences we obtain different members of the good family.

[67] If G_ρ is a partition of a set K, the Borel field generated by G_ρ is the smallest set of subsets of K which (1) includes K, (2) includes every element of G_ρ, and (3) is closed under countable union and complementation.

[68] See result (3.7) in van Fraassen 1979, p. 357. The strong law of large numbers implies, in effect, that for each $G \in \mathbf{G}$, the complement of E in $\{G\} \times K^\infty$ has measure 0 in terms of the canonic measure on K^∞ relative to the probability function P. (See note 11.)

[69] For the reader's comfort, here and elsewhere I describe the elements of a good family of ideal experiments as sequences of outcomes in a sample space; but—as I have already noted—van Fraassen defines them as *ordered pairs*, formed by (i) one such sequence and (ii) a countable partition of the sample space. Thus, in the text above, the reader who wishes to be rigorous ought to replace σ by $\langle\sigma, \{+,-\}\rangle$.

[70] In accordance with standard mathematical usage, the expresion 'almost always' is here taken to mean always except at most on a set of measure 0 (which may well not be empty, or even finite). The expression is therefore relative to a suitably defined measure function apart from which it is meaningless.

[71] Popper does not give an authority for this proposal, which is akin to but not identical with the following definition offered by Cramér: "Whenever we say that the probability of an event E with respect to an experiment $\mathfrak{E}$ is equal to P, the concrete meaning of this assertion will [. . .] be the following: In a *long* series of repetitions of $\mathfrak{E}$, it is practically certain that the frequency of E will be *approximately* equal to P" (1946, p. 148). The two words I have italicized indicate that for connecting the long run with the single case Cramér's definition relies on Bernoulli's Theorem, the weak law of large numbers, and not on the strong law, as does the proposal considered by Popper.

[72] Note also that our everyday ideas of cause and effect must be—and indeed are—capable of coping with the many orderly but multibranched natural processes which nobody has yet been able to analyze into deterministic single-track developments. Thus, for example, if a student dies right after being clubbed by a policeman, we judge that the policeman killed him, although on that very day the same policeman and his colleagues have clubbed thousands of students who survived. Being imbued with a scientific outlook we expect that biology will one day produce a deterministic theory capable of accounting for the difference between one case and the others. Our judgment, however, does not depend on such expectations but relies on the prescientific idea of causality without determination. See, for instance, Anscombe 1971.

[73] The importance of keeping 'propensity' separate from 'chance' is clearly brought out in the following example, which throws much light on Mellor's thinking:

> Suppose a healthy man has, from January to June, a propensity to die in a year that would be displayed directly by a chance of death of 0.05. Suppose now that in June he contracts a disease which changes his propensity to that which would be displayed by a chance of 0.3. It is clear that the chance of the man dying in the

> whole year, January to December, is between 0.05 and 0.3 and that at no time does he have a propensity of which this chance is a direct display. [. . .] The chance of the man dying during the given year is not a property of the man, ascribable to him at a particular time in the year and capable of changing from time to time, as the propensity is. It is a property of the trial of waiting a year—during which the man's propensity changes. It makes no more sense to locate it temporally within the location of the trial than it would to ask how the chance of heads changes during the toss of a coin, or what the length of a rod is half way along it.
> (Mellor 1971, p. 73)

[74] Inner product spaces were defined in Section 2.8.4. Remember that in the case of complex vector spaces condition (i*) replaces condition (i). In other words, if (**u**,**v**) denotes the inner product of the vectors **u** and **v** and a^* stands for the complex conjugate of the complex number *a*, then $(\mathbf{u},\mathbf{v}) = (\mathbf{v},\mathbf{u})^*$. A linear operator Λ on a vector space *S* is just a linear mapping of *S* into itself. If **a** and **b** are vectors, and *a* and *b* are scalars, the image by Λ of the linear combination $a\mathbf{a} + b\mathbf{b}$ satisfies the linearity condition: $\Lambda(a\mathbf{a} + b\mathbf{b}) = a\Lambda(\mathbf{a}) + b\Lambda(\mathbf{b})$. A vector **v** is said to be an eigenvector of Λ if there is a scalar *a* such that $\Lambda\mathbf{v} = a\mathbf{v}$, so that the operation of Λ on **v** merely rescales **v** (or leaves it unchanged). The scalar *a* is then said to be an eigenvalue of Λ. If *S* is a complex vector space with an inner product, any linear operator Λ on *S* has an adjoint Λ^*, that is, a linear operator which satisfies, for every pair of vectors **a** and **b**, the condition $(\Lambda(\mathbf{a}),\mathbf{b}) = (\mathbf{a},\Lambda^*(\mathbf{b}))$. Λ is self-adjoint if $\Lambda = \Lambda^*$. From the definition of adjointness and condition (i*) on inner products it readily follows that if *a* is an eigenvalue of a self-adjoint operator, *a* is always equal to its complex conjugate and therefore lacks an imaginary part. Thus the eigenvalues of self-adjoint operators are fit for representing the values of physical quantities.

[75] According to the *Oxford English Dictionary,* 'world' stems from the Germanic roots *wer* = 'man' and *ald* = 'age', the original etymological meaning being, therefore, 'age' or 'life of man'. This is also the primary meaning of αἰών (cf. *Iliad* 16.453, 22.58; but also Plato, *Gorgias* 448c6; *Leges* III, 701c4), a word occurring in the Greek New Testament in several passages (e.g., Matthew 13:22, Romans 12:2) where the King James version has 'world'.

[76] Almost three centuries after disposition predicates were ridiculed in the mock examination in Molière's *Le malade imaginaire* and Continental Cartesians fighting for clarity and distinctness gallantly resisted the onslaught of Newtonian force (*virtus*), another great French poet ruefully conceded: "Mais rendre la lumière/suppose d'ombre une morne moitié." Ironically, by that time scotophobia was taking hold of English-speaking philosophy.

[77] Levi (1967, pp. 193–94) gives the following example:

> Salt dissolves in water when it is immersed; wood does not. This difference in the behavior of salt and water is to be attributed—so it is suspected—to some as yet inadequately characterized property of salt. Moreover, sugar and salt seem to share this characteristic (a conjecture that may prove to be mistaken in the light of subsequent inquiry). [. . .] Whatever is water-soluble dissolves in water when immersed. 'Is water-soluble' is a place-holder for certain conditions that are inadequately characterized by available theories. Wood and salt differ in that the latter is, and the former is not water-soluble. 'Is water-soluble' is used here to mark that difference in a manner that will hopefully be replaced by a more adequate characterization in the light of subsequent inquiry.

[78] Derived, through 'creed', from the Latin verb *credo,* 'I believe'; see the *Oxford English Dictionary,* s.v. 'creedal, credal'. As far as I know, the adjective **credalis* is not attested in

classical Latin.

[79] In sampling with replacement the total size of the sampled population is irrelevant. But even in the more familiar case in which the sample is drawn once and for all, and no item has a chance of being chosen twice, the size of the sampled population does not significantly affect the results, provided that it is much larger than the sample.

[80] In what looks to me like an ill-advised attempt to make subjective previsions a little more objective, some authors have sought to express them in terms of a fancied real-valued scale of "utility"—thus avoiding the notorious dependence of a person's monetary valuations on her net worth and cash flow. In this approach the quantity called 'prevision' is defined as a ratio between "utilities."

[81] By the same token one may, indeed, say that a tautology T has more than just a good chance of being satisfied and that a contradiction $\neg T$ stands no chance at all. And yet, when T is built from propositions describing events in a probability space, probabilities equal to 1 and 0 must be assigned, respectively, to the event that T, and to the event that $\neg T$. This assignment of probabilities to such determinate but not contingent events is, I submit, an inevitable but innocuous artifact of the mathematical representation. I am grateful to John Norton for prompting me to express my view on this issue.

[82] As indicated in note 21, the sample need not be strictly random in the sense that each member of the population had the same chance of being chosen. But the sampling procedure should have no bias for or against individuals with any features relevant to the matter at hand.

[83] Note, however, that the condition "without any further qualification" is hardly ever met. Let C stand for the narrowest concept under which a given object is grasped at a given moment. Any further consideration of the object will produce further specification of the concept under which it is grasped. Hence, if the object is grasped under C *for the sake* of such further consideration, C can only do its job while *in the course* of being qualified further. Indeed, the idea that one can stop thought at an instant t and determine the narrowest concept under which some object was being grasped at t is sheer philosophical fiction. The concept C under which, at a particular time, a particular object is understood in scientific inquiry is surrounded by flickering, mostly unverbalized, tentative specifications that a realistic epistemology must regard as pertaining to C itself.

[84] Robert Leslie Ellis wrote in 1854 that "the fundamental principle of the Theory of Probabilities may be regarded as included in the following statement, — 'The conception of a genus implies that of numerical relations among the species subordinated to it'" (1856, p. 606). John Venn quoted this passage approvingly in the second and third editions of his classic *Logic of Chance.* But because Venn was intent on conceiving the numerical relations in question as limits of empirical relative frequencies he disparaged Ellis' use of 'genus' and 'species', and favored the term 'series' instead (Venn 1888, p. 9).

[85] Of course, if S_C is infinite, the foregoing definition does not generally make sense unless, with de Finetti, we require $\mathbf{p}$ to be just finitely additive, or, with Solovay, we give up the Axiom of Choice in its standard form (see note 13). However, such austerity is not called for if C is the concept of a physical quantity whose subordinate concepts can be represented by points and sets of points in $\mathbf{R}^n$ (for some positive integer n). For all practical purposes, it is sufficient to have the probability function $\mathbf{p}$ defined, not on $\mathcal{B}(S_C)$, but on the smallest Borel field generated by the open balls of $\mathbf{R}^n$.

[86] If the particle has integral spin eigenvalues it is a boson; if it has half-odd-integral spin eigenvalues it is a fermion. A boson gas obeys the Bose-Einstein statistics, and a fermion gas obeys the Fermi-Dirac statistics.

[87] On the other hand, if such a grounding is lacking, talk of cosmic chances borders on nonsense, as in the following quotation from a recent philosophical book: "The probability

concept that should be used in analyzing counterfactuals [. . .] is an extension of the notion of a chance setup, when the whole world up to a certain time is taken as a setup" (Kvart 1986, p. 103).

[88] The paragraph to which this note is appended was written a few months before I had occasion to see Michael Tooley's *Causation* (1987). After reading it, I am under the impression that my description of ideal chances agrees with—or at least comes very close to—his conception of statistical or probabilistic laws of nature. The simplest such laws are expressed by statements of the following form:

> It is a law that the probability that something with the occurrent property *P* has the occurrent property *Q* equal to *k*.
>
> (Tooley 1987, p. 148)

There are, however, some important differences between Tooley's handling of this matter and mine. I am ready to welcome 'probability' in the above statement-form as a primitive term, which signifies a new, creative extension to the realm of universals of Galileo's quantitative concept of facility of occurrence (originally introduced for particular physical setups; see Section 4.3). But Tooley rejects probability as propensity—and, hence, Galileo's concept—on semantic grounds which I do not share. Apparently, it was this rejection that led him to his novel conception of probabilistic laws of nature (while I lighted on ideal chances simply by reflecting on Classical Statistical Mechanics). On the same grounds, he must find a definition for the relation between property *P* and property *Q* expressed in the above statement-form, and which he calls 'probabilification to degree *k*'. A 17-line definition is finally offered after a 10-page search. It turns on the use of 'logical probability' (= 'degree of confirmation') as a primitive term. I regard this concept as a stillborn creature of philosophy, and therefore I cannot countenance Tooley's proposal. On the other hand, if he ever reads me, he will probably dismiss my own attitude as irresponsible. Of course, Tooley, for all his subtlety and inventiveness, remains committed, as an Australian realist, to the hard and fast preconceptions of being and truth which—before reading him—I had dubbed 'sclerotic'.

[89] There are other plausible ways of measuring the probability that a line λ through a circle *K* meets a certain region of *K*, given that any position of $\lambda \cap K$ in *K* is just as probable as any other (see J. Bertrand 1907, pp. 4–5).

[90] A mechanical system satisfying this condition is currently said to be *ergodic*. This designation, mysteriously stemming from the Greek ἔργον, 'work', and ὅδος, 'path', was imposed by Paul and Tatiana Ehrenfest (1912, p. 30), who attributed it to Boltzmann. But Boltzmann actually used the term *Ergode* to denote what is now called, after Gibbs, a microcanonical ensemble, i.e., a *collection* of systems whose representative trajectories in phase space lie within the same hypersurface of constant energy (see Brush 1976, pp. 368–69; Boltzmann 1884 in Boltzmann 1968, III, 134). An ergodic system in the now current sense cannot exist, because a particular solution of Hamilton's equations for a mechanical system is a continuous injective mapping of a time interval—which is unidimensional—into a multidimensional region of phase space, and such a mapping cannot be surjective. Brush suggests that what Boltzmann really wished to say when he made the claim mentioned in the text was that a typical mechanical system would behave quasi-ergodically, i.e., that its representative point would eventually come *arbitrarily close* to every point in *C*. But—as the Ehrenfests noted—quasi-ergodicity does not have the same implications as ergodicity. The point set covered by the trajectory of a quasi-ergodic system has volume 0 in *C*, and there is no reason why the lengths of its several parts should stand to one another in the same ratios as the volumes of the regions of *C* through which they respectively meander. For an inspiring

and up-to-date discussion of this matter, with references to recent literature, see Earman 1986, Chapter IX.

Chapter 5

[1] A very useful survey of such attempts and their failings is contained in Michael J. Loux's anthology *The Possible and the Actual* (1979). My own views come closest to those expressed in the article by Mondadori and Morton on pp. 235–52. William Lycan's "The Trouble with Possible Worlds" (pp. 274–316) offers a more considerate, but no less devastating, criticism of polycosmism.

[2] At any rate, in philosophy classrooms. In ordinary conversation, it would certainly be very silly to say on a summer afternoon under a bright blue sky that it is possible that the sun is shining.

[3] Thus, for the polycosmist, the modal operators are quantifiers that bind a tacit variable which ranges over the set of worlds. In opposition to Leibniz, current polycosmist semantics understands by 'world' an aggregate not of things but of facts (after Wittgenstein, *Tractatus*, 1.1). Thus, Plantinga (1974, p. 45) defines a 'possible world' as being 'simply a possible state of affairs that is maximal', i.e., which is such that for every state of affairs *S*, it either includes *S* (i.e., is consistent with *S*) or it precludes *S* (i.e., is inconsistent with *S*).

[4] On this question, see Lakoff 1987, passim. For those who think they see the world as an aggregate, the best antidote is, to my mind, Chapter 3 of Heidegger's *Sein und Zeit*; see also Eugen Fink, *Spiel als Weltsymbol* (1960).

[5] David Lewis (1986) comments with his usual brilliance on several such projects and the advantages they may derive from polycosmism. The chapter is entitled "A Philosopher's Paradise."

[6] Michael Tooley 1987, p. 25, makes it very clear why he—who, by the way, is not a polycosmist—would not take 'possible' or 'necessary' as primitives. According to him, no primitive predicate of scientific and philosophical discourse "can be such that elementary statements containing that predicate necessarily involve nonextensional contexts" (i.e., contexts in which pairs of coextensive terms or pairs of logically equivalent sentences cannot be substituted for one another *salva veritate*). And, of course, the terms 'possible' and 'necessary' do not meet this condition. Tooley's justification for requiring it is that, in his considered view, if *p* is an elementary statement that is free of quantifiers and contains only primitive terms, the state of affairs that makes *p* true must also be the truth-maker of an atomic statement of the first-order predicate calculus (Tooley 1987, p. 26). This claim, reminiscent of Russell's notorious Axiom of Reducibility, is far from being evident to me—indeed, I am unable to see that it has any force.

[7] To put it in a way that may find favor with the professors: My awareness that I exist (*sum*) contains the awareness that I can (*possum*). But since to be for me is to-be-with-other-persons-among-things-in-the-world, I may just as well generalize and go on to say in kitchen Latin, *nullum esse sine posse*; or in Plato's Greek, ἱκανὸν . . . ὅρον που τῶν ὄντων, ὅταν τῳ παρῇ ἡ τοῦ πάσχειν ἢ δρᾶν καὶ πρὸς τὸ σμικρότατον δύναμις (*Sophista* 248c; note that the—presumably manifest—presence of possibilities is here put forward as the mark of reality; cf. *Sophista* 247e, *Phaidrus* 270d).

[8] For, as Aristotle neatly put it, "obviously change is the actuality of the possible *as such*" (ἡ τοῦ δυνατοῦ, ᾗ δυνατόν, ἐντελέχεια, φανερὸν ὅτι κίνησίς ἐστιν—*Physica* III, 1, 201^{b}5). Note that even Descartes, who, blinded by the light of geometry, could not make head or tail of Aristotle's saying, knew a lump of wax to be flexible and changeable. "But what is this being

flexible and changeable? Is it just that I imagine that this wax can turn from round into square, and from square into triangular? By no means; for I grasp it as being capable of innumerable changes of this kind" (Descartes, *Meditatio Secunda*; AT, VII, 32). I would never achieve such a grasp by merely comparing the initial and final states I actually perceive in the wax.

[9] 'Power' is, of course—like its French cousin, 'pouvoir'—the direct descendant of 'potere', the vulgar Latin equivalent of 'posse'.

[10] τῷ μὲν [scil. τῷ δυνατὸν εἶναι—R.T.] γὰρ ἄμφω ἐνδέχεσθαι συμβαίνειν. ἅμα γὰρ δυνατὸν εἶναι καὶ μὴ εἶναι (*De Int.* 13; 22^b19ff.). ὅσα γὰρ κατὰ τὸ δύνασθαι λέγεσθαι, ταὐτόν ἐστι δυνατὸν τἀναντία (*Metaph.* Θ, 9; 1051^a6). Cf. also Aquinas, *Summa c. Gent.* 1.67: "Contingens a necessario differt secundum quod unumquodque in sua causa est: contingens enim sic in sua causa est ut non esse ex ea possit et esse; necessarium vero non potest ex sua causa nisi esse." Note that Hintikka, like Aquinas, uses 'contingent' for 'possible' in its ordinary sense and reserves 'possibility proper' for the modern philosophical sense, which—as we shall see below—is already prominent in Aquinas.

[11] τὸ γὰρ ἀναγκαῖον ὁμωνύμως ἐνδέχεσθαι λέγομεν (*An.Pr.* I, 13, 32^a21; cf. I, 3, 25^a37).

[12] I call this view of possibility Diodoran because it was put forward by Aristotle's younger contemporary Diodorus Cronus (fl. ca. 300 B.C) in the conclusion of his celebrated Master Argument: "Nothing is possible which neither is nor will be true" (Epictetus, *Diss.* ii.19.1). As Hintikka (1973, p. 182) rightly notes, the key ingredients of the Master Argument can all be found in Aristotle's writings. For a forceful criticism of Hintikka's approach see Sorabji 1980, especially Chapter 8.

[13] See, e.g., Aquinas's *Summa contra Gentiles*, 1.8, where—under the caption "That God's substance is His power"—we read three times, as an incantation, "Est igitur ipse sua potentia" ("He Himself is therefore His power").

[14] "Non autem potest dici quod Deus sit omnipotens, quia potest omnia quae sunt possibilia naturae creatae: quia divina potentia in plura extenditur" (Aquinas, *Summa Theol.* 1, qu.25, a.3).

[15] *Metaphysica*, Δ, 12, 1019^b34. The passage in question is part of Aristotle's "philosophical dictionary," s.v. δύναμις. He mentions there several ways in which one speaks of the possible (τὸ δυνατός) without referring it to a given power (δύναμις). Aquinas seizes on one of them. In this particular sense, "impossible is that whose contrary is necessarily false" (ἀδύνατον μὲν οὗ τὸ ἐναντίον ἐξ ἀνάγκης ἀληθές); therefore, "the possible is when it is not necessary that the contrary be false" (τὸ δυνατόν, ὅταν μὴ ἀναγκαῖον ᾖ τὸ ἐναντίον ψεῦδος εἶναι). In his Commentary on the Metaphysics of Aristotle, §971, Aquinas remarks that 'posse' is always referred to 'esse', 'being'. Since one of the several meanings of 'being' concerns not the nature of things but truth and falsehood in the composition of propositions, there must be a matching sense of 'impossible', based not on the lack of a particular power but on the incompatibility of terms in propositions ("non propter privationem alicuius potentiae, sed propter repugnantiam terminorum in propositionibus"). See also *De Pot.*, qu.1, a.3: "Impossibile quod dicitur secundum nullam potentiam, sed secundum se ipsum, dicitur ratione discoherentiae terminorum." I find none of this in Aristotle. However, his example of impossibility in the sense here considered is the incommensurability of the diagonal, which indeed is proved by deriving a contradiction from its negation.

[16] *Gorgias* 483e. In Plato's dialogue, the phrase is used by Callicles, an unidentifiable, possibly fictional character. He is voicing the view that superior men stand above our petty human justice, "according to the *nomos* of nature, though not indeed according to the one that we establish for us."

[17] A *graph* can be informally described as a finite collection of points (*vertices*) joined by

lines (*edges*). Every edge joins an unordered pair of vertices. There may be vertices to which no edge at all is attached. But no edge may join more than two vertices, and there may be no more than one edge between two given vertices or between a vertex and itself. (If multiple edges between two vertices are allowed, we speak of a *pseudograph*). In an *oriented* graph every edge joins an ordered pair of vertices. The same two vertices can then be joined, in reverse order, by another edge. Graphs and oriented graphs can be readily characterized as Bourbakian species of structure:

(1) A graph **G** is a pair $\langle V,E \rangle$, where V—the set of vertices of **G**—is a finite set and E—the set of edges of **G**—is a subset of the set $\{\{a,b\} \mid a,b \in V\}$.

(2) An oriented graph **G*** is a pair $\langle V,E \rangle$, where V is a finite set and E is a subset of V^2.

A *path* in **G** is an n-tuple $\langle a_1, \ldots, a_n \rangle$ (or an infinite sequence $a_1, a_2, \ldots$) in which all odd-numbered terms are vertices and all even-numbered terms are edges of **G**, and, for any $i \geq 1$, $a_{2i} = \{a_{2i-1}, a_{2i+1}\}$. A path in **G*** is defined in the same way, except that the last condition is replaced by the following: for any $i \geq 1$, $a_{2i} = \langle a_{2i-1}, a_{2i+1} \rangle$.

[18] The graph of chess is too large to be figured out in full even by our largest computer, but a characterization of it can be readily sketched using the definition of **G*** in note 17 and our acquaintance with the rules of chess. To readers with little or no mathematical training, the following hints will provide a useful exercise in mathematical representation. Each position in chess can be viewed as an injective mapping (see Section 2.8.2) of all or some of the chesspieces into the set **B** of squares on the chessboard. (For this approach to work one must, of course, label the individual pieces when there are two or more of a kind, e.g., by numbering the white and black pawns from 1 to 8, and by marking each rook, knight, or bishop to distinguish it from the other one of the same color). By supplementing the 64 elements of **B** with a box β, to which we assign the captured pieces, we eliminate the tiresome distinction between positions involving all 32 pieces and positions involving only some of them. Still, we must complete the set Π of the normal chesspieces with sufficient doublets to substitute for pawns reaching rows 1 or 8. Denote the enriched set of pieces by Π^*. Let **C*** $= \langle V,E \rangle$ be the graph of chess. We take V to be the collection of all the mappings $\Phi: \Pi^* \rightarrow \mathbf{B} \cup \{\beta\}$ which are injective on $\Phi^{-1}(\mathbf{B})$ and do not send a king to β. This collection probably contains many positions that can never be reached in a game of chess; we let them stand as isolated vertices in our graph. Any $\Phi \in V$ which assigns the white and black pieces of Π to their customary initial places in rows 1, 2, 7, and 8 of **B**, while confining the doublets to β, would normally be accepted as an initial position of the game. Let V_0 be the set of all elements of V which can be built from such a position Φ by permuting pairs of equivalent pieces of the same color. To simplify matters, we single out a particular member of V_0 as *the* initial position of chess and denote it by Φ_I. (Most of the other members of V_0 cannot be reached from Φ_I and will therefore stand as isolated vertices in **C***; however, the following is a crazy but perfectly legal chess opening leading from Φ_I to another position in V_0: 1. Nc3, Nc6; 2. Ne4, Nf6; 3. Nf3, Ne5; 4. Neg5, Neg4; 5. Nh3, Nh6; 6. Nhg1, Nhg8; 7. Ne5, Ne4; 8. Nc4, Nc5; 9. Na3, Na6; 10. Nb1, Nb8.) The set E of moves can now be defined recursively as follows:

(i) The set E_1 of first moves contains the twenty pairs $\langle \Phi_I, \Phi^1{}_1 \rangle, \ldots, \langle \Phi_I, \Phi^1{}_{20} \rangle$, where $\Phi^1{}_r$ $(1 \leq r \leq 20)$ ranges over the 20 positions that differ from Φ_I only at one point, namely, in the value (location) assigned to one of White's eight pawns (which is displaced, along the same column, from row 2 to row 3, or from row 2 to row 4) or in the value assigned to one of White's knights (which is displaced from row 1 to 3 rook or to 3 bishop, on the same side).

(ii) The set E_{n+1} of $(n+1)$th moves $(n \geq 1)$ contains all the pairs $\langle \Phi^n{}_r, \Phi^{n+1}{}_s \rangle$ such that $\Phi^n{}_r$ is the endpoint of a move in E_n and $\Phi^{n+1}{}_s$ is a position legally attainable from $\Phi^n{}_r$ by White if n is even, and by Black if n is odd.

(iii) $E = \bigcup_{n=1}^{\infty} E_n$.

Consider a Black move $\langle \Phi^n{}_r, \Phi^{n+1}{}_s \rangle$, with odd n. $\Phi^{n+1}{}_s$ typically differs from $\Phi^n{}_r$ either (1) in the value assigned to a single black piece X, or (2) in the values assigned to a single black piece X and a single white piece Y. In case (1), $\Phi^{n+1}{}_s(\mathrm{X})$ is a square which is not assigned by $\Phi^n{}_r$ to any piece ($\Phi^{n+1}{}_s(\mathrm{X}) \notin \Phi^n{}_r(\Pi)$), and which is legally accessible to X from $\Phi^n{}_r(\mathrm{X})$. In case (2), $\Phi^{n+1}{}_s(\mathrm{X})$ is again legally accessible to X from $\Phi^n{}_r(\mathrm{X})$; moreover, $\Phi^{n+1}{}_s(\mathrm{X}) = \Phi^n{}_r(\mathrm{Y})$ and $\Phi^{n+1}{}_s(\mathrm{Y}) = \beta$ (Y is captured by Black on the $(n+1)$th move). There are, however, two interesting exceptions. One of them is (3) *pawn conversion*: X is a black pawn, $\Phi^n{}_r(\mathrm{X})$ lies on row 2, and a square Q in row 1 is legally accessible to X from $\Phi^n{}_r(\mathrm{X})$; X′ is a black piece (usually a queen), such that $\Phi^n{}_r(\mathrm{X}') = \beta$; $\Phi^{n+1}{}_s(\mathrm{X}) = \beta$; and $\Phi^{n+1}{}_s(\mathrm{X}') = \mathrm{Q}$. The other is (4) *castling*: $\Phi^{n+1}{}_s$ differs from $\Phi^n{}_r$ in the values assigned to the black king K and one of the black rooks, R_σ; $\Phi^n{}_r(\mathrm{K}) = \Phi_{\mathrm{I}}(\mathrm{K})$ and $\Phi^n{}_r(\mathrm{R}_\sigma) = \Phi_{\mathrm{I}}(\mathrm{R}_\sigma)$; $\Phi^{n+1}{}_s(\mathrm{K})$ and $\Phi^{n+1}{}_s(\mathrm{R}_\sigma)$ are the appropriate final positions for castling to side σ; moreover, every square Q in row 8 placed between $\Phi_{\mathrm{I}}(\mathrm{K})$ and $\Phi_{\mathrm{I}}(\mathrm{R}_\sigma)$ must be empty and unmenaced in position $\Phi^n{}_r$ (that is to say, $\mathrm{Q} \notin \Phi^n{}_r(\Pi)$ and Q is not legally accessible for any white piece Y from $\Phi^{n+1}{}_s(\mathrm{Y}) = \Phi^n{}_r(\mathrm{Y})$). A further complication consists in this: Black can castle to side σ at the $(n+1)$th move only if K and R_σ have not been moved at any earlier stage of the game, that is, if $\Phi(\mathrm{K})$ and $\Phi(\mathrm{R}_\sigma)$ have remained constant through all the first n moves of the game; hence, if $\langle \Phi^n{}_r, \Phi^{n+1}{}_s \rangle$ is a castling move meeting the aforesaid conditions, and $\langle \Phi_1, \ldots, \Phi^n{}_r \rangle$ is a path in **C*** which contains a position Φ such that $\Phi_{\mathrm{I}}(\mathrm{K}) \neq \Phi(\mathrm{K})$ or $\Phi_{\mathrm{I}}(\mathrm{R}_\sigma) \neq \Phi(\mathrm{R}_\sigma)$, then the path $\langle \Phi_{\mathrm{I}}, \ldots, \Phi^n{}_r, \langle \Phi^n{}_r, \Phi^{n+1}{}_s \rangle, \Phi^{n+1}{}_s \rangle$ does not represent a possible game of chess.

A White move $\langle \Phi^n{}_r, \Phi^{n+1}{}_s \rangle$ (n even) can be handled in essentially the same way.

[19] We are told that while Plato stayed with Dionysius of Syracuse "the tyrant's palace was filled with dust, from the multitude of people doing geometry" (Plutarch, *Dion* xiii). The Greeks drew their geometrical figures on strewn sand.

[20] Friedrich Ueberweg (1851) attempted to build Euclid's system of geometry upon a set of axioms centered on the concept of rigid motion. The same goal was later pursued by Hermann von Helmholtz (1866, 1868), and achieved by Mario Pieri (1899). A different approach to the axiomatization of geometry, first tried by Moritz Pasch (1882), was carried to completion by David Hilbert (1899). Other axiomatizations have been given by Veblen (1904, 1911), Huntington (1912), Bachmann (1951), etc.

[21] A region of the plane is said to be simply connected if its boundary can be gradually contracted to a point inside it without leaving the region. Thus, the interiors of a circle and a polygon are simply connected; but an annular region or a punctured region (from which one or more points have been deleted) are not.

[22] From the denial of a proposition equivalent to Postulate 5, J. H. Lambert (1786) inferred that the two sides of a right angle have a common parallel at a fixed distance from its vertex. He concluded that he had thereby proved Postulate 5 by reductio ad absurdum. The said distance—appropriately described as the upper bound of the height of an isosceles

right triangle—was gleefully acknowledged by the lawyer and amateur geometrician F. K. Schweikart as a characteristic Constant of the *Astralgeometrie* he announced to Gauss in a private communication of December 1818. Euclidian geometry was just the limiting case in which the Constant is infinite. "Astral Geometry" is, of course, just a fancy name for what is now known as Bolyai-Lobachevsky geometry.

[23] If two coplanar straights λ and μ meet a third one τ at different points P and Q, they form with it eight angles. The four angles that lie between λ and μ—i.e., the angles adjacent to the segment PQ—are known as *internal* angles and are grouped into two pairs, located on either side of τ. Euclid's Postulate 5 says that if the internal angles on one side of τ add up to less than two right angles, λ and μ will meet on that side of τ.

[24] Torretti 1978b discusses Dingler's philosophy of geometry with ample references to his writings. I should mention that, in my view, Dingler never managed to pinpoint so neatly as Lorenzen (1984) the reason why physics is supposedly committed to Euclidian geometry.

[25] A reconstructed Dinglerian may still contend that a scientific tradition which, like ours, feeds on measurements taken with Euclidian instruments is bound with necessity to employ geometrical systems which agree locally (to a good approximation) with Euclidian geometry. Given the unmitigated success of Euclidian tools in every branch of industry, such a claim, though philosophically interesting, is surely otiose.

[26] Henri Bergson, who first perceived this conception as a product of evolution, not a self-evident truth, described it as the distinctive mark of human intelligence (as opposed, say, to the "instinct" of insects). I think this is quite exaggerated and misleading, for many civilizations have remained foreign to the understanding of time on the analogy of a straight line, and still today, when it permeates not just science and engineering, but the entire fabric of social life, some of our contemporaries continue to be put off by it.

[27] I am thinking of the explication of the linear continuum as a structured set, such as the set of points in a straight line or the set of instants in time. See, e.g., E. V. Huntington 1905 (reprinted as Huntington 1955). This explication, based on the work of Georg Cantor, was rejected by a small but vocal minority of mathematicians, led by L. E. J. Brouwer. Note, however, that Brouwer does not oppose the assimilation of time to the linear continuum, but only the crumbling of the latter into points (and of time into instants). His "neo-intuitionism considers the falling apart of moments of life into qualitatively different parts, to be reunited only while remaining separated by time[,] as the fundamental phenomenon of the human intellect, [which] by abstracting from its emotional content [becomes] the fundamental phenomenon of mathematical thinking, the intuition of the bare two-oneness. [. . .] This basal intuition of mathematics [. . .] gives rise immediately to the intuition of the linear continuum, i.e., of the 'between', which is not exhaustible by the interposition of new units and which therefore can never be thought of as a mere collection of units" (Brouwer 1913, pp. 85–86).

[28] A second lemma follows: If two points move with constant speed on different lines, and two lengths are taken on each line, such that the first point traverses each length on its line in the same time as the other point traverses one of the lengths on the second line, then the two pairs of lengths are proportional to each other. This proposition equates the ratios between two pairs of lengths, in the standard manner of Euclid's Book V, but the middle term of its proof is the ratio between a pair of times.

[29] Τίνων ὑποτεθέντων δι' ὁμαλῶν καὶ τεταγμένων καὶ ἐγκυκλίων κινήσεων δυνήσεται **διασωθῆναι τὰ** περὶ τὰς κινήσεις τῶν πλανᾶσθαι λεγομένων **φαινόμενα** (Simplicius, *In Aristotelis de Caelo*, 488.16–18; my emphasis). Simplicius ascribes this program to Plato himself. The boldfaced words are the historical source of the slogan "To save the phenomena," which according to Pierre Duhem and Bas van Fraassen epitomizes the task of mathematical physics. Plato initially distrusted the possibility of accurately predicting the

motion of heavenly bodies, not because of any limitation of the human mind, but due to the inherent inability of matter to keep a schedule (*Republic*, 529b–530d). Later in life, however, he cited the exact periodicity of the planets as evidence that they were piloted by intelligent beings (*Laws*, 821e–822a, 897c, 967b).

[30] This is not such a strong demand as it may seem at first sight. Should a planet exactly comply with the dictates of a given Eudoxian model, then, inevitably, its trajectory will actually be the resultant of the combination of uniform circular motions foreseen in that model. It will, of course, be also the resultant of many other such combinations. (The problem of constructing a Eudoxian model for the motion of a point on a spherical surface generally admits more than one solution.) But I frankly see no difficulty in identifying the adequate concept of the observed trajectory with the entire class of *equivalent* Eudoxian models—i.e., the set of all the Eudoxian models which yield that trajectory as a resultant. After all, Eudoxian spheres are not hard bodies one might run against in interplanetary flight. The sole effect of their existence and their motions is to produce the motion of the respective planet. (The solid aether balls that Aristotle substituted for Eudoxus' spheres, like the pegs and cogs of some 19th century models of the electromagnetic field, were spawned by premathematical thinking and bear witness to its great resilience.) A greater difficulty for a realistic interpretation of Eudoxian astronomy is the following: Eudoxian models account for the motion of a planet's projection on the heavenly vault but not for its radial motion to and from the Earth. Hence, if planets do have such a radial motion, Eudoxian astronomy is simply inadequate. (Greek astronomers reached this conclusion fairly soon.) But a similar difficulty besets every physical theory. If there were a whole new dimension to the behavior of so-called elementary particles, which our physicists have hitherto failed to read in—or into—the recorded observations, we could hardly expect our current theories to account for it.

[31] As a matter of fact, the differential and integral calculus that Newton invented in the 1660s was independently reinvented by Leibniz in the 1670s, and Leibniz's friend and follower Jacques Bernoulli was one of the first to employ it, publicly and explicitly, to solve a differential equation.

[32] Clearly, in the present context a point $t \in \mathbf{R}$ is suitable whenever $t \in \pi_1(U)$, where π_1 denotes the projection of $\mathbf{R}^{n+1}$ into $\mathbf{R}$ by $\langle u_1, \ldots, u_{n+1} \rangle \mapsto u_1$.

[33] Proofs may be found in H. Cartan 1967, pp. 116–20 (§§ II.1.5–II.1.7) and also in Brauer and Nobel 1969, an excellent textbook in English which has been recently reprinted by Dover. The above presentation closely follows Cartan's book. The two theorems on the existence and uniqueness of solutions quoted above are proved by Cartan for the more general case in which the continuous mapping f on the r.h.s. of eqn. (1) is defined on a subset U of $\mathbf{R} \times \mathbf{E}$, where $\mathbf{E}$ is an arbitrary real Banach space (see Section 2.8.4). In that case, of course, the theorems assume that f is κ-Lipschitzian in $\mathbf{r} \in \mathbf{E}$. A solution of eqn. (1) is then a mapping φ of a real interval into $\mathbf{E}$ meeting conditions (i)–(iii). Its first derivative φ' is, strictly speaking, a point in the vector space $\mathcal{L}(\mathbf{R},\mathbf{E})$ of linear mappings of $\mathbf{R}$ into $\mathbf{E}$. But $\mathcal{L}(\mathbf{R},\mathbf{E})$ is canonically isomorphic with $\mathbf{E}$ and may therefore be identified with the latter. (One must also substitute the norm $\| \ \|$ of space $\mathbf{E}$ for the $\mathbf{R}^n$ distance function in inequality (2), condition (iiiε), etc.)

[34] Let $\mathbf{E}$ be a Banach space over $\mathbf{R}$. Let U be an arbitrary subset of $\mathbf{R} \times \mathbf{E}^n$. Let f be a continuous mapping of U into $\mathbf{E}$. An *ordinary differential equation of the nth order* is written thus:

$$d^n\mathbf{r}/dt^n = f(t, \mathbf{r}, d\mathbf{r}/dt, \ldots, d^{n-1}\mathbf{r}/dt^{n-1}) \qquad (*)$$

with $t \in \mathbf{R}$, $\mathbf{r} \in \mathbf{E}$. A solution is any mapping φ: $\mathbf{I} \to \mathbf{E}$ (where $\mathbf{I}$ is an arbitrary interval in $\mathbf{R}$) which meets the following three conditions:

(i′) All derivatives of the first n orders, $\varphi', \ldots, \varphi^{(n)}$, are defined and continuous on the interior of **I**.

(ii′) For every $t \in \mathbf{I}$, $\langle t, \varphi(t), \varphi'(t), \ldots, \varphi^{(n-1)}(t)\rangle \in U$.

(iii′) For every $t \in \mathbf{I}$, $\varphi^{(n)}(t) = f(t, \varphi(t), \varphi'(t), \ldots, \varphi^{(n-1)}(t))$.

Searching for a solution to eqn. (*) is then tantamount to searching for a solution of a system of n equations of the first order, viz.,

$$d\mathbf{r}/dt = \mathbf{r}_1, \qquad d\mathbf{r}_1/dt = \mathbf{r}_2, \ldots,$$
$$d\mathbf{r}_{n-2}/dt = \mathbf{r}_{n-1}, \qquad d\mathbf{r}_{n-1}/dt = f(t, \mathbf{r}, \mathbf{r}_1, \ldots, \mathbf{r}_{n-1})$$

Instead of looking for a single unknown function φ meeting conditions (i′)–(iii′), one looks for a system of n unknown functions $\varphi, \varphi_1, \ldots, \varphi_{n-1}$, meeting conditions (i) and (ii) and such that

(iii″) $\varphi'(t) = \varphi_1(t), \varphi_1'(t) = \varphi_2(t) \ldots, \varphi_{n-2}'(t) = \varphi_{n-1}(t),$
$\varphi_{n-1}'(t) = f(t, \varphi_1(t), \ldots, \varphi_{n-1}(t))$.

35 The following rough-and-ready characterization of partial differential equations is given at the beginning of a classical treatise on the subject:

> A partial differential equation is given as a relation of the form
>
> (1) $$F(x,y, \ldots, u, u_x, u_y, \ldots, u_{xx}, u_{xy}, \ldots) = 0,$$
>
> where F is a function of the variables $x,y, \ldots, u, u_x, u_y, \ldots, u_{xx}, u_{xy}, \ldots$; a function $u(x,y,\ldots)$ of the independent variables $x,y, \ldots$ is sought such that equation (1) is identically satisfied in these independent variables if $u(x,y, \ldots)$ and its partial derivatives
>
> $$u_x = \frac{\partial u}{\partial x}, \quad u_y = \frac{\partial u}{\partial y}, \quad \ldots,$$
> $$u_{xx} = \frac{\partial^2 u}{\partial x^2}, \quad u_{xy} = \frac{\partial^2 u}{\partial x \partial y}, \quad \ldots,$$
> $$\cdots\cdots \quad \cdots\cdots \quad \cdots\cdots$$
>
> are substituted in F. Such a function $u(x,y,\ldots)$ is called a *solution of the partial differential equation* (1).
>
> (Courant and Hilbert, MMP, vol. II, p. 1)

36 The reader is free to call it “logical necessity” if he so desires, provided that he takes ‘logical’ simply as Greek for ‘conceptual’ and not as referring to the theory of deductive inference in a first-order language. The latter falls within the compass of the theory of recursive functions and recursive enumerability, but mathematics extends well beyond it.

[37] As a matter of historical fact, the Hamiltonian formulation of mechanics was found after Laplace wrote this passage, but he was well acquainted with the equivalent Lagrangian formulation.

[38] Let me give these phrases in context. Here is the continuation of the passage translated above:

> The human mind displays, in the perfection achieved by astronomy, a feeble sketch of such an intelligence. Its discoveries in mechanics and geometry, plus those concerning universal gravitation, have enabled it to comprehend the past and future states of the system of the world [i.e., the Solar System] in the same analytic expressions. Applying the same method to some other objects within its purview it has succeeded in bringing the observed phenomena under general laws and in forecasting those that given circumstances should bring to pass. All its efforts in the quest for truth tend to make it incessantly approach the intelligence we have just conceived, from which, however, it will always remain infinitely far.
>
> (Laplace, OC, vol. VIII, p. vii)

[39] Einstein 1949, p. 64. On Einstein's way from Special to General Relativity see Stachel 1989, Norton 1989, and my own little *divertissement*, Torretti 1989.

[40] The so-called singularities of General Relativity are not, like the singularities of Classical Electrodynamics, points of spacetime where the physical quantities of interest become undefined. In General Relativity, the physical quantities of interest define the spacetime itself, so the "singularities" are rather like lacunae in the latter.

[41] At some point in time the tendency grew so strong that a statement to the effect that differential equations are "the natural expression of the principle of causality" could find its way into a college textbook. (The quotation is from Hopf 1948, p. 1.) In a paper published in the second issue of *Philosophy of Science* Henry Margenau emphatically warned the reader: "There is no law of connection between cause and effect known to science; moreover, these concepts are foreign to physical analysis. Nor is it of any avail to inject them externally, for the meaning usually conveyed by the words in question is expressed more adequately and precisely by technical terms like boundary condition, initial and final state" (Margenau 1934, in Danto and Morgenbesser 1960, p. 437). And yet, throughout that same paper Margenau uses the adjective 'causal' as equivalent with my 'GDE' (restricted to differential equations which do not contain explicit functions of the time).

[42] Nancy Cartwright has repeatedly stressed the contrast between causal storytelling and the subsumption of particular facts under a general law (i.e., between etiology and nomology, as I propose to say for short). It is a pity that I did not get a copy of *Nature's Capacities and Their Measurement* (Cartwright 1989) until after the manuscript of *Creative Understanding* had been sent off to the publisher, for there is much in that rich and challenging book that is directly relevant to the present section, and I might have articulated some of my ideas differently if I had had a chance to see it. This is not the proper place for reviewing it, but for the sake of making my own position clearer to the reader, I will record here my reaction to some of its claims. Here is a sample of statements which I unreservedly endorse: "The domain and limitations on the domain [of a physical theory] can be constructed only by already using the theory and the concepts of the theory. Theories come before observables, or are at least born together with them" (p. 162). "There is no going from pure theory to causes, no matter how powerful the theory" (p. 39). "New causal knowledge can be built only from old causal knowledge. There is no way to get it from equations and associations by themselves" (p. 54). "Hume's picture [of causation] is exactly upside down"

(p. 91). I gleefully admit that "nature is complex through and through" (p. 72); indeed, I would add that it is only through the exercise of mythopoetic freedom that a referent for the subject of this statement can be found—in effect, founded—amidst all the complexity. I will also grant that "laws—in the conventional empiricist sense—have no fundamental role to play in scientific theory" (p. 185), provided that the parenthetical phrase is not overlooked. But when it comes to genuine physical laws—which, as Erhard Scheibe (1989) ably notes, do not generalize over a variety of empirical cases but rather over the elements (points, particles, etc.) of a singular instance (application, model) of the theory they belong to—I should be loath to spurn them with Cartwright (1989, p. 218) as "pieces of science fiction." I rather see the literature of mathematical physics as a paradigm of nonfiction (while not forgetting, of course, that nonfiction too is a form of discourse—a literary genre, if you wish). Finally, I do not believe that "causality is at the core of scientific explanation" (p. 218)—at any rate not in mathematical physics, as bequeathed to us by Galileo, Newton, Fourier, Maxwell, Einstein, Dirac. Nor do I buy Cartwright's suggestion that "Aristotelian concepts" provide "the natural way to describe the explanatory structures of physics" (p. 226).

[43] Zeno Vendler, who "rules out persons or objects from the ranks of causes," believes that "counter-examples like 'John caused the disturbance' can be made harmless by pointing to the possibility of inserting a verb-nominalization: 'John's action caused the disturbance'" (1966, p. 13). But, compared with the former sentence, the latter sounds pretty unnatural and pedantic and can only be justified in the eyes of someone bent on making a philosophical point. (On the other hand, 'John's inaction caused the disturbance', if apposite, will sound quite natural; presumably because John's causing anything typically involves some form of action, and if we wish blame him for something that he merely let happen, we must explicitly report his inaction.) G. H. von Wright has argued that "the concepts of cause and nomic necessity [. . .] presuppose, are dependent upon [. . .] the concepts of action and agency" (1974, p. 48). Nevertheless, "for the sake of clarity," he is "anxious to *separate* agency from causation" (p. 49; his italics), presumably because he insists in conflating the latter with nomic necessity. He achieves such separation by making nice yet artificial verbal distinctions.

[44] The primary meaning of the Latin word *causa* is difficult to ascertain, because the first extant texts that use it display the word in a rich variety of meanings and were written after the Roman intelligentsia was conquered by the Greek. It is perhaps worth noting that while both αἰτία and *causa* have technical legal meanings, they point in different directions: whereas αἰτία specifically designates the *accusation* submitted to a court, *causa* means a legal case—whence a good case, a claim—and also an alleged reason or extenuating plea. This may be related to the fact that in its extralegal sense *causa* means 'motive', 'reason', and not only a cause understood as a physically active and productive factor. See *Oxford Latin Dictionary*, s.v. CAUSA.

[45] Liddell, Scott, and Jones quote line 35 of Pindar's first Olympian Ode as the earliest occurrence of αἰτία. The word there means 'fault', 'blame'. In Homer the adjective αἴτιος is used for saying that someone is at fault (οὔ τί μοι αἴτιοί εἰσιν, "they are in no way at fault in my regard"—*Il.*, 1.153). Aristotle uses αἰτία and the nominalization τὸ αἴτιον as equivalent, but—as Michael Frede 1980, p. 222, reports—the Stoic philosopher Chrysippus made the following distinction between these two expressions: An αἴτιον, a cause, is an entity, but an αἰτία is an account of the αἴτιον, or the account about the αἴτιον as an αἴτιον. The Hippocratic treatise *On Ancient Medicine*, probably written ca. 430 B.C., contains a passage which nicely contrasts αἰτία and τὸ αἴτιον (though not in the manner proposed by Chrysippus). The author speaks about relapses that may happen during convalescence, after the patient has done something out of the ordinary, e.g., taken a bath or a walk, or eaten something unusual. In such cases, many physicians, being ignorant of the factor actually responsible for the relapse (τὸ μὲν αἴτιον ἀγνοεῦντας), place the blame on the unusual action (τὴν αἰτίην τούτων τινὶ ἀνατιθέντας), even though it may in fact be beneficial (*De Vet. Med.*, 21).

[46] "Much unjustified criticism of Aristotle's doctrine would have been avoided if the word 'cause' had not been used in translations, but it has become traditional and no other single word does better. In reading what follows, therefore, remember that the four so-called 'causes' are *types of explanatory factor*" (Ackrill 1981, p. 36).

[47] Post-Aristotelian philosophers returned to the pre-Aristotelian notion of cause. Frede (1980, p. 219, n. 2) quotes the following passage by Clement of Alexandria (2nd century A.D.): φάμεν [. . .] τὸ αἴτιον ἐν τῷ ποιεῖν καὶ ἐνεργεῖν καὶ δρᾶν νοεῖσθαι ("we say . . . that the cause is understood [to lie] in the producing, the action, the doing"—*Strom.* I, 17.82.3). This agrees well with a statement in Plato's *Banquet:* ἡ γάρ τοι ἐκ τοῦ μὴ ὄντος εἰς τὸ ὄν ἰόντι ὁτῳοῦν αἰτία πᾶσά ἐστι ποίησις ("the cause that something goes from not being into being is in every case a making"—*Symp.* 205b).

[48] If the structure of reality, as it is manifested through the English language, were such that only events can operate as causes, we, who are not events but persons, would be hard put to do what we ought to do.

[49] The symmetric nature of the relation between GDE-states must not be confused with the time symmetry of the more important differential equations of physics. Time symmetry means simply that whenever $f(t)$ is a solution, $f(-t)$ is also a solution (t being the time variable). It holds for the equations of classical and quantum mechanics and electrodynamics and for the Einstein field equations of General Relativity but not, for instance, for the heat equation: $\partial u/\partial t = k(\partial^2 u/\partial x^2 + \partial^2 u/\partial y^2 + \partial^2 u/\partial z^2)$. The symmetry described in the text above holds, of course, for every GDE-system.

[50] This point was made by Mark Steiner 1986. In her recent proposal for explicating causal *propagation,* Cartwright (1989, p. 245) introduces "the convenient fiction that time is discrete," for the sake of avoiding "a more cumbersome notation." I contend that without this fiction Cartwright's explication would not preserve the semblance of causality. On the other hand, the use of differential equations in physics presupposes that time is continuous.

[51] A like view was put forward by T. S. Kuhn in his essay "Concepts of Cause in the Development of Physics" (1971). Kuhn relates nomological explanation by differential equations to Aristotle's explanation by a formal cause. He stresses, however, that "the resemblance to Aristotelian explanation displayed by explanations in [mechanics, electricity, etc.] is only structural" (1977, p. 27). In my opinion, even this guarded claim is excessive.

[52] This note is addressed to readers acquainted with the doctrine of causation put forward by Wesley Salmon in his book *Scientific Explanation and the Causal Structure of the World* (1984). I studied that book shortly after its publication, five years before writing the present chapter on necessity. On rereading it after the chapter was written, I noticed that the above duality of GDE-processes ("nomology") and cause-effect relations ("etiology") parallels Salmon's distinction between "causal processes" and "causal interactions." By a "causal process" Salmon means any distinct spatio-temporal process that is liable to interventions which mildly modify ("mark") its further development. E.g., the transmission of light from a searchlight to a wall is a causal process, for if you make the beam go through a piece of red glass, the segment between the glass and the wall gets reddened; but the quick displacement of the beam's projection on the wall as the searchlight moves is not a causal process, for even if the moving spot of light takes on a red color while it falls, say, on a red portion of the wall, it recovers its normal color as soon as it leaves that portion behind. I shall not question here these ideas, which originated with Reichenbach. If we give them the benefit of every doubt, we may grant (i) that a GDE-process stripped of its crisp mathematical representation would amount to a causal process in Salmon's sense, at any rate in those tame cases where a mild outside intervention on the process does but mildly alter its subsequent course (how Salmon would deal with GDE-governed chaos is, as far as I know, a moot question); and (ii) that causal processes in Salmon's sense are among the natural candidates for scientific representation

as GDE-processes. Whether the adjective 'causal' is a good one to use in this context is apparently a question of taste. I think that to do so merely adds confusion to an irking subject, but I acknowledge Salmon's right to think otherwise. "Causal interaction" is Salmon's substitute for the plain, ordinary, unabashedly anthropomorphic concept of causation that I favor. He conceives a causal interaction as the spatio-temporal intersection of two distinct "causal processes" after which each is "marked" by the other (cf. the definition in Salmon 1984, p. 171). This notion of mutual interference between distinct and independent natural processes is most congenial to Aristotelian science but I doubt that it will go well with modern mathematical physics. The GDE-processes to which the latter resorts for the representation of nature are self-contained, holistic affairs. Hence, if two such processes come across each other, they will fuse into one and ought presumably to be reconceived as constituting a single GDE-process from the outset—unless their coupling is quite weak, in which case each can still be handled on its own, with a small perturbation added but without properly paying heed to the other.

REFERENCES

Abraham, R., and J. E. Marsden (1978). *Foundations of Mechanics.* Reading, MA: Benjamin/Cummings.

Achinstein, P., and O. Hannaway, eds. (1985). *Observation, Experiment, and Hypothesis in Modern Physical Science.* Cambridge: MIT Press.

Ackermann, R. J. (1985). *Data, Instruments and Theory: A Dialectical Approach to Understanding Science.* Princeton: Princeton University Press.

Ackrill, J. L. (1981). *Aristotle the Philosopher.* Oxford: Oxford University Press.

Adams, E. W. (1955). "Axiomatic foundations of rigid body mechanics." Ph.D. dissertation. Stanford University.

Adams, E. W. (1959). "The foundations of rigid body mechanics and the derivation of its laws from those of particle mechanics." In Henkin et al., eds. (1959). *The Axiomatic Method with Special Reference to Geometry and Physics. Proceedings of an International Symposium held at the University of California, Berkeley, December 26, 1957–January 4, 1958.* Amsterdam: North-Holland. Pp. 250–265.

Adams, E. W. (1962). "On rational betting systems." *Archiv für mathematische Logik und Grundlagenforschung.* **6**: 7–29; 112–128.

Alexandri Aphrodisiensis *De Anima.* Edidit I. Bruns. Berlin: G. Reimer, 1887. (Commentaria in Aristotelem Græca.)

Anscombe, G. E. M. (1971). *Causality and Determination: An Inaugural Lecture.* Cambridge: Cambridge University Press.

Aquinas, T. (*S.c.Gent.*). *Summa contra Gentiles.* Torino: Marietti, 1946.

Aquinas, T. (*In Metaph.*). *In Duodecim Libros Metaphysicorum Aristotelis Expositio.* Cura et studio Raymundi M. Spiazzi. Torino: Marietti, 1950.

Aquinas, T. (*De Pot.*). "De Potentia." In Aquinas, *Quaestiones Disputatae.* Cura et studio Pauli M. Pession. Torino: Marietti, 1953. Vol. II, pp. 7–276.

Aquinas, T. (*S. Theol.*). *Summa Theologiae.* Cura Fratrum eiusdem Ordinis. Madrid: Biblioteca de Autores Cristianos, 1955. 5 vols.

Arbuthnot, J. (1710). "An argument for divine providence taken from the constant regularity observed in the births of both sexes." *Royal Society of London Philosophical Transactions.* **27**: 186–190.

Archimedes (1972). *Opera omnia cum commentariis Eutocii.* Editio stereotypa editionis anni MCMXIII. Iterum edidit Iohan Ludvig Heiberg and corrigenda adiecit Evangelos S. Stamatis. Stuttgart: B. G. Teubner. 3 vols.

Aristoteles. *Opera.* Ex recognitione I. Bekkeris edidit Academia Regia Borussica. Berlin: G. Reimer, 1831. 2 vols. (My quotations refer, as is customary, to the pages, columns, and lines of this edition; however, they are taken from various more recent editions and may not always agree with Bekker's text.)

Bachmann, F. (1951). "Zur Begründung der Geometrie aus dem Spiegelungsbegriff." *Mathematische Annalen.* **123**: 341–344.

Ball, W. E. (1982). "Programming languages and systems." In S. V. Pollack, ed., *Studies in Computer Science.* The Mathematical Association of America. Pp. 52–94.

Balzer, W. (1978). *Empirische Geometrie und Raum-Zeit-Theorie in mengentheoretischer Darstellung.* Kronberg/Ts.: Scriptor.

Balzer, W. (1985). "On a new definition of theoreticity." *Dialectica.* **39**: 127–145.

Balzer, W. (1986). "Theoretical terms: A new perspective." *Journal of Philosophy.* **83**: 71–90.

Balzer, W., and C. U. Moulines (1980). "On theoreticity." *Synthese.* **44**: 467–494.

Balzer, W., C. U. Moulines, and J. Sneed (1986). "The structure of empirical science: Local and global." In R. Barcan Marcus et al., eds., *Logic, Methodology and Philosophy of Science VII: Proceedings of the Seventh International Congress of Logic, Methodology and Philosophy of Science.* Amsterdam: North-Holland. Pp. 291–306.

Balzer, W., C. U. Moulines, and J. D. Sneed (1987). *An Architectonic for Science: The Structuralist Program.* Dordrecht: D. Reidel.

Balzer, W., D. A. Pearce, and H. J. Schmidt, eds. (1984). *Reduction in Science: Structure, Examples, Philosophical Problems.* Dordrecht: D. Reidel.

Balzer, W., and J. Sneed (1977/78). "Generalized net structures of empirical theories." *Studia Logica.* **36**: 195–211; **37**: 167–194.

Bartelborth, T. (1988). *Eine logische Rekonstruktion der klassischen Electrodynamik.* Frankfurt a.M.: Peter Lang.

Barwise, J., and J. Etchemendy (1986). *Turing's World.* Santa Barbara, CA: Kinko's Academic Courseware Exchange. (Computer program.)

Bayes, T. (1763). "An essay towards solving a problem in the doctrine of chances." *Royal Society of London Philosophical Transactions.* **53**: 370–418.

Bell, J. L. (1988). *Toposes and Local Set Theories: An Introduction.* Oxford: Clarendon Press.

Bell, J. L., and M. Machover (1977). *A Course in Mathematical Logic.* Amsterdam: North-Holland.

Benacerraf, P., and H. Putnam, eds. (1983). *Philosophy of Mathematics: Selected Readings.* Second edition. Cambridge: Cambridge University Press.

Bernoulli, J. (1713). *Ars conjectandi: Opus posthumum.* Basilea: Impensis Thurnisiorum, fratrum.

Bertrand, J. (1907). *Calcul des probabilités.* New York: Chelsea (n/d; reprint of the second edition, published in Paris).

Beth, E. W. (1948/49). "Analyse sémantique des théories physiques." *Synthese.* **7**: 206–207.

Boltzmann, L. (1884). "Über die Eigenschaften monocyklischer und anderer damit verwandter Systeme." *Sitzungsberichte der Akademie der Wissenschaften zu Wien, math.-naturwiss. Klasse.* **90**: 231–245.

Boltzmann, L. (1968). *Wissenschaftliche Abhandlungen.* Herausgegeben von F. Hasenöhrl. New York: Chelsea. 3 vols.

Bondi, H. (1961). *Cosmology.* Second edition. Cambridge: Cambridge University Press.

Boole, G. (1854). *An Investigation of the Laws of Thought on Which Are Founded the Mathematical Theories of Logic and Probabilities.* London: Macmillan.

Bourbaki, N. (1970). *Théorie des ensembles.* Paris: Hermann.

Boyle, R. (WW). *Works.* London, 1744.

Braithwaite, R. B. (1953). *Scientific Explanation: A Study of the Function of Theory, Probability and Law in Science, Based upon the Tarner Lectures, 1946.* Cambridge: Cambridge University Press.

Braithwaite, R. B. (1957). "On unknown probabilities." In Körner 1957, pp. 3–11.

Brauer, F., and J. A. Nobel (1969). *The Qualitative Theory of Ordinary Differential Equations.* New York: W. A. Benjamin.

Brody, B. A., ed. (1970). *Readings in the Philosophy of Science.* Englewood Cliffs: Prentice-Hall.

Bromberger, S. (1966). "Why-questions." In Brody 1970, pp. 66–87 (originally published in Colodny 1966).

Brouwer, L. E. J. (1913). "Intuitionism and formalism." *Bulletin of the American Mathematical Society.* **20**: 81–96.

Brown, H. I. (1987). *Observation and Objectivity.* New York: Oxford University Press.

Brush, S. G. (1976). *The Kind of Motion We Call Heat: A History of the Kinetic Theory of Gases in the 19th Century.* Amsterdam: North-Holland.

Bunge, M. (1967). *Foundations of Physics.* Berlin: Springer.

Bunge, M. (1979). *Causality and Modern Science.* Third revised edition. New York: Dover. (First edition published as *Causality.* Cambridge: Harvard University Press, 1959.)

Bunge, M., et al. (1971). *Les théories de la causalité.* Paris: Presses Universitaires de France.

Butts, R. E., and J. Hintikka, eds. (1977). *Basic Problems in Methodology and Linguistics.* Dordrecht: D. Reidel.

Campbell, N. R. (1920). *Physics: The Elements.* Cambridge: Cambridge University Press. (Republished as *Foundations of Science: The Philosophy of Theory and Experiment.* New York: Dover, 1957.)

Carnap, R. (1928). *Der logische Aufbau der Welt.* Berlin.

Carnap, R. (1932). "Die physikalische Sprache als Universalsprache der Wissenschaft." *Erkenntnis.* **2**.

Carnap, R. (1936/37). "Testability and meaning." *Philosophy of Science.* **3**: 419–471; **4**: 1–40.

Carnap, R. (1937). *The Logical Syntax of Language.* Translated by A. Smeaton. London: Routledge & Kegan Paul.

Carnap, R. (1939). "Foundations of logic and mathematics." In Neurath, Carnap, and Morris 1971, pp. 139–212.

Carnap, R. (1950). *The Logical Foundations of Probability.* Chicago: University of Chicago Press.

Carnap, R. (1956). "The methodological character of theoretical concepts." *Minnesota Studies in the Philosophy of Science.* **1**: 38–76.

Carnap, R. (1958). *Introduction to Symbolic Logic and Its Applications.* New York: Dover.

Carnap, R. (1961). *Der logische Aufbau der Welt/Scheinprobleme der Philosophie.* Second edition. Hamburg: Felix Meiner.

Carnap, R. (1962). *The Logical Foundations of Probability.* Second edition. Chicago: University of Chicago Press.

Carnap, R. (1966). *Philosophical Foundations of Physics: An Introduction to the Philosophy of Science.* Edited by M. Gardner. New York: Basic Books.

Carnap, R., and R. C. Jeffrey, eds. (1971). *Studies in Inductive Logic and Probability.* Vol. I. Berkeley: University of California Press.

Carroll, L. (AA). *The Annotated Alice: Alice's Adventures in Wonderland & Through the Looking Glass.* With an Introduction and Notes by M. Gardner. New York: Bramhall House, 1960.

Cartan, E. (1923). "Sur les variétés à connexion affine et la théorie de la relativité généralisée (première partie)." *Annales de l'École Normale Supérieure.* **40**: 325–412.

Cartan, H. (1967). *Calcul différentiel.* Paris: Hermann.

Cartwright, N. (1983). *How the Laws of Physics Lie.* Oxford: Clarendon Press.

Cartwright, N. (1989). *Nature's Capacities and Their Measurement.* Oxford: Clarendon Press.

Church, A. (1940). "On the concept of a random sequence." *Bulletin of the American Mathematical Society.* **46**: 130–135.

Churchland, P. M. (1979). *Scientific Realism and the Plasticity of Mind.* Cambridge: Cambridge University Press.

Churchland, P. M. (1988). "Perceptual plasticity and theoretical neutrality: A reply to Jerry Fodor." *Philosophy of Science.* **55**: 167–187.

Clagett, M. (1959). *The Science of Mechanics in the Middle Ages.* Madison: University of Wisconsin Press.

Clagett, M. (1968). *Nicole Oresme and the Medieval Geometry of Qualities and Motions: A Treatise on the Uniformity and Difformity of Intensities Known as Tractatus de configurationibus qualitatum et motuum.* Edited with an introduction, English translation, and commentary. Madison: University of Wisconsin Press.

Clavelin, M. (1968). *La philosophie naturelle de Galilée.* Paris: Vrin.

Clavelin, M. (1983). "Conceptual and technical aspects of the Galilean geometrization of the motion of heavy bodies." In W. R. Shea, ed., *Nature Mathematized.* Dordrecht: D. Reidel. Pp. 23–50.

Cohen, I. B. (1967). "Newton's attribution of the first two laws of motion to Galileo." In *Atti del Symposium Internazionale di Storia, Metodologia, Logica e Filosofia della Scienza "Galileo nella Storia e nella Filosofia della Scienza."* Pp. xxv–xliv.

Cohen, I. B. (1980). *The Newtonian Revolution with Illustrations of the Transformation of Scientific Ideas.* Cambridge: Cambridge University Press.

Cohen, P. J. (1963/64). "The independence of the continuum hypothesis." *Proceedings of the National Academy of the U.S.A.* **50**: 1143–1148; **51**:105–110.

Colodny, R. G., ed. (1965). *Beyond the Edge of Certainty.* Englewood Cliffs: Prentice-Hall.

Colodny, R. G., ed. (1966). *Mind and Cosmos.* Pittsburgh: Pittsburgh University Press.

Colodny, R. G., ed. (1970). *The Nature and Function of Scientific Theories: Essays in Contemporary Science and Philosophy.* Pittsburgh: Pittsburgh University Press.

Colodny, R. G., ed. (1972). *Paradigms and Paradoxes: The Philosophical Challenge of the Quantum Domain.* Pittsburgh: Pittsburgh University Press.

Colodny, R. G., ed. (1977). *Logic, Laws, & Life: Some Philosophical Complications.* Pittsburgh: Pittsburgh University Press.

Colodny, R. G., ed. (1986). *From Quarks to Quasars: Philosophical Problems of Modern Physics.* Pittsburgh: Pittsburgh University Press.

Costa de Beauregard, O. (1989). *Time, the Physical Magnitude.* Dordrecht: D. Reidel.

Courant, R., and D. Hilbert (MMP). *Methods of Mathematical Physics.* Vol. II. *Partial Differential Equations.* By R. Courant. New York: Interscience.

Cournot, A. A. (1843). *Exposition de la théorie des chances et des probabilités.* Paris: Hachette.

Cramér, H. (1946). *Mathematical Methods of Statistics.* Princeton: Princeton University Press.

Danto, A., and S. Morgenbesser, eds. (1960). *Philosophy of Science.* Cleveland: World Publishing Co.

David, F. N. (1962). *Games, Gods and Gambling: The Origins and History of Probability and Statistical Ideas from the Earliest Times to the Newtonian Era.* London: Charles Griffin.

Davidson, D. (1974). "On the very idea of a conceptual scheme." *Proceedings and Addresses of the American Philosophical Society.* **47**: 5–20.

Davidson, D. (1984). *Inquiries into Truth and Interpretation.* Oxford: Clarendon Press.

Davis, M. (1958). *Computability and Unsolvability.* New York: McGraw-Hill.

Dawid, A. P. (1985). "Probability, symmetry and frequency." *British Journal for the Philosophy of Science.* **36**: 107–128.

Descartes, R. (AT). *Œuvres.* Edited by C. Adam and P. Tannery. Paris: Cerf, 1897–1912. 12 vols.

Dewey, J. (1938). *Logic: The Theory of Inquiry.* New York: Holt.

Diederich, W., A. Ibarra, and T. Mormann (1989). "Bibliography of structuralism." *Erkenntnis.* **30**: 387–407.

Diels, H., and W. Kranz (DK). *Die Fragmente der Vorsokratiker.* Siebente Auflage. Berlin: Weidmannsche Verlagsbuchhandlung, 1954. 3 vols.

Dijksterhuis, E. J. (1987). *Archimedes.* Translated by C. Dikshoorn. Princeton: Princeton University Press.

DiSalle, R. (1988). "Space, time, and inertia in the foundations of Newtonian physics, 1870–1905." Ph.D. dissertation. University of Chicago.

Dixon, W. G. (1978). *Special Relativity: The Foundations of Macroscopic Physics.* Cambridge: Cambridge University Press.

Donnellan, K. S. (1966). "Reference and definite descriptions." *Philosophical Review.* **75**: 281–304.

Drake, S. (1975). "The role of music in Galileo's experiments." *Scientific American.* **232** 6: 98–104.

Drake, S. (1978). *Galileo at Work: His Scientific Biography.* Chicago: University of Chicago Press.

Dreyer, J. L. E. (1953). *History of Astronomy from Thales to Kepler.* New York: Dover. (Originally published as *History of the Planetary Systems from Thales to Kepler.* Cambridge: Cambridge University Press, 1901.)

Droste, J. (1916). "The field of a single centre in Einstein's theory of gravitation, and the motion of a particle in that field." *K. Nederlandse Akademie van Wetenschappen, Proceedings.* **19**: 197–215.

Ducasse, C. J. (1924). *Causation and the Types of Necessity.* Seattle: University of Washington Press.

Ducasse, C. J. (1969). *Causation and the Types of Necessity.* New York: Dover. (Originally published by the University of Washington Press in 1924. This reprint includes four articles on causality published by Ducasse in 1930, 1957, 1961, and 1965.)

Duhem, P. (1914). *La théorie physique: son objet – sa structure.* Deuxième édition revue et augmentée. Paris: Marcel Rivière.

Earman, J. (1986). *A Primer on Determinism.* Dordrecht: D. Reidel.

Edwards, P., ed. (1967). *The Encyclopedia of Philosophy.* New York: Macmillan. 8 vols.

Ehrenfest, P., and T. Ehrenfest (1912). "Begriffliche Grundlage der statistischen Auffassung in der Mechanik." In *Enzyklopädie der mathematischen Wissenschaften.* Leipzig: B. G. Teubner. Vol. IV 2, II, Heft 6.

Eilenberg, S., and S. Mac Lane (1945). "General theory of natural equivalences." *American Mathematical Society Transactions.* **58**: 231–294.

Einstein, A. (1905a). "Über eine die Erzeugung und Verwandlung des Lichts betreffenden heuristischen Gesichtspunk." *Annalen der Physik.* (4) **17**: 132–148.

Einstein, A. (1905b). "Zur Elektrodynamik bewegter Körper." *Annalen der Physik.* (4) **17**: 891–921.

Einstein, A. (1907). "Plancksche Theorie der Strahlung und die Theorie der spezifischen Wärme." *Annalen der Physik.* (4) **22**: 180–190. (Correction on p. 800.)

Einstein, A. (1915). "Erklärung der Perihelbewegung des Merkur aus der allgemeinen Relativitätstheorie." *K. Preußische Akademie der Wissenschaften Sitzungsberichte.* Pp. 831–839.

Einstein, A. (1934). *Mein Weltbild.* Zweite Auflage. Amsterdam: Querido Verlag.

Einstein, A. (1949). "Autobiographisches." In P. A. Schilpp, ed., *Albert Einstein Philosopher-Scientist*. Evanston: Open Court. Vol. I, pp. 2–94.

Ellis, B. (1963/64). "Universal and differential forces." *British Journal for the Philosophy of Science*. **14**: 177–194.

Ellis, R. L. (1849). "On the foundations of the theory of probabilities." *Transactions of the Cambridge Philosophical Society*. **8**: 1–6.

Ellis, R. L. (1856). "Remarks on the fundamental principle of the theory of probabilities." *Transactions of the Cambridge Philosophical Society*. **9**: 605–607.

Enderton, H. B. (1977). "Elements of recursion theory." In J. Barwise, ed., *Handbook of Mathematical Logic*. Amsterdam: North-Holland. Pp. 527–566.

Epictetus. *The Discourses as Reported by Arrian, the Manual, and the Fragments*. With an English translation by W. A. Oldfather. London: Heinemann, 1956. 2 vols. (Loeb Classical Library.)

Feller, W. (1971). *An Introduction to Probability Theory and Its Applications*. Vol. II. Second edition. New York: Wiley.

Fermi, E. (1934). "Versuch einer Theorie der β-Strahlen." *Zeitschrift für Physik*. **88**: 161–171.

Feyerabend, P. K. (1958). "An attempt at a realistic interpretation of experience." *Proceedings of the Aristotelean Society*. **58**: 143ff. (Reprinted in Feyerabend 1981, vol. I, pp. 17–36.)

Feyerabend, P. K. (1960). "Das Problem der Existenz theoretischer Entitäten." In E. Töpitsch, ed., *Probleme der Wissenschaftstheorie: Festschrift für Viktor Kraft*. Wien. Pp. 35–72.

Feyerabend, P. K. (1962). "Explanation, reduction and empiricism." *Minnesota Studies in the Philosophy of Science*. **3**: 28–97. (Reprinted in Feyerabend 1981, vol. I, pp. 44–96.)

Feyerabend, P. K. (1965). "Problems of empiricism." In Colodny 1965, pp. 145–260.

Feyerabend, P. K. (1981). *Philosophical Papers*. Cambridge: Cambridge University Press. 2 vols.

Feynman, R. P., and M. Gell-Mann (1958). "Theory of the Fermi interaction." *Physical Review*. **109**: 193–198.

Feynman, R. P., R. B. Leighton, and M. Sands (1963). *The Feynman Lectures on Physics*. Vol. I. Reading, MA: Addison-Wesley.

Field, H. (1973). "Theory change and the indeterminacy of reference." *Journal of Philosophy*. **1973**: 462–481.

Fine, T. L. (1973). *Theories of Probability*. New York: Academic Press.

Finetti, B. de (1930). "Funzione caratteristica di un fenomeno aleatorio." *Memorie della Reale Accademia dei Lincei*. **IV** 5: 86–133.

Finetti, B. de (1931). "Sul significato soggettivo della probabilità." *Fundamenta Mathematicae*. **17**: 298–329.

Finetti, B. de (1937). "La prévision: Ses lois logiques, ses sources subjectives." *Annales de l'Institut Henri Poincaré*. **7**: 1–68. (Translated as "Foresight: Its logical laws, its subjective sources" in Kyburg and Smokler 1980, pp. 53–118.)

Finetti, B. de (1970). *Teoria della probabilità*. Torino: Einaudi.

Finetti, B. de (1972). *Probability, Statistics and Truth: The Art of Guessing*. New York: Wiley.

Finetti, B. de (1974). *Theory of Probability: A Critical Introductory Treatment*. Translated by A. Machi and A. Smith. New York: Wiley. 2 vols.

Fink, E. (1960). *Spiel als Weltsymbol*. Stuttgart: Kohlhammer.

Fodor, J. (1984). "Observation reconsidered." *Philosophy of Science*. **51**: 23–43.

Fodor, J. A. (1988). "A reply to Churchland's 'Perceptual plasticity and theoretical neutrality'." *Philosophy of Science*. **55**: 188–198.

Fotinis, A. P. (1979). *The De Anima of Alexander of Aphrodisias: A Translation and Commentary.* Washington, DC: University Press of America.

Fraassen, B. C. van (1970). "On the extension of Beth's semantics to physical theories." *Philosophy of Science.* **37**: 325–339.

Fraassen, B. C. van (1972). "A formal approach to the philosophy of science." In Colodny 1972, pp. 303–366.

Fraassen, B. C. van (1977). "Relative frequencies." *Synthese.* **34**: 133–166.

Fraassen, B. C. van (1979). "Foundations of probability: A modal frequency interpretation." In G. Toraldo di Francia, ed., *Problemi dei fondamenti della fisica.* Bologna: Società Italiana di Fisica. Pp. 344–394. (Rendiconti della Scuola Internazionale di Fisica Enrico Fermi, LXXII corso.)

Fraassen, B. C. van (1980). *The Scientific Image.* Oxford: Oxford University Press.

Fraenkel, A. (1922). "Axiomatische Begründung der transfiniten Kardinalzahlen, I." *Mathematische Zeitschrift.* **13**: 153–188.

Frank, P. (1946). "Foundations of Physics." In Neurath, Carnap, and Morris 1971, pp. 423–565.

Frede, M. (1980). "The original notion of cause." In M. Schofield, M. Burnyeat, and J. Barnes, eds. *Doubt and Dogmatism: Studies in Hellenistic Epistemology.* Oxford: Clarendon Press. Pp. 217–241.

Frege, G. (NS). *Nachgelassene Schriften.* Hamburg: Felix Meiner.

Friedman, M. (1983). *Foundations of Space-Time Theories: Relativistic Physics and Philosophy of Science.* Princeton: Princeton University Press.

Friedrichs, K. (1927). "Eine invariante Formulierung des Newtonschen Gravitationsgesetzes und des Grenzüberganges vom Einsteinschen zum Newtonschen Gesetz." *Mathematische Annalen.* **98**: 566–575.

Fries, J. F. (1842). *Versuch einer Kritik der Principien der Wahrscheinlichkeitsrechnung.* Braunschweig: Vieweg.

Furley, D. (1987). *The Greek Cosmologists.* Vol. I. *The Formation of the Atomic Theory and Its Earliest Critics.* Cambridge: Cambridge University Press.

Gähde, U. (1983). *T-Theorizität und Holismus.* Frankfurt a. M./Bern: Peter Lang.

Galileo Galilei (EN). *Le Opere.* Nuova ristampa della Edizione Nazionale. Firenze: G. Barbera, 1964–1966. 20 vols.

Galileo Galilei (1638). *Discorsi e dimostrazioni matematiche intorno à due nuove scienze attenenti alla mecanica & i movimenti locali.* In Leida, appresso gli Elsevirii.

Galileo Galilei (1974). *Two New Sciences, Including Centers of Gravity and Force of Percussion.* Translated by S. Drake. Madison: University of Wisconsin Press.

Galison, P. (1987). *How Experiments End.* Chicago: University of Chicago Press.

Garber, D. (1983). "Old evidence and logical omniscience in Bayesian confirmation theory." *Minnesota Studies in the Philosophy of Science.* **10:** 99–131.

Garber, D., and S. Zabell (1979). "On the emergence of probability." *Archive for History of Exact Sciences.* **21**: 33–53.

Gasking, D. (1955). "Causation and recipes." *Mind.* **64**: 479–487.

Gibbs, J. W. (1902). *Elementary Principles in Statistical Mechanics Developed with Especial Reference to the Rational Foundation of Thermodynamics.* New York: Scribner.

Giere, R. N. (1973). "Objective single-case probabilities and the foundations of statistics." In P. Suppes et al., eds., *Logic, Methodology and Philosophy of Science IV.* Amsterdam: North-Holland. Pp. 467–483.

Giere, R. N. (1976a). "A Laplacean formal semantics for single case probabilities." *Journal of Philosophical Logic.* **5**: 321–353.

Giere, R. N. (1976b). "Empirical probability, objective statistical methods and scientific inquiry." In W. Harper and C. A. Hooker, eds., *Foundations of Probability Theory, Statistical Inference and Statistical Theories in Science.* Dordrecht: D. Reidel. Vol. 2, pp. 63–101.

Giere, R. N. (1988). *Explaining Science: A Cognitive Approach.* Chicago: University of Chicago Press.

Gillies, D. A. (1973). *An Objective Theory of Probability.* London: Methuen.

Glymour, C. (1977). "Indistinguishable space-times and the fundamental group." *Minnesota Studies in the Philosophy of Science.* **8**: 50–60.

Glymour, C. (1982). "Conceptual scheming, or, Confessions of a metaphysical realist." *Synthese.* **51**: 169–180.

Gödel, K. (1938). "The consistency of the axiom of choice and of the generalized continuum hypothesis." *Proceedings of the National Academy of Sciences of the U.S.A.* **24**: 556–557.

Gödel, K. (1940). *The Consistency of the Axiom of Choice and of the Generalized Continuum Hypothesis with the Axioms of Set Theory.* Princeton: Princeton University Press.

Goodman, N. (1951). *The Structure of Appearance.* Cambridge: Harvard University Press.

Goodman, N. (1966). *The Structure of Appearance.* Second edition. Indianapolis: Bobbs-Merrill.

Graunt, J. (1662). *Natural and Political Observations Mentioned in a Following Index, and Made upon the Bills of Mortality . . . with Reference to the Government, Religion, Trade, Growth, Ayre, Diseases, and the Several Changes of the Said City . . .* . London: J. Martin, J. Allestry & T. Dicas.

Graunt, J. (1676). *Natural and Political Observations Mentioned in a Following Index, and Made upon the Bills of Mortality . . . with Reference to the Government, Religion, Trade, Growth, Air, Diseases, and the Several Changes of the Said City.* Fifth edition, much enlarged. London: John Martyn.

Gregory, R. L., ed. (1987). *The Oxford Companion to the Mind.* Oxford: Oxford University Press.

Hacker, P. M. S. (1987). *Appearance and Reality: A Philosophical Investigation into Perception and Perceptual Qualities.* Oxford: Basil Blackwell.

Hacking, I. (1965). *Logic of Statistical Inference.* Cambridge: Cambridge University Press.

Hacking, I. (1975). *The Emergence of Probability: A Philosophical Study of Early Ideas about Probability, Induction and Statistical Inference.* Cambridge: Cambridge University Press.

Hacking, I. (1983). *Representing and Intervening: Topics in the Philosophy of Natural Science.* Cambridge: Cambridge University Press.

Hacking, I. (1988). "On the stability of the laboratory sciences." *Journal of Philosophy.* **85**: 507–514.

Hall, A. R., and M. B. Hall, eds. (1978). *Unpublished Scientific Papers of Isaac Newton: A Selection from the Portsmouth Collection in the University Library, Cambridge.* Cambridge: Cambridge University Press. (First published 1962.)

Hanson, N. R. (1958). *Patterns of Discovery.* Cambridge: Cambridge University Press.

Hardin, C. L. (1988). *Color for Philosophers: Unweaving the Rainbow.* Indianapolis: Hackett.

Harré, R. (1970). *The Principles of Scientific Thinking.* London: Macmillan.

Harrison, E. R. (1965). "Olbers' paradox and the background radiation density in an isotropic homogeneous universe." *Monthly Notices of the Royal Astronomical Society.* **131**: 1–12.

Harrison, E. R. (1977). "The dark night sky paradox." *American Journal of Physics.* **45**: 119–124.

Hartkämper, A., and H. J. Schmidt, eds. (1981). *Structure and Approximation in Physical Theories.* Proceedings of a Colloquium on Structure and Approximation in Physical

Theories, held at Osnabrück, FRG, in June 1980. New York: Plenum Press.

Hasert, F. J., et al. (1973a). "Search for elastic muon-neutrino electron scattering." *Physics Letters.* **46B**: 121–124.

Hasert, F. J., et al. (1973b). "Observation of neutrino-like interactions without muon or electron in the Gargamelle neutrino experiment." *Physics Letters.* **46B**: 138–140.

Hasert, F. J., et al. (1974). "Observation of neutrino-like interactions without muon or electron in the Gargamelle neutrino experiment." *Nuclear Physics.* **B73**: 1–22.

Havas, P. (1964). "Four-dimensional formulations of Newtonian mechanics and their relation to the special and the general theory of relativity." *Reviews of Modern Physics.* **36**: 938–965.

Havas, P. (1987). "Simultaneity, conventionalism, general covariance, and the special theory of relativity." *General Relativity and Gravitation.* **19**: 435–453.

Heath, T. L. (1953). *The Works of Archimedes Edited in Modern Notation with Introductory Chapters.* With a supplement, "The Method of Archimedes." New York: Dover. (Reprint of the book published by the Cambridge University Press in 1912; first published in 1897, without the supplement.)

Heidegger, M. (SZ). *Sein und Zeit.* Siebente unveränderte Auflage. Tübingen: Max Niemeyer.

Heijenoort, J. van (1967). *From Frege to Gödel: A Source Book in Mathematical Logic, 1879–1931.* Cambridge: Harvard University Press.

Heilig, K. (1978). "Carnap and de Finetti on bets and the probability of singular events: The Dutch Book argument reconsidered." *British Journal for the Philosophy of Science.* **29**: 325–346.

Helmholtz, H. von (1866). "Über die tatsächlichen Grundlagen der Geometrie." *Verhandlungen des naturhistorisch-medicinischen Vereins zu Heidelberg.* **4**: 197–202.

Helmholtz, H. von (1868). "Über die Tatsachen, die der Geometrie zum Grunde liegen." *Nachrichten von der Kgl. Gesellschaft der Wissenschaften und der Georg-Augusts-Universität aus dem Jahre 1868.* Nr. 9: 193–221.

Helmholtz, H. von (1887). "Zählen und Messen, erkenntnistheoretisch betrachtet." In *Philosophische Aufsätze Eduard Zeller gewidmet.* Leipzig: Verlag Fues. (Reprinted in H. von Helmholtz, *Die Tatsachen in der Wahrnehmung/Zählen und Messen.* Darmstadt: Wissenschaftliche Buchgesellschaft, 1959. Pp. 77–112.)

Hempel, C. G. (1965). *Aspects of Scientific Explanation and Other Essays in the Philosophy of Science.* New York: The Free Press.

Hempel, C. G., and P. Oppenheim (1948). "Studies in the logic of explanation." *Philosophy of Science.* **15**: 135–175.

Herodotus. With an English translation by A. D. Godley. Cambridge: Harvard University Press, 1920–1925. 4 vols. (Loeb Classical Library.)

Hesse, M. B. (1970). "Is there an independent observation language?" In Colodny 1970, pp. 35–77. (Reprinted as "Theory and observation" in Hesse 1974, pp. 9–44, and in Hesse 1980, pp. 63–110.)

Hesse, M. B. (1974). *The Structure of Scientific Inference.* Berkeley: University of California Press.

Hesse, M. B. (1980). *Revolutions and Reconstructions in the Philosophy of Science.* Bloomington: Indiana University Press.

Hilbert, D. (1899). "Die Grundlagen der Geometrie." In *Festschrift zur Feier der Enthüllung des Gauss-Weber Denkmals.* Leipzig. B. G. Teubner. Pp. 3–92.

Hintikka, J. (1957). "Necessity, universality and time in Aristotle." *Ajatus.* **20**: 65–90. (Reprinted in J. Barnes et al., eds., *Articles on Aristotle.* London: Routledge, 1979. Vol. 3, pp.

108–124.)

Hintikka, J. (1960). "Aristotle's different possibilities." *Inquiry.* **3**: 17–28.

Hintikka, J. (1973). *Time and Necessity: Studies in Aristotle's Theory of Modality.* Oxford: Clarendon Press.

Hippocrates. Vol. I. With an English translation by W. S. J. Jones. Cambridge: Harvard University Press, 1923. (Loeb Classical Library.)

Hölder, O. (1901). "Die Axiome der Quantität und die Lehre vom Mass." *Verhandlungen der K. Sächsischen Gesellschaft der Wissenschaften zu Leipzig, Math.-Phys. Classe.* **53**: 1–64.

Hopf, L. (1948). *Introduction to the Differential Equations of Physics.* Translated by W. Nef. New York: Dover.

Hume, D. (THN). *A Treatise of Human Nature.* Edited by L. A. Selby-Bigge. Oxford: Clarendon Press, 1888.

Huntington, E. V. (1905). "The continuum as a type of order: An exposition of the modern theory; with an appendix on the transfinite numbers." *Annals of Mathematics.* **6**:151–184; **7**: 15–43.

Huntington, E. V. (1912). "A set of postulates for abstract geometry, expressed in terms of the simple relation of inclusion." *Mathematische Annalen.* **73**: 522–559.

Huntington, E. V. (1955). *The Continuum and Other Types of Serial Order.* New York: Dover. (This is a reprint of a book published in 1917 by Harvard University Press, which in turn was a reprint of Huntington 1905.)

Huygens, C. (1657). "De ratiociniis in aleae ludo." In *Francisci a Schooten Exercitationum Mathematicarum libri quinque... Quibus accedit C. Hugenii tractatus de ratiociniis in aleae ludo.* Lugd. Batav.: J. Elsevir.

Jackson, F. (1977). *Perception: A Representative Theory.* Cambridge: Cambridge University Press.

Jeffrey, R. C. (1971). "Probability measures and integrals." In Carnap and Jeffrey 1971, vol. I, pp. 169–223.

Jeffrey, R. C. (1977). "Mises redux." In Butts and Hintikka 1977, pp. 213–222.

Kamke, E. (1932). *Einführung in die Wahrscheinlichkeitstheorie.* Leipzig: S. Hirzel.

Kamlah, A. (1981). "G. Ludwig's positivistic reconstruction of theoretical concepts." In Hartkämper and Schmidt 1981, pp. 71–90.

Kant, I. (Ak.). *Gesammelte Schriften.* Herausgegeben von der K. Preußischen, bzw. Deutschen Akademie der Wissenschaften. Berlin, 1902ff.

Kant, I. (1781). *Critik der reinen Vernunft.* Riga: Johann Friedrich Hartknoch.

Kant, I. (1787). *Critik der reinen Vernunft.* Zweyte hin und wieder verbesserte Auflage. Riga: Johann Friedrich Hartknoch.

Kant, I. (CPR). *Critique of Pure Reason.* Translated by N. Kemp Smith. London: Macmillan, 1958.

Kemeny, J. G., and P. Oppenheim (1956). "On reduction." *Philosophical Studies.* **7**: 6–19. (Reprinted in Brody 1970, pp. 307–318.)

Kendall, M. G. (1956). "The beginnings of a probability calculus." *Biometrika.* **43**: 1–14.

Keynes, J. M. (1921). *A Treatise on Probability.* London: Macmillan.

Koch, G., and F. Spizzichino, eds. (1982). *Exchangeability in Probability and Statistics.* Proceedings of the International Conference on Exchangeability in Probability and Statistics, Rome, 6–9 April 1981, in honor of Professor Bruno de Finetti. Amsterdam: North-Holland.

Kolmogoroff, A. N. (1933). *Grundbegriffe der Wahrscheinlichkeitsrechnung.* Berlin: Springer.

König, J. (1905). "Über die Grundlagen der Mengenlehre und das Kontinuumsproblem."

Mathematische Annalen. **61**: 156–160.

Körner, S., ed. (1957). *Observation and Interpretation: A Symposium of Philosophers and Physicists.* Proceedings of the Ninth Symposium of the Colston Research Society held at the University of Bristol, 2–4 April, 1957. London: Butterworth.

Krajewski, W. (1977). *Correspondence Principle and Growth of Science.* Dordrecht: D. Reidel.

Krantz, D. H., R. D. Luce, P. Suppes, and A. Tversky (1971). *Foundations of Measurement.* Vol. I. *Additive and Polynomial Representations.* New York: Academic Press.

Kripke, S. A. (1972). "Naming and necessity." In D. Davidson and G. Harman, eds., *Semantics of Natural Languages.* Dordrecht: D. Reidel. Pp. 253–355.

Kripke, S. A. (1980). *Naming and Necessity.* Oxford: Basil Blackwell. (This is a revised edition of Kripke 1972.)

Kuhn, T. S. (1962). *The Structure of Scientific Revolutions.* Chicago: University of Chicago Press.

Kuhn, T. S. (1964). "A function for thought experiments." In *L'aventure de la science: Mélanges Alexandre Koyré.* Paris: Hermann. Pp. 307–334. (Reprinted in Kuhn 1977, pp. 240–265.)

Kuhn, T. S. (1970). *The Structure of Scientific Revolutions.* Second edition. Chicago: University of Chicago Press.

Kuhn, T. S. (1971). "Concepts of cause in the development of physics." In Kuhn 1977, pp. 21–30. (Originally published as "Les notions de causalité dans le developpement de la physique" in Bunge et al. 1971, pp. 8–18.)

Kuhn, T. S. (1977). *The Essential Tension: Selected Studies in Scientific Tradition and Change.* Chicago: University of Chicago Press.

Kuhn, T. S. (1978). *Black-Body Theory and the Quantum Discontinuity, 1894–1912.* Oxford: Clarendon Press.

Kuratowski, K. (1921). "Sur la notion d'ordre dans la théorie des ensembles." *Fundamenta Mathematicae.* **2**: 161–171.

Kvart, I. (1986). *A Theory of Counterfactuals.* Indianapolis: Hackett.

Kyburg H. E., Jr., and H. E. Smokler, eds. (1980). *Studies in Subjective Probability.* Huntington, NY: Krieger.

Lakoff, G. (1987). *Women, Fire, and Dangerous Things: What Categories Reveal about the Mind.* Chicago: University of Chicago Press.

Lambert, J. H. (1786). "Theorie der Parallellinien." *Magazin für die reine und angewandte Mathematik.* Nr. 2: 134–164; Nr. 3: 325–358.

Landau, L. D., and E. M. Lifshitz (1960). *Course of Theoretical Physics.* Vol. I. *Mechanics.* Translated from the Russian by J. B. Sykes and J. S. Bell. Oxford: Pergamon Press.

Lange, L. (1885). "Über das Beharrungsgesetz." *K. Sächsische Gesellschaft der Wissenschaften zu Leipzig. Mathematisch-Physische Classe. Berichte über die Verhandlungen.* **37**: 333–351.

Laplace, P. S. de (OC). *Œuvres complètes.* Paris: Gauthier-Villars, 1878–1912. 14 vols.

Laue, M. von (1911). *Das Relativitätsprinzip.* Braunschweig: Vieweg.

Laue, M. von (1955). *Die Relativitätstheorie.* 1. Band. *Die spezielle Relativitätstheorie.* 6. durchgesehene Auflage. Braunschweig: Vieweg.

Leibniz, G. W. (SS). *Sämtliche Schriften und Briefe.* Herausgegeben von der Preußischen, bzw. Deutschen Akademie der Wissenschaften zu Berlin. Darmstadt/Leipzig/Berlin, 1923ff.

Leibniz, G. W. (GP). *Die philosophischen Schriften.* Herausgegeben von C. J. Gerhardt. Hildesheim: Olms, 1965. 7 vols.

Levi, I. (1967). *Gambling with Truth: An Essay on Induction and the Aims of Science.* New York: Knopf.

Levi, I. (1977a). "Subjunctives, dispositions and chances." *Synthese.* **34**: 423–455.

Levi, I. (1977b). "Direct inference." *Journal of Philosophy.* **74**: 5–29.

Levi, I. (1980). *The Enterprise of Knowledge: An Essay on Knowledge, Credal Probability and Chance.* Cambridge: MIT Press.

Levi, I. (1984). *Decisions and Revisions: Philosophical Essays on Knowledge and Value.* Cambridge: Cambridge University Press.

Levi, I., and S. Morgenbesser (1964). "Belief and disposition." *American Philosophical Quarterly.* **1**: 221–232.

Lewis, C. I. (1946). *An Analysis of Knowledge and Valuation.* La Salle: Open Court.

Lewis, D. (1984). "Putnam's paradox." *Australasian Journal of Philosophy.* **62**: 221–236.

Lewis, D. (1986). *On the Plurality of Worlds.* Oxford: Basil Blackwell.

Lindberg, D. C. (1970). *John Pecham and the Science of Optics: Perspectiva Communis.* Edited with an introduction, English translation, and critical notes. Madison: University of Wisconsin Press.

Lindley, D. V. (1965). *Introduction to Probability and Statistics from a Bayesian Viewpoint.* Cambridge: Cambridge University Press. 2 vols.

Locke, J. (*Essay*). *An Essay concerning Human Understanding.* Collated and annotated by A. C. Fraser. New York: Dover, 1959. 2 vols.

Loève, M. (1978). *Probability Theory.* Fourth edition. New York: Springer. 2 vols.

Long, A. A., and D. N. Sedley (1987). *The Hellenistic Philosophers.* Vol. 1. *Translations of the Principal Sources with Philosophical Commentary.* Vol. 2. *Greek and Latin Texts with Notes and Bibliography.* Cambridge: Cambridge University Press.

Lorenzen, P. (1984). *Elementargeometrie: Das Fundament der Analytischen Geometrie.* Mannheim: Bibliographisches Institut.

Loux, M. J., ed. (1979). *The Possible and the Actual: Readings in the Metaphysics of Modality.* Ithaca: Cornell University Press.

Ludwig, G. (1970). *Deutung des Begriffs "physikalische Theorie" und axiomatische Grundlegung der Hilbertraumstruktur der Quantenmechanik durch Hauptsätze des Messens.* Heidelberg: Springer. (Lecture Notes on Physics, Nº 4.)

Ludwig, G. (1974). *Einführung in die Grundlagen der theoretischen Physik.* Band 2. *Elektrodynamik, Zeit, Raum, Kosmos.* Düsseldorf: Bertelsmann.

Ludwig, G. (1978). *Die Grundstrukturen einer physikalischen Theorie.* Berlin: Springer.

Ludwig, G. (1983). *Foundations of Quantum Mechanics, I.* Translated by C. A. Hain. New York: Springer.

Ludwig, G. (1984). "Restriction and embedding." In Balzer, Pearce, and Schmidt 1984, pp. 17–31.

Lycan, W. (1979). "The trouble with possible worlds." In Loux 1979, pp. 274–316.

Mac Lane, S. (1971). *Categories for the Working Mathematician.* New York: Springer.

Mac Lane, S. (1986). *Mathematics: Form and Function.* New York: Springer.

Maier, A. (1949). *Die Vorläufer Galileis im 14. Jahrhundert.* Roma: Edizioni di "Storia e Letteratura."

Maier, A. (1966). *Die Vorläufer Galileis im 14. Jahrhundert: Studien zur Naturphilosophie der Spätscholastik.* 2., erweiterte Auflage. Roma: Edizioni di "Storia e Letteratura."

Malament, D. (1977a). "Causal theories of time and the conventionality of simultaneity." *Noûs.* **11**: 293–300.

Malament, D. (1977b). "Observationally indistinguishable space-times." *Minnesota Studies in the Philosophy of Science.* **8**: 61–80.

Margenau, H. (1934). "Meaning and scientific status of causality." In A. Danto and S. Morgenbesser 1960, pp. 435–449. (Originally published in *Philosophy of Science.* **1**: 133–

148 (1934).)

Martin-Löf, P. (1966). "The definition of a random sequence." *Information and Control.* **9**: 602–619.

Martin-Löf, P. (1969). "The literature on von Mises' Kollektivs revisited." *Theoria.* **35**: 12–37.

Maxwell, G. (1962). "The ontological status of theoretical entities." *Minnesota Studies in the Philosophy of Science.* **3**: 3–27.

Maxwell, J. C. (1860). "Illustrations of the dynamical theory of gases." *Philosophical Magazine.* (4) **19**: 19–32; **20**: 21–37.

Maxwell, J. C. (1879). "On Boltzmann's Theorem on the average distribution of energy in a system of material points." *Transactions of the Cambridge Philosophical Society.* **12**: 547–570.

Maxwell, J. C. (1890). *The Scientific Papers.* Edited by W. D. Niven. Cambridge: Cambridge University Press. 2 vols.

McKinsey, J. C. C., A. C. Sugar, and P. C. Suppes (1953). "Axiomatic foundations of classical particle mechanics." *Journal of Rational Mechanics and Analysis.* **2**: 253–272.

Mellor, D. H. (1971). *The Matter of Chance.* Cambridge: Cambridge University Press.

Merrill, G. H. (1980). "The model-theoretic argument against realism." *Philosophy of Science.* **47**: 69–81.

Mill, J. S. (1843). *A System of Logic Ratiocinative and Inductive.* London: John W. Parker.

Mill, J. S. (SL). *A System of Logic Ratiocinative and Inductive.* Toronto: University of Toronto Press, 1973. (*Collected Works,* vols. 7 & 8.)

Miller, R. W. (1987). *Fact and Method: Explanation, Confirmation and Reality in the Natural and Social Sciences.* Princeton: Princeton University Press.

Minkowski, H. (1908). "Die Grundgleichungen für die elektromagnetischen Vorgänge in bewegten Körper." *Göttinger Nachrichten.* Pp. 53–111. (Reprinted in *Mathematische Annalen.* **68**: 472–525 (1910).)

Minkowski, H. (1909). "Raum und Zeit." *Physikalische Zeitschrift.* **10**: 104–111.

Mises, R. von (1928). *Wahrscheinlichkeit, Statistik und Wahrheit.* Wien: Springer.

Mises, R. von (1931). *Wahrscheinlichkeitsrechnung und ihre Anwendung in der Statistik und theoretischen Physik.* Leipzig: Deuticke.

Mises, R. von (1957). *Probability, Statistics and Truth.* Second revised English edition, prepared by H. Geiringer. London: Allen & Unwin.

Mises, R. von (1964). *Mathematical Theory of Probability and Statistics.* New York: Academic Press.

Moivre, A. de (1718). *The Doctrine of Chances; or, A Method of Calculating the Probability of Events in Play.* London: Printed by W. Pearson, for the author.

Moivre, A. de (1756). *The Doctrine of Chances; or, A Method of Calculating the Probabilities of Events in Play.* Third edition. London: A. Millar.

Mondadori, F., and A. Morton (1976). "Modal realism: The poisoned pawn." *Philosophical Review.* **85**: 3–20. (Reprinted in Loux 1979, pp. 235–252.)

Montmort, P. R. de (1723). *Essay d'analyse sur les jeux de hazard.* Seconde édition revûe et augmentée de plusieurs lettres. Paris: Jacque Quillau.

Moulines, C. U. (1973). *La estructura del mundo sensible (sistemas fenomenalistas).* Barcelona: Ariel.

Moulines, C. U. (1976). "Approximate application of empirical theories." *Erkenntnis.* **10**. (Revised Spanish version in Moulines 1982, pp. 164–190.)

Moulines, C. U. (1981). "A general scheme for intertheoretic approximation." In Hartkämper and Schmidt 1981, pp. 123–146.

Moulines, C. U. (1982). *Exploraciones metacientíficas: Estructura, contenido y desarrollo de la ciencia.* Madrid: Alianza.

Moulines, C. U. (1984). "Links, loops, and the global structure of science." *Philosophia Naturalis.* **21**: 254–265.

Moulines, C. U. (1987). "Referencia de términos científicos e inconmensurabilidad." In J. J. Acero and T. Calvo Martínez, eds., *Symposium Quine: Actas del Symposium Internacional sobre el Pensamiento Filosófico de Willard V. Quine, celebrado en Granada del 18 al 21 de marzo de 1986.* Granada: Universidad de Granada.

Moulines, C. U., and I. Jané (1981). "Aproximaciones admisibles dentro de las teorías empíricas." *Crítica.* **38**: 53–75.

Moulines, C. U., and R. Torretti (1989). "Extractos de una correspondencia." *Diálogos.* **53**: 123–137.

Myatt, G. (1969). "Background problems in a bubble chamber neutrino experiment." *CERN Yellow Report* 69–28, pp. 145–158.

Nagel, E. (1949). "The meaning of reduction in the natural sciences." In Stauffer, ed., *Science and Civilization.* Madison: University of Wisconsin Press. Pp. 99–145. (Reprinted in Danto and Morgenbesser 1960, pp. 288–312.)

Nagel, E. (1961). *The Structure of Science.* New York: Harcourt, Brace & World.

Nagel, E., P. Suppes, and A. Tarski, eds. (1962). *Logic, Methodology and Philosophy of Science.* Stanford: Stanford University Press.

Narens, L. (1985). *Abstract Measurement Theory.* Cambridge: MIT Press.

Neumann, C. (1870). *Über die Principien der Galilei-Newton'schen Theorie: Akademische Antrittsvorlesung gehalten in der Aula der Universität Leipzig am 3. November 1869.* Leipzig: B. G. Teubner.

Neurath, O., R. Carnap, and C. Morris, eds. (1971). *Foundations of the Unity of Science: Towards an International Encyclopedia of Unified Science.* Chicago: University of Chicago Press. 2 vols.

Newton, I. (*Principia*). *Philosophiæ naturalis principia mathematica.* Third edition (1726) with variant readings. Assembled and edited by A. Koyré and I. B. Cohen. Cambridge: Harvard University Press, 1972. 2 vols.

Newton, I. (Cajori). *Mathematical Principles of Natural Philosophy and His System of the World.* Translated into English by A. Motte in 1729. Translation revised and supplied with historical and explanatory appendix by F. Cajori. Berkeley: University of California Press, 1934. 2 vols.

Newton-Smith, W. H. (1981). *The Rationality of Science.* London: Routledge & Kegan Paul.

Nickles, T. (1973a). "Two concepts of intertheoretic reduction." *Journal of Philosophy.* **70**: 181–201.

Nickles, T. (1973b). "Heuristics and justification in scientific research: comments on Shapere." In F. Suppe 1977, pp. 571–589.

Norton, J. (1989). "The chaos of possibilities: On Einstein's heuristic methods." Paper read at the Conference on the Logic of Scientific Discovery held at the University of North Carolina, Greensboro, in March 1989. (To be published in the proceedings of that conference.)

Pasch, M. (1882). *Vorlesungen über neuere Geometrie.* Leipzig: B. G. Teubner.

Pearce, D., and V. Rantala (1982a). "Realism and formal semantics." *Synthese.* **52**: 39–53.

Pearce, D., and V. Rantala (1982b). "Realism and reference: Some comments on Putnam." *Synthese.* **52**: 439–448.

Pearson, E. S., and M. G. Kendall, eds. (1970). *Studies in the History of Statistics and Probability.* London: Charles Griffin.

Peirce, C. S. (CP). *Collected Papers.* Edited by C. Hartshorne, P. Weiss, and A. W. Burks. Cambridge: Belknap Press of Harvard University Press. 8 vols.

Penrose, R. (1989). *The Emperor's New Mind: Concerning Computers, Minds and the Laws of Physics.* Oxford: Oxford University Press.

Pickering, A. (1984a). "Against putting the phenomena first." *Studies in the History and Philosophy of Science.* **15**: 85–117.

Pickering, A. (1984b). *Constructing Quarks: A Sociological History of Particle Physics.* Chicago: University of Chicago Press.

Pieri, M. (1899). "Della geometria elementare come sistema ipotetico-deduttivo; monografia del punto e del moto." *Memorie della Reale Accademia delle Scienze di Torino, Classe di Sc. Fisiche, Matematiche e Naturali, Serie 2.* **48**: 1–62.

Plantinga, A. (1974). *The Nature of Necessity.* Oxford: Clarendon Press.

Platon. *Œuvres Complètes.* Paris: Société d'Édition Les Belles Lettres, 1920ff. 13 vols. (I quote Plato from this edition. As is customary, reference is made to the pages of Henri Estienne's edition, Paris 1578.)

Plutarch. *Lives.* VI. *Dion and Brutus, Timoleon and Aemilius Paulus.* With an English translation by B. Perrin. London: Heinemann, 1970. (Loeb Classical Library.)

Poincaré, H. (SM). *Science et méthode.* Paris: Flammarion, 1922.

Poincaré, H. (FS). *The Foundations of Science: Science and Hypothesis; The Value of Science; Science and Method.* Translated by G. B. Halsted. New York: Science Press, 1929.

Poisson, S. D. (1837). *Recherches sur la probabilité des jugements en matière criminelle et en matière civile, précedées des règles générales du calcul des probabilités.* Paris: Bachelier.

Popper, K. R. (1935). *Logik der Forschung: Zur Erkenntnistheorie der modernen Naturwissenschaft.* Wien: Springer.

Popper, K. R. (1950). "Indeterminism in Quantum Physics and in Classical Physics." *British Journal for the Philosophy of Science.* **1**: 117–133, 173–195.

Popper, K. R. (1957). "The propensity interpretation of the calculus of probability and the quantum theory." In Körner 1957, pp. 65–70.

Popper, K. R. (1959a). *The Logic of Scientific Discovery.* New York: Basic Books.

Popper, K. R. (1959b). "The propensity interpretation of probability." *British Journal for the Philosophy of Science.* **10**: 25–42.

Popper, K. R. (1963). *Conjectures and Refutations: The Growth of Scientific Knowledge.* London: Routledge and Kegan Paul.

Popper, K. R. (1972). *Objective Knowledge: An Evolutionary Approach.* Oxford: Oxford University Press.

Popper, K. R. (1983). *Realism and the Aim of Science.* From the *Postscript to the Logic of Scientific Discovery.* Edited by W. W. Bartley III. Totowa, NJ: Rowman and Littlefield.

Putnam, H. (PP). *Philosophical Papers.* Cambridge: Cambridge University Press, 1975–83. 3 vols.

Putnam, H. (1962). "What theories are not." In Putnam PP, vol. 1, pp. 215–227. (First published in Nagel, Suppes, and Tarski 1962.)

Putnam, H. (1970). "Is semantics possible?" In Putnam PP, vol. 2, pp. 139–152. (First published in H. Kiefer and M. Munitz, eds. *Language, Belief and Metaphysics.* Albany: State University of New York Press.)

Putnam, H. (1973). "Explanation and reference." In Putnam PP, vol. 2, pp. 196–214. (First published in G. Pearce and P. Maynard, *Conceptual Change.* Dordrecht: D. Reidel. Pp. 199–221.)

Putnam, H. (1975). "The meaning of 'meaning'." In Putnam PP, vol. 2, pp. 215–271. (First published in *Minnesota Studies in the Philosophy of Science.* **7**.)

Putnam, H. (1978). *Meaning and the Moral Sciences.* London: Routledge & Kegan Paul.

Putnam, H. (1978). "Realism and reason." In Putnam, *Meaning and the Moral Sciences.* London: Routledge and Kegan Paul. Pp. 123–138.

Putnam, H. (1980). "Models and reality." In Putnam PP, vol. 3, pp. 1–25. (First published in *Journal of Symbolic Logic.* **45**: 464–482 (1980).)

Putnam, H. (1981). *Reason, Truth and History.* Cambridge: Cambridge University Press.

Putnam, H. (1982a). "Why there isn't a ready-made world." In Putnam PP, vol. 3, pp. 205–228. (First published in *Synthese.* **51**: 141–167 (1982).)

Putnam, H. (1982b). "Why reason can't be naturalized." In Putnam PP, vol. 3, pp. 229–247. (First published in *Synthese.* **52**: 3–23 (1982).)

Putnam, H. (1988). *Representation and Reality.* Cambridge: MIT Press.

Quine, W. V. (1948). "On what there is." In W. V. Quine, *From a Logical Point of View.* Cambridge: Harvard University Press. Pp. 1–19. (First published in *Review of Metaphysics.* **2**: 21–28 (1948).)

Quine, W. V. (1951). "Two dogmas of empiricism." In W. V. Quine, *From a Logical Point of View.* Cambridge: Harvard University Press. Pp. 20–46. (First published in *Philosophical Review.* **60**: 20–43 (1951).)

Quine, W. V. (1960). *Word and Object.* New York: Wiley.

Quine, W. V. (1961). *From a Logical Point of View: Nine Logico-Philosophical Essays.* Second edition, revised. Cambridge: Harvard University Press.

Quine, W. V. (1969). *Ontological Relativity and Other Essays.* New York: Columbia University Press.

Railton, P. (1978). "A deductive-nomological model of probabilistic explanation." *Philosophy of Science.* **45**: 206–226.

Ramsey, F. P. (1926). "Truth and probability." In Ramsey 1978, pp. 58–100. (First published in Ramsey 1931, pp.156–198.)

Ramsey, F. P. (1931). *The Foundations of Mathematics and Other Logical Essays.* London: Routledge & Kegan Paul.

Ramsey, F. P. (1978). *Foundations.* London: Routledge and Kegan Paul.

Raychaudhuri, A. K. (1979). *Theoretical Cosmology.* Oxford: Clarendon Press.

Reichenbach, H. (1924). *Axiomatik der relativistischen Raum-Zeit-Lehre.* Braunschweig: Vieweg. (Quotations refer to reprint of 1965.)

Reichenbach, H. (1928). *Philosophie der Raum-Zeit-Lehre.* Berlin: W. de Gruyter.

Reichenbach, H. (1935). *Wahrscheinlichkeitslehre: Eine Untersuchung über die logischen und mathematischen Grundlagen der Wahrscheinlichkeitsrechnung.* Leiden: A. W. Sijthoff.

Reichenbach, H. (1949). *The Theory of Probability: An Inquiry into the Logical and Mathematical Foundations of the Calculus of Probability.* Translated by E. H. Hutten and M. Reichenbach. Berkeley: University of California Press.

Reichenbach, H. (1971). *The Theory of Probability: An Inquiry into the Logical and Mathematical Foundations of the Calculus of Probability.* Second edition. Translated by E. H. Hutten and M. Reichenbach. Berkeley: University of California Press.

Renyi, A. (1970). *Foundations of Probability.* San Francisco: Holden-Day.

Révész, P. (1968). *The Laws of Large Numbers.* New York: Academic Press.

Robb, A. A. (1914). *A Theory of Time and Space.* Cambridge: Cambridge University Press.

Robertson, H. P., and T. W. Noonan (1968). *Relativity and Cosmology*. Philadelphia: W. B. Saunders.

Rohrlich, F., and L. Hardin (1983). "Established theories." *Philosophy of Science.* **50**: 603–617.

Rosenberg, J. F. (1980). *One World and Our Knowledge of It.* Dordrecht: D. Reidel.

Russell, B. (1908). "Mathematical logic as based on the theory of types." *American Journal of Mathematics.* **30**: 222–262.

Saccheri, G. (1733). *Euclides ab omni nævo vindicatus sive conatus geometricus quo stabiliuntur prima ipsa universæ geometriæ principia.* Mediolani: Ex Typographia Pauli Antonii Montani.

Salam, A. (1968). "Weak and electromagnetic interactions." In N. Svartholm, ed., *Proceedings of the Eighth Nobel Symposium.* Stockholm: Almqvist & Wicksell. Pp. 367–377.

Salmon, W. (1984). *Scientific Explanation and the Causal Structure of the World.* Princeton: Princeton University Press.

Savage, L. J. (1954). *The Foundations of Statistics.* New York: Wiley.

Schaffner, K. F. (1967). "Approaches to reduction." *Philosophy of Science.* **34**: 137–147.

Scheffler, I. (1963). *The Anatomy of Inquiry: Philosophical Studies in the Theory of Science.* New York: Knopf.

Scheibe, E. (1989). "General laws of nature and the uniqueness of the universe." Lecture delivered at the Conference on the Origin and Evolution of the Universe held by the Académie Internationale de Philosophie des Sciences in Lima, Perú, in August 1989. (To be published in the proceedings of that conference).

Schnorr, C. P. (1970a). "Über die Definition von effektiven Zufallstests." *Zeitschrift für Wahrscheinlichkeit und verwandten Gebieten.* **15**: 297–312, 313–328.

Schnorr, C. P. (1970b). *Zufälligkeit und Wahrscheinlichkeit: Eine algoritmische Begründung der Wahrscheinlichkeitstheorie.* Berlin: Springer. (Lecture Notes in Mathematics Nº 218.)

Schnorr, C. P. (1977). "A survey of the theory of random sequences." In Butts and Hintikka 1977, pp. 193–211.

Schrödinger, E. (1926). "Über das Verhältnis der Heisenberg-Born-Jordanschen Quantenmechanik zu der meinen." *Annalen der Physik.* **79**: 734–756.

Schwartz, S. P., ed. (1977). *Naming, Necessity and Natural Kinds.* Ithaca: Cornell University Press.

Schwarzschild, K. (1916). "Über das Gravitationsfeld eines Massenpunktes nach der Einsteinschen Theorie." *K. Preußische Akademie der Wissenschaften Sitzungsberichte.* Pp. 189–216.

Scriven, M. (1958). "Definitions, explanations and theories." *Minnesota Studies in the Philosophy of Science.* **2**: 99–195.

Scriven, M. (1962). "Explanations, predictions and laws." *Minnesota Studies in the Philosophy of Science.* **3**: 130–230.

Searle, J. R. (1983). *Intentionality: An Essay in the Philosophy of Mind.* Cambridge: Cambridge University Press.

Shapere, D. (1964). "The structure of scientific revolutions." *Philosophical Review.* **73**: 382–394. (Reprinted in Shapere 1984, pp. 37–48.)

Shapere, D. (1966). "Meaning and scientific change." In Colodny 1966, pp. 41–85. (Reprinted in Shapere 1984, pp. 58–101.)

Shapere, D. (1982). "The concept of observation in science and philosophy." *Philosophy of Science.* **49**: 485–525.

Shapere, D. (1984). *Reason and the Search for Knowledge: Investigations in the Philosophy of Science.* Dordrecht: D. Reidel.

Shapere, D. (1985). "Obervation and the Scientific Enterprise." In Achinstein and Hannaway 1985, pp. 21–45.

Shapiro, S. (1985). "Second-order languages and mathematical practice." *Journal of Symbolic Logic*. **50**: 714–742.

Shea, W. R., ed. (1983). *Nature Mathematized: Historical and Philosophical Case Studies in Classical Modern Natural Philosophy*. Dordrecht: D. Reidel.

Shimony, A. (1977). "Is observation theory-laden? A problem in naturalistic epistemology." In Colodny 1977, pp. 185–208.

Simplicius. *In Aristotelis de Caelo Commentaria*. Edidit H. Diels. Berlin: G. Reimer, 1894. (Commentaria in Aristotelem Græca.)

Sklar, L. (1967). "Types of inter-theoretic reduction." *British Journal for the Philosophy of Science*. **18**: 109–124.

Skyrms, B. (1980). *Causal Necessity*. New Haven: Yale University Press.

Skyrms, B. (1984). *Pragmatics and Empiricism*. New Haven: Yale University Press.

Smart, J. J. C. (1963). *Philosophy and Scientific Realism*. London: Routledge & Kegan Paul.

Smith, P. (1981). *Realism and the Progress of Science*. Cambridge: Cambridge University Press.

Sneed, J. D. (1971). *The Logical Structure of Mathematical Physics*. Dordrecht: D. Reidel.

Solovay, R. M. (1970). "A model of set theory in which every set of reals is Lebesgue-measurable." *Annals of Mathematics*. **92**: 1–56.

Sorabji, R. (1980). *Necessity, Cause, and Blame: Perspectives on Aristotle's Theory*. Ithaca: Cornell University Press.

Spector, M. (1975). "Russell's Maxim and reduction as replacement." *Synthese*. **32**: 135–176.

Spector, M. (1978). *Concepts of Reduction in Physical Science*. Philadelphia: Temple University Press.

Stachel, J. (1989). "Einstein's search for general covariance, 1912–1915." In D. Howard and J. Stachel, eds., *Einstein and the History of General Relativity: Proceedings of the 1986 Osgood Hill Conference, North Andover, Massachusetts, May 8–11, 1986*. Boston: Birkhäuser. Pp. 63–100.

Stegmüller, W. (1970). *Theorie und Erfahrung*. Berlin: Springer.

Stegmüller, W. (1973a). *Theorie und Erfahrung*. Zweiter Halbband. *Theorienstrukturen und Theoriendynamik*. Berlin: Springer.

Stegmüller, W. (1973b). *Personelle und Statistische Wahrscheinlichkeit*. Zweiter Halbband. *Statistisches Schließen, statistische Begründung, statistische Analyse*. Berlin: Springer.

Stegmüller, W. (1976). *The Structure and Dynamics of Theories*. New York: Springer.

Stegmüller, W. (1979). *The Structuralist View of Theories. A Possible Analogue of the Bourbaki Programme in Physical Science*. Berlin: Springer.

Stegmüller, W. (1986). *Theorie und Erfahrung*. Dritter Teilband. *Die Entwicklung des neuen Strukturalismus seit 1973*. Berlin: Springer.

Stein, H. (1977). "Some philosophical prehistory of General Relativity." *Minnesota Studies in the Philosophy of Science*. **8**: 3–49.

Steiner, M. (1986). "Events and causality." *Journal of Philosophy*. **65**: 249–264.

Stevenson, L. (1982). *The Metaphysics of Experience*. Oxford: Oxford University Press.

Stigler, S. M. (1986). *The History of Statistics: The Measurement of Uncertainty before 1900*. Cambridge: Belknap Press of Harvard University Press.

Strawson, P. F. (1959). *Individuals: An Essay in Descriptive Metaphysics*. London: Methuen.

Streintz, H. (1883). *Die physikalischen Grundlagen der Mechanik*. Leipzig: Teubner.

Sudarshan, C. G., and R. E. Marshak (1958). "Chirality noninvariance and the universal Fermi interaction." *Physical Review.* **109**: 1860–1862.

Suppe, F., ed. (1977). *The Structure of Scientific Theories.* Second edition. Urbana: University of Illinois Press.

Suppes, P. (1960). "A comparison of the meaning and uses of models in mathematics and the empirical sciences." *Synthese.* **12**: 287–301.

Suppes, P. (1962). "Models of data." In Nagel, Suppes, and Tarski 1962, pp. 252–261.

Suppes, P. (1967). "What is a scientific theory?" In S. Morgenbesser, ed., *Philosophy of Science Today.* New York: Basic Books. Pp. 55–67.

Suppes, P. (1969). "The structure of theories and the analysis of data." In Suppe 1977, pp. 266–307.

Suppes, P. (1984). *Probabilistic Metaphysics.* Oxford: Basil Blackwell.

Teller, P. (1973). "Conditionalization and observation." *Synthese.* **26**: 218–258.

Thomson, J. (1884). "On the Law of Inertia, the Principle of Chronometry and the Principle of Absolute Clinural Rest, and of Absolute Rotation." *Royal Society of Edinburgh Proceedings.* **12**: 568–578.

Thomson, W., and P. G. Tait (1867). *Treatise on Natural Philosophy.* Cambridge: Cambridge University Press.

Todhunter, I. (1865). *A History of the Mathematical Theory of Probability from the Time of Pascal to That of Laplace.* London: Macmillan.

Tooley, M. (1987). *Causation: A Realist Approach.* Oxford: Clarendon Press.

Torretti, R. (1978a). *Philosophy of Geometry from Riemann to Poincaré.* Dordrecht: D. Reidel.

Torretti, R. (1978b). "Hugo Dingler's Philosophy of Geometry." *Diálogos.* **32**: 85–128. (Correction in *Diálogos.* **50**: 191–192 (1987).)

Torretti, R. (1983). *Relativity and Geometry.* Oxford: Pergamon Press.

Torretti, R. (1984). "Kosmologie als ein Zweig der Physik." In B. Kanitscheider, ed., *Moderne Naturphilosophie.* Würzburg: Königshausen & Neumann. Pp. 183–200.

Torretti, R. (1986a). "Observation." *British Journal for the Philosophy of Science.* **37**: 1–23.

Torretti, R. (1986b). "Physical theories, I." *Diálogos.* **48**: 183–212.

Torretti, R. (1989). "Einstein's luckiest thought." Paper read at the Conference on the Logic of Scientific Discovery held at the University of North Carolina, Greensboro, in March 1989. (To be published in the proceedings of that conference.)

Toulmin, S. (1961). *Foresight and Understanding: An Enquiry into the Aims of Science.* Bloomington: Indiana University Press.

Truesdell, C. (1984). *An Idiot's Fugitive Essays on Science: Methods, Criticism, Training, Circumstances.* New York: Springer.

Ueberweg, F. (1851). "Die Prinzipien der Geometrie wissenschaftlich dargestellt." In M. Brasch, ed., *Die Welt- und Lebensanschauung Friedrich Ueberwegs.* Leipzig: Gustav Engel, 1889. Pp. 263–316. (Originally published in *Archiv für Philosophie und Pädagogik.* **17**).

Veblen, O. (1904). "A system of axioms for geometry." *American Mathematical Society Transactions.* **5**: 343–384.

Veblen, O. (1911). "The foundations of geometry." In J. W. A. Young, ed., *Monographs on Topics of Modern Mathematics Relevant to the Elementary Field.* New York: Longmans Green. Pp. 3–51.

Vendler, Z. (1962). "Effects, results and consequences." In R. J. Butler, ed., *Analytical Philosophy.* First series. Oxford: Basil Blackwell. Pp. 1–15.

Venn, J. (1866). *The Logic of Chance.* London: Macmillan.

Venn, J. (1888). *The Logic of Chance.* Third edition. London: Macmillan.

Ville, J. (1939). *Étude critique de la notion collectif.* Paris: Gauthier-Villars.

Wagon, S. (1985). *The Banach-Tarski Paradox.* Cambridge: Cambridge University Press.

Wald, A. (1937). "Die Widerspruchsfreiheit des Kollektivbegriffs in der Wahrscheinlichkeitsrechnung." *Ergebnisse eines mathematischen Kolloquiums.* **8**: 38–72.

Wallis, J. (1693). "De Postulato Quinto et Definitione Quinta, Lib. 6 Euclidis disceptatio geometrica." In Wallis, *Opera mathematica.* Oxonii: E theatro Sheldoniano. Vol. II, pp. 669–678.

Weil, A. (1937). *Sur les espaces à structure uniforme et sur la topologie générale.* Paris: Hermann.

Weinberg, S. (1967). "A model of leptons." *Physical Review Letters.* **19**: 1264–1266.

Weinberg, S. (1972). *Gravitation and Cosmology: Principles and Applications of the Theory of Relativity.* New York: Wiley.

Wiggins, D. (1980). *Sameness and Substance.* Cambridge: Harvard University Press.

Will, C. M. (1981). *Theory and Experiment in Gravitational Physics.* Cambridge: Cambridge University Press.

Winnie, J. W. (1986). "Invariants and objectivity: A theory with applications to relativity and geometry." In Colodny 1986, pp. 71–180.

Wittgenstein, L. (1960). *Schriften: Tractatus logico-philosophicus. Tagebücher 1914–1916. Philosophische Untersuchungen.* Frankfurt a. M.: Suhrkamp.

Wittgenstein, L. (PU). *Philosophical Investigations.* Second edition. Oxford: Basil Blackwell, 1958.

Wittgenstein, L. (BB). *Preliminary Studies for the "Philosophical Investigations," Generally Known as the Blue and Brown Books.* Second edition. Oxford: Basil Blackwell, 1969.

Wright, G. H. von (1974). *Causality and Determinism.* New York: Columbia University Press.

Young, E. C. M. (1967) "High energy neutrino interactions." *CERN Yellow Report.* 67–12.

Zermelo, E. (1904). "Beweis, daß jede Menge wohlgeordnet werden kann." *Mathematische Annalen.* **59**: 514–516.

Zermelo, E. (1908a). "Neuer Beweis für die Möglichkeit einer Wohlordnung." *Mathematische Annalen.* **65**: 101–128.

Zermelo, E. (1908b). "Untersuchungen über die Grundlagen der Mengenlehre I." *Mathematische Annalen.* **65**: 261–281.

INDEX

SOUTHEASTERN MASSACHUSETTS UNIVERSITY
3 2922 00284 333 9